Implementation of Chemometrics and Other Techniques as Means of Authenticity and Traceability to Detect Adulteration in Foods for the Protection of Human Health

Implementation of Chemometrics and Other Techniques as Means of Authenticity and Traceability to Detect Adulteration in Foods for the Protection of Human Health

Editor

Theodoros Varzakas

MDPI • Basel • Beijing • Wuhan • Barcelona • Belgrade • Manchester • Tokyo • Cluj • Tianjin

MDPI

Editor
Theodoros Varzakas
Food Science and Technology
University of Peloponnese
Kalamata
Greece

Editorial Office
MDPI
St. Alban-Anlage 66
4052 Basel, Switzerland

This is a reprint of articles from the Special Issue published online in the open access journal *Foods* (ISSN 2304-8158) (available at: www.mdpi.com/journal/foods/special_issues/Chemometrics_Authenticity_Traceability_Food).

For citation purposes, cite each article independently as indicated on the article page online and as indicated below:

LastName, A.A.; LastName, B.B.; LastName, C.C. Article Title. *Journal Name* **Year**, *Volume Number*, Page Range.

ISBN 978-3-0365-6661-0 (Hbk)
ISBN 978-3-0365-6660-3 (PDF)

Contents

About the Editor

Theodoros Varzakas

Theodoros Varzakas has a Bachelor (Honours) in Microbiology and Biochemistry (1992), a Ph.D. in Food Science and Technology, and an MBA in Food and Agricultural Management from Reading University, UK (1998). He has also worked as a postdoctoral research staff at the same university. He has worked in large pharmaceutical and multinational food companies in Greece for 5 years and has at least 22 years of experience in the public sector. Since 2005, he has served as Assistant, Associate, and Full Professor at the Department of Food Science and Technology, University of Peloponnese, ex Technological Educational Institute of Peloponnese, Greece, specializing in issues of food technology, food processing/engineering, and food quality and safety. He is also a Section Editor in Chief for the Journal *Foods* in Food Security and Sustainability (2020–), was an ex Editor in Chief for Current Research in Nutrition and Food Science (2015–2019), and is a reviewer and member of the editorial board in many international journals. He has written more than 250 research papers and chapters in books and has presented more than 160 papers and posters at national and international conferences. He has written and edited 16 books in Greek and 14 in English on sweeteners, biosensors, food engineering, and food processing, chemometrics and authenticity published by CRC. He has participated in many European and national research programs as a coordinator or scientific member. His work has been cited in over 4000 citations with an h-index of 31.

Ex EFSA advisor/scientific expert to panel of biological hazards, visiting professor at Ghent University global campus, South Korea, Research Fellow at University Technology Malaysia and Expert in JEMRA of FAO/WHO.

According to September 2022 data-update for "Updated science-wide author databases of standardized citation indicators" Varzakas Theodoros is in the 2% of citations worldwide.

Preface to "Implementation of Chemometrics and Other Techniques as Means of Authenticity and Traceability to Detect Adulteration in Foods for the Protection of Human Health"

Authenticity and traceability are crucial in order to overcome frauds in the international food trade.

Classification of foods, such as olive oils, according to their variety and/or geographical origin is of great importance for producers, importers, and consumers. Toward this target of food classification, different multivariate statistical procedures are employed, such as cluster analysis, factor analysis, multidimensional scaling, discriminant analysis, correspondence analysis, canonical analysis, and Procrustes analysis.

Recently, artificial intelligence has also been applied to solve food characterization problems.

Different analytical approaches have been employed for the adulteration of foods, such as gas chromatography–mass spectrometry (GC/MS), compound-specific isotope analysis (CSIA), isotope ratio mass spectrometry (IRMS), NMR spectroscopy, Fourier transform mid-infrared (FTIR), near-infrared (FT-NIR), and Raman (FT-Raman) spectroscopy. Moreover, chemometric methods have been used to process experimental data, such as linear discriminant analysis (LDA) and artificial neural networks (ANN). The aim of this e-book is to describe chemometrics and authenticity in a range of foods of plant and animal origin.

Theodoros Varzakas
Editor

foods MDPI

Editorial

Implementation of Chemometrics and Other Techniques as Means of Authenticity and Traceability to Detect Adulteration in Foods for the Protection of Human Health

Theodoros Varzakas

Department of Food Science and Technology, University of the Peloponnese, 24100 Kalamata, Greece; t.varzakas@uop.gr

Citation: Varzakas, T. Implementation of Chemometrics and Other Techniques as Means of Authenticity and Traceability to Detect Adulteration in Foods for the Protection of Human Health. *Foods* **2023**, *12*, 652. https://doi.org/10.3390/foods12030652

Received: 22 January 2023
Revised: 30 January 2023
Accepted: 31 January 2023
Published: 2 February 2023

The authenticity of foods of plant and animal origin is key to safeguarding both quality and safety aspects without jeopardizing consumers' health. Hence, this Special Issue brings advances in the area of the authentication and chemometrics of foods, in attempt to prevent fraud, a phenomenon which has increased recently. Different food materials have been investigated and the results have been reported.

Tocopherols (α-, β-, γ-, and δ-) are part of the commonly known vitamin E complex, along with the corresponding tocotrienols, which protect against the non-enzymatic peroxidation of polyunsaturated fatty acids, are lipid-soluble, and can be found in high-fat plant foods, such as walnuts. In the first paper, Mitsikaris et al. [1] determined the levels of tocopherols in walnut seed oils by HPLC-UV. The levels of the samples varied from four European countries and showed that the Ukrainian walnut seed oils exhibited significantly higher total concentrations. Moreover, a higher mean concentration of α-tocopherol was reported for Greek walnuts compared to the French and Bulgarian walnuts.

Zhao et al. [2] discriminated the lamb meat quality of Protected Geographical Indication (PGI) Sunite lamb, from two other banners in the Inner Mongolia autonomous region by stable isotopes and modeling. They found that this methodology, in conjunction with chemometric approaches such as the data-driven soft independent modeling of class analogy (DD–SIMCA), can be used as an effective indicator for protecting PGI Sunite lamb, taking into account geographical origin, feeding system, age, and gender.

Rana et al. [3] discriminated the physico-functional properties and chemometric techniques (principal component analysis (PCA) and multiclass discriminant analysis (MDA)) of four *Cinnamomum* species. They found a clear separation of the different *Cinnamomum* species using the above techniques, thus showing that this combination is effective against food fraud.

Siddiqui et al. [4] employed Fourier infrared spectroscopy (FTIR) to determine adulteration in meat mixtures of beef, lamb, and chicken. They also used PCA and multiclass support vector machine (M-SVM) and found that the highest classification accuracy value of 85% was presented in beef and lamb samples for both adulterated and non-adulterated classes. This method could be a rapid quality control tool in the meat industry and might be employed in halal authentication.

Shomaji et al. [5] reported the use of low-field ^{1}H-nuclear magnetic resonance (NMR) relaxometry to determine dye contamination on vegetables. It is very well known that non-food-grade dyes could be very toxic. It was concluded that the proposed low-cost detection approach can be used to generate warning flags if the detected dye concentrations are over the limit of the accepted standards for food dyes.

The article by Agriopoulou et al. [6] discussed Greek table olive varieties from protected designation of origin (PDO) areas. The authors used orthogonal partial least square discriminant analysis (OPLS-DA) for the discrimination and classification of table olives and found the model to be effective in olive fruit authentication.

Zhao et al. [7] employed multivariate data analysis applied to elemental analysis, stable isotope analysis, and fatty acid analysis in combination with orthogonal partial least squares discriminant analysis (OPLS-DA) to determine the geographical origin of milk in four neighboring provinces. The discrimination of milk took place in farms with different distances of less than 11 km in each province, and the discriminant distance was successfully reduced to 0.7 km.

They found that for a relatively close sample origin distance, a single technique, such as the fatty acid chemical parameter analysis, could be completely superior compared to the combination multiple technologies. These results could be used to improve milk traceability in China.

The paper by Tarapoulouzi et al. [8] reported on stable isotope ratio analysis and orthogonal projections to geographically discriminate Greek olive cultivars. This combination was very good and they showed that the most important isotope markers for the discrimination of olive oil samples were $\delta^{18}O$ and $\delta^{2}H$.

Tarapoulouzi et al. [9] discussed and reviewed the evolution of the use of chemometrics on honey composition/the physico-chemical parameters during processing and storage in order to determine the authenticity of honey. They verified it as an effective tool to optimizing quality control and the safety protection of consumers' health.

Another review paper by Grassi et al. [10] compared the different chemometric techniques (from clustering to classification and regression) along with spectroscopy, chromatography, electrochemical sensors, and other on-site detection devices against milk adulteration. They also presented the steps which should be followed to develop a chemometric model to face adulteration issues.

Finally, Avila-Sosa et al. [11] determined the specific chemical markers coupled with chemometric methods to discriminate the adulterated samples of saffron. Saffron is an important colorant, antioxidant, and source of phytochemicals aromatic spice due to the large number of chemical compounds found in the by-products (flower parts) of saffron (catechin, quercetin, delphinidin, etc.). They found that the geographical origin and harvest/postharvest characteristics of saffron could play a key role in chemical characterization.

Data Availability Statement: Data is contained within the article.

Acknowledgments: The editor would like to acknowledge the support from *Foods* and the help provided from his colleagues in publishing these quality articles.

Conflicts of Interest: The author declares no conflict of interest.

References

1. Mitsikaris, P.D.; Kokokiris, L.; Pritsa, A.; Papadopoulos, A.N.; Kalogiouri, N.P. Investigating the Tocopherol Contents of Walnut Seed Oils Produced in Different European Countries Analyzed by HPLC-UV: A Comparative Study on the Basis of Geographical Origin. *Foods* **2022**, *11*, 3719. [CrossRef] [PubMed]
2. Zhao, R.; Liu, X.; Wang, J.; Wang, Y.; Chen, A.-L.; Zhao, Y.; Yang, S. Proposing Two Local Modeling Approaches for Discriminating PGI Sunite Lamb from Other Origins Using Stable Isotopes and Machine Learning. *Foods* **2022**, *11*, 846. [CrossRef] [PubMed]
3. Rana, P.; Liaw, S.-Y.; Lee, M.-S.; Sheu, S.-C. Discrimination of Four *Cinnamomum* Species with Physico-Functional Properties and Chemometric Techniques: Application of PCA and MDA Models. *Foods* **2021**, *10*, 2871. [CrossRef]
4. Siddiqui, M.A.; Khir, M.H.M.; Witjaksono, G.; Ghumman, A.S.M.; Junaid, M.; Magsi, S.A.; Saboor, A. Multivariate Analysis Coupled with M-SVM Classification for Lard Adulteration Detection in Meat Mixtures of Beef, Lamb, and Chicken Using FTIR Spectroscopy. *Foods* **2021**, *10*, 2405. [CrossRef] [PubMed]
5. Shomaji, S.; Masna, N.V.R.; Ariando, D.; Paul, S.D.; Horace-Herron, K.; Forte, D.; Mandal, S.; Bhunia, S. Detecting Dye-Contaminated Vegetables Using Low-Field NMR Relaxometry. *Foods* **2021**, *10*, 2232. [CrossRef] [PubMed]
6. Agriopoulou, S.; Tarapoulouzi, M.; Bedine Boat, M.A.; Rébufa, C.; Dupuy, N.; Theocharis, C.R.; Varzakas, T.; Roussos, S.; Artaud, J. Authentication and Chemometric Discrimination of Six Greek PDO Table Olive Varieties through Morphological Characteristics of Their Stones. *Foods* **2021**, *10*, 1829. [CrossRef] [PubMed]
7. Zhao, R.; Su, M.; Zhao, Y.; Chen, G.; Chen, A.; Yang, S. Chemical Analysis Combined with Multivariate Statistical Methods to Determine the Geographical Origin of Milk from Four Regions in China. *Foods* **2021**, *10*, 1119. [CrossRef] [PubMed]

8. Tarapoulouzi, M.; Skiada, V.; Agriopoulou, S.; Psomiadis, D.; Rébufa, C.; Roussos, S.; Theocharis, C.R.; Katsaris, P.; Varzakas, T. Chemometric Discrimination of the Geographical Origin of Three Greek Cultivars of Olive Oils by Stable Isotope Ratio Analysis. *Foods* **2021**, *10*, 336. [CrossRef] [PubMed]
9. Tarapoulouzi, M.; Mironescu, M.; Drouza, C.; Mironescu, I.D.; Agriopoulou, S. Insight into the Recent Application of Chemometrics in Quality Analysis and Characterization of Bee Honey during Processing and Storage. *Foods* **2023**, *12*, 473. [CrossRef]
10. Grassi, S.; Tarapoulouzi, M.; D'Alessandro, A.; Agriopoulou, S.; Strani, L.; Varzakas, T. How Chemometrics Can Fight Milk Adulteration. *Foods* **2023**, *12*, 139. [CrossRef] [PubMed]
11. Avila-Sosa, R.; Nevárez-Moorillón, G.V.; Ochoa-Velasco, C.E.; Navarro-Cruz, A.R.; Hernández-Carranza, P.; Cid-Pérez, T.S. Detection of Saffron's Main Bioactive Compounds and Their Relationship with Commercial Quality. *Foods* **2022**, *11*, 3245. [CrossRef]

foods
MDPI

Review

Insight into the Recent Application of Chemometrics in Quality Analysis and Characterization of Bee Honey during Processing and Storage

Maria Tarapoulouzi [1,*,†], Monica Mironescu [2,*], Chryssoula Drouza [3,†], Ion Dan Mironescu [2] and Sofia Agriopoulou [4]

1 Department of Chemistry, Faculty of Pure and Applied Science, University of Cyprus, P.O. Box 20537, Nicosia 1678, Cyprus
2 Faculty of Agricultural Sciences Food Industry and Environmental Protection, Lucian Blaga University of Sibiu, Bv. Victoriei 10, 550024 Sibiu, Romania
3 Department of Agricultural Production, Biotechnology and Food Science, Cyprus University of Technology, P.O. Box 50329, Limassol 3036, Cyprus
4 Department of Food Science and Technology, University of the Peloponnese, Antikalamos, 24100 Kalamata, Greece
* Correspondence: tarapoulouzi.maria@ucy.ac.cy (M.T.); monica.mironescu@ulbsibiu.ro (M.M.)
† These authors contributed equally to this work.

Abstract: The application of chemometrics, a widely used science in food studies (and not only food studies) has begun to increase in importance with chemometrics being a very powerful tool in analyzing large numbers of results. In the case of honey, chemometrics is usually used for assessing honey authenticity and quality control, combined with well-established analytical methods. Research related to investigation of the quality changes in honey due to modifications after processing and storage is rare, with a visibly increasing tendency in the last decade (and concentrated on investigating novel methods to preserve the honey quality, such as ultrasound or high-pressure treatment). This review presents the evolution in the last few years in using chemometrics in analyzing honey quality during processing and storage. The advantages of using chemometrics in assessing honey quality during storage and processing are presented, together with the main characteristics of some well-known chemometric methods. Chemometrics prove to be a successful tool to differentiate honey samples based on changes of characteristics during storage and processing.

Keywords: honey; quality parameters; chemometrics; storage; processing; thermal treatment

Citation: Tarapoulouzi, M.; Mironescu, M.; Drouza, C.; Mironescu, I.D.; Agriopoulou, S. Insight into the Recent Application of Chemometrics in Quality Analysis and Characterization of Bee Honey during Processing and Storage. *Foods* **2023**, *12*, 473. https://doi.org/10.3390/foods12030473

Academic Editor: Ricard Boqué

Received: 23 November 2022
Revised: 30 December 2022
Accepted: 16 January 2023
Published: 19 January 2023

1. Introduction

Bee honey is a natural food extracted from honeycombs, together with bee wax, pollen and propolis, royal jelly and venom [1,2]. Honey, also known as "the food of the Gods" [3], was always used as food by humans [4]; its success is represented by the very high content in simple sugars (honey is a supersaturated sugar solution with about 25–35% glucose and 35–45% fructose, together with 1–2% sucrose) [5]. Honey is considered to be the most energy-dense food in nature [6], with an appreciable energy value (100 g of bee honey provides 310 kcal) [7]. It has a glycemic index in the range 32–87 [8], seeming to show a hypoglycemic effect [9].

Honey is a complex food, a valuable gift of nature [10,11]; research in the field shows that, in addition to the sugars, bee honey also contains valuable compounds in small quantities, namely: vitamins (B1, B2, B6, C, K, A, D, E, K, etc.), proteins (an average of 0.5% in the case of floral honey and higher doses in the composition of manna honey), dextrins, coloring and odorous substances, organic acids (malic, pantothenic, citric, oxalic, lactic, succinic, etc.), trace elements (calcium, iron, potassium, silver, nickel, beryllium, etc.) and others [9,12,13]. Because of this very complex composition and valuable compound richness, bee honey is used in its natural state [14,15] or incorporated in food matrices to improve sensorial [16], functional and physicochemical properties [17]. The antimicrobial [18,19], antioxidant [20–22], antiviral [23] and anti-inflammatory action [24] of different types of honey is well known; the research has also demonstrated the anticarcinogenic activity of honey [25].

The addition of food ingredients is prohibited in honey, including food additives, as well as any other addition. Similarly, the removal of any of the natural components of honey is prohibited, including pollen, unless such removal cannot be avoided [26]. The authenticity of honey can be defined as the precise identification of its botanical and geographic origin in relation to its unique composition and properties [27]. Honey adulteration has generally been identified with trained sensory evaluation based on viscosity, color, taste and aroma The testers in such sensory evaluations require long-term professional training to ensure the reliability of the results. However, these methods are mostly subjective and prone to human error. Beyond the sensory assessment, methods described in the literature include microscopy, immunoassays, physico-chemical analyses, chromatography, mass spectrometry, near-infrared spectroscopy, Raman spectroscopy, DNA transcoding and enzyme-linked immunosorbent assay [28].

In general, the quality of honey and its chemical composition are related to many factors, such as its geographic, botanical or plant origin, climate and seasonality; other factors could be external, such as environmental factors, processing methods of honey by beekeepers, storage conditions and deliberate adulteration by producers [29]. The quality of honey can also be characterized by its purity; the purity of honey can be determined by its physico-chemical properties, i.e., moisture content, pH, free acidity, total soluble solids, sugar content, color intensity, 5-hydroxymethylfurfural (5-HMF) content and amino acid content [30]. Heavy metals can accumulate in the honey, from the soil [31] or from the air [32]. Therefore, the evaluation of physico-chemical properties and certification of botanical origin is vital to determine the quality and authenticity of honey. These factors may affect market price and consumer acceptance. Consumers' safety and their protection from fraud is the overall goal [33].

For production on an industrial scale, raw bee honey has to be extracted from the honeycombs and then processed. Due to their natural characteristics, all varieties of flower honey, except acacia honey, crystallize shortly after extraction from honeycombs; crystallization of flower honey is a natural process and is due to the fact that most varieties of honey are supersaturated solutions of sugars (at an average temperature of 20 °C) [34]. The honey crystallization degree is dependent on the absolute pollen count [35]. Crystallization does not alter the biological and medicinal properties of the product and should not be considered as a defect, but rather as a guarantee of its authenticity [36]. However, many consumers prefer fluid honey and have reservations about buying and consuming crystallized honey [37]. In order to meet the demands of consumers, many solutions are used to obtain decrystallized honey and to keep it in this state for as long as possible; on the other hand, it is very important for the technologists to bring honey in a liquid state through liquefaction for further processing [34]. After liquefaction, technologies such as ultrasonication, microwave and infrared irradiation or heating are used to eliminate water in excess and to sterilize the raw bee honey, to continue honey decrystallisation and to temper it for further processing steps such as filtration or mixing with other compounds [38].

Many research papers and reviews indicate the difficulties in analyzing the great amount of results generated by the literature, but also the disparity of the results, due to different methods used to appreciate honey quality [9,19,22,39–42]. The progress of methods and techniques used, the multivariate methods applied to analyze honey, indicates chemometrics as one of the most adequate solutions for large quantities of data evaluation and interpretation [43]. Nowadays, the field of chemometrics is well-applied in the fields of honey authenticity and quality control. There are numerous studies published in the last few years, e.g., between 2017 and 2021 combining chemometrics with well-established analytical methods in honey authentication and quality control [44–64]. Based on chemometrics,

Bruker has developed the largest database containing the spectroscopic ^{1}H NMR signatures of 28,000 reference honeys, monofloral and polyfloral types from more than 50 countries, aiming at the botanical and geographical determination or adulteration of honeys [65]. This review aims to present the state-of-the-art in using chemometrics in analyzing honey quality during processing and storage; the advantages of using chemometrics are presented extensively in the next sections, and also the main characteristics of some well-known chemometric methods. From our knowledge, no review on the chemometric methods used for assessing honey quality during processing and storage has been published and this is the most original contribution of this publication.

2. Methodology and Design

The literature search was conducted in order to identify the most relevant research articles and reviews illustrating the state-of-the art in using chemometrics in analyzing bee honey quality during processing and storage. The online databases queried were Web of Science, ScienceDirect, Google scholar, MDPI, ResearchGate, PubMed, Scopus and the Wiley Online Library. The keywords used individually and combined on the searching engines were: honey, quality parameters, chemometrics, storage, processing, thermal treatment. By selecting the publications to be included, a similar pathway as that in [66] was used. The criteria were: (i) research that focused on the use of chemometric methods in analyzing honey quality and changes/adulteration during storage and processing; (ii) papers written in English; (iii) papers with accessible full text (in some cases after requests from the authors).

After a short introductory discussion on honey as food and its properties, we discuss the aspects related to storage and the two main processing steps—dewatering and thermal treatment—and their influence on the honey quality (Section 3). Then, we present the main chemometric methods and the quality parameters of honey mostly investigated by using chemometrics and we intensively analyze the main publications applying chemometrics in honey processing and storage (Section 4).

The review includes mainly publications from the period 2010–2022 (the last 12 years), together with a few older publications considered as being very important for this research.

3. Honey Processing and Storage

After the extraction from the honeycomb, honey contains pollen, beeswax and other materials, and these impurities have to be removed from the preheated honey by straining and coarse filtration. Dewatering, liquefaction and filtration are the main operations in processing honey [67] to assure its commercial quality and these processes will be discussed here, together with storage before or after processing. The product can be bottled as it is or can be enriched with other different ingredients, including: (a) bioactive compounds such as propolis or bee bread [68,69], coumarin [70] or spirulina [71]; (b) dried fruits [72–74]; (c) flavored materials usually accepted by the consumers, such as cocoa or cinnamon [75,76], to obtain novel foods. Figure 1 presents the main industrial treatments of the raw honey received from the primary suppliers and/or beekeepers.

According to the EU Regulations, the compositional criteria of honey are: moisture content (M), simple sugars (fructose + glucose and sucrose), water-insoluble solids, electrical conductivity, free acids, 5-hydroxymethylfurfural (HMF) content and diastase value (the last two criteria in the processed honey) [39,77] together with melissopalynological analysis and other physical–chemical parameters (such as pH, proline content, enzymatic activity and ash content) and with sensorial characteristics (such as color and volatile profile), they are the classical tools to provide important information on the honey's origin and authenticity [78]. Other important parameters, especially for honey storage and processing, are: water activity (a_w) and rheological behavior [79,80]. HMF and M can be used to classify blended honey samples according to their crystallization degrees [35]. According to the European and international norms, a special importance must be given to the knowledge of these quality parameters, but also to the causes that determine changes in their initial values [36,81,82].

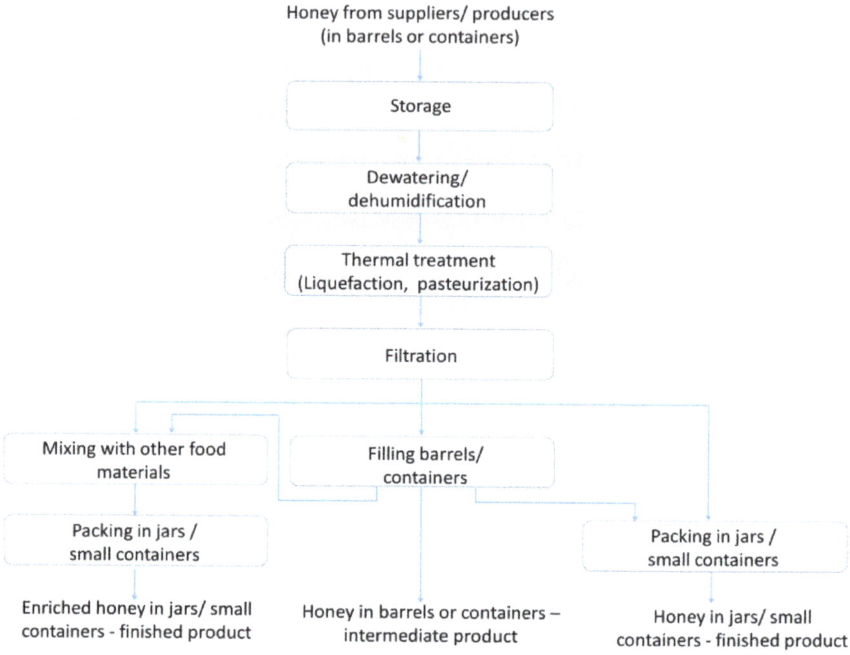

Figure 1. Steps in processing honey from suppliers or producers, highlighting the main operations where modifications can occur.

3.1. Influence of Storage Conditions on Raw Honey Quality

After the extraction from the comb, bee honey could be preserved for a long time (from months to years). The storage conditions will influence honey quality; refrigeration temperatures lead to a darkening of the color, whereas storage at room temperatures and higher (28 °C) results in modification of the acidity and an increase in HMF and the microbiota in honeys, together with a small modification of the composition in monosaccharides [83] and phenols [84]. Although low temperatures do not favor honey crystallization, its refrigeration determines the internal reorganization of the water molecules with increased fermentation risk when the honey returns to room temperature [85].

There is a negative correlation between HMF content and diastase (amylase) activity during storage; a longer storage time increases the HMF content, with a reduction in the enzymatic activity [83]. The storage of honey at higher temperatures (35 °C) for a longer time (from 6 to 9 months) increases the HMF content to values exceeding those accepted by the legislation [86]. A deeper discussion on these parameters is given in Section 4.2. of this review.

The storage conditions influence the antibacterial properties of honey. Besides propolis, two antibacterial compounds are the most investigated in honey, methylglyoxal, the compound responsible for the antimicrobial action of Manuka honey [87] and hydrogen peroxide obtained from glucose degradation by glucose oxidase in bee honey [88,89]. It should be mentioned that the peroxidase activity of honey is lost during storage, together with the exposure to light and heating [90–92].

The aroma compounds in honey are a complex mix of volatile components with various functions and relatively low molecular weight [93]. The storage results in reduction in its odorous compounds, especially at higher temperatures; temperatures near to refrigeration maintain the volatile compounds in honey [94].

The honey properties during storage seem to be more dependent on the honey type than the storage conditions [95].

3.2. Influence of Storage Conditions on Raw Honey Quality

Dewatering and dehumidification refer both to the Moisture (M); M can be removed from honey through dewatering or through dehumidification from air in contact with honey. The acceptable range of moisture content is 16.4–20.0% [96]; higher moisture can cause honey fermentation by osmotolerant yeasts [97] and favors faster crystallization [98]. Table 1 presents some methods used for M removal.

Table 1. Honey dewatering and dehumidification methods used in the last 12 years.

Method	Device and Conditions	Reference
Use of dry air	Heating in a desiccant honey dehydrator (with silica gel desiccant bed) with dehumidified air at 35 °C or 45 °C	[99]
	Heating of containers (having hot-water jacket) combined with treatment with dried air (until 40 °C)	[100]
	Desiccant-bed silica gel heating and drying the air, with recirculation at 40–55 °C up to 36 h	[101]
	Dehydrator system with control of temperature, drying air speed, relative humidity and honey exposure surface	[102]
	Vertical centrifugal honey-dehydrator with an external electric heat source and a closed air circuit and heat pump	[103]
Vacuum drying	Use of Low Temperature Vacuum Drying (LTVD) (30 °C) with induced nucleation technique	[104]
	Ultrasonic vacuum drying at 40 kHz 80 W	[105]

3.3. Honey Thermal Treatment

The thermal treatment of honey involves two main operations, liquefaction and pasteurization. Liquefaction is realized for the honey decrystallization, by using different methods. The classical liquefaction is conducted by heating honey at temperatures around 50 °C (45 to maximum 60 °C) for a long time (12 h), with sensorial changes such as: darkening, contracting the taste of caramel and weakening or even the disappearance of the specific flavor [106]; the antioxidant activity and the phenolic compounds seem to also become modified [106,107].

Another method used is microwave heating, with the increase in HMF concentration and decrease in antioxidant properties as disadvantages [108]. Novel methods such as ultrasounds or high-pressure treatments seem to be effective in destroying/minimizing the crystals or in increasing the period before starting crystallization and they have the advantage of being shorter [109–111]. Higher ultrasound power input lowers the liquefaction time (in minutes) [112]. Ultrasounds are beneficial for the extraction of volatile and semi-volatile compounds [113], and they do not affect the physical–chemical parameters (moisture, pH, diastase activity, HMF content) and sensorial characteristics (color) [114,115]; ultrasounds have antimicrobial activity, too, depending on the pathogens and honey types [116–118]. Studies have shown that the high-pressure treatment improves the nutritional and antimicrobial characteristics of honey [110,119].

Liquefied honey can be pasteurized at higher temperatures (around 80–90 °C) for seconds to destroy microorganisms (especially yeasts, responsible for fermentation); a secondary effect is the moisture reduction and the delaying of crystallization [120]. The non-thermal treatments such as ultrasounds [121] or irradiation [122] are promising techniques for destroying the microorganisms in honey. In addition to the classical way of heating the water for obtaining the temperatures required for the thermal treatment (liquefaction or pasteurization), a new trend is to use geothermal water as a sustainable resource [123,124].

4. Chemometrics Used in Honey Quality Analysis during Storage and Processing

4.1. Introduction to Chemometrics in Honey Quality Analysis

Chemometrics means the use of statistical and in recent times Artificial Intelligence (AI) methods to characterize and classify samples based on large quantities of analytical data. The most frequent applications are the identification of the class of a sample and the prediction of its properties not covered by the analysis [125]. The methods can be divided into unsupervised and supervised ones.

4.1.1. Unsupervised Chemometric Methods

The unsupervised methods do not require a training set, and they group the samples based on the similarities between them; a new sample will be placed in the existing groups or possibly will create a new group. These methods are suited for preliminary, exploratory and qualitative analysis especially when no prior classifying information is available. Examples of such methods are:

- ANOVA or ANalysis Of VAriance is used to compare statistical populations in order to decide if there are statistically significant differences between them. Its use has become a standard requirement for proving the soundness and validity of a research hypothesis. In the context of chemometrics, ANOVA is used to investigate the effect of independent variables on the dependent variable. If multiple dependent variables are of interest then a Multivariate ANOVA (MANOVA) is performed [126];
- Cluster Analysis (CA) which groups samples in clusters with the most used being:
  - o HCA or Hierarchical Cluster Analysis [127], which uses distance-based methods to group the data in hierarchical clusters and to place a new sample in this hierarchy;
  - o K-means clustering, which is a non-hierarchical clustering of data in k clusters.

Examples of the use of clustering techniques in honey chemometrics can be found in [128,129]. In these individual pieces of research, polycyclic aromatic hydrocarbons (PAHs) were identified in honey by using ultrasound-vortex-assisted dispersive liquid–liquid micro-extraction followed by a triple quadrupole gas chromatograph/mass spectrometer (DLLME-GC-MS). The dataset was then grouped using k-Mean cluster analysis and PCA in order to identify geographic specific pollution [128] or human activity specific pollution patterns [129].

- Principal Component Analysis (PCA) is used to reduce the dimensionality of a sample space when many features are investigated for many samples; they are plotted in a reduced space where the axes are combinations of the features chosen so that the relations between them (distances) are preserved [130]. PCA principal use is for visualization and qualitative analysis and it needs the use of a secondary method—usually a supervised Discriminant Analysis (DA) method—for classification. In [128,129], PCA is used in combination with k-Mean cluster analysis to visualize the grouping of pollutants based on geographical location [127] or human activity [128]. PCA was used by [131] to cluster honey types based on the data expressing the content in vi-tamin B2 and Cu and the antioxidant activity measured by 2,2′-azino-bis(3-ethylbenzothiazoline-6-sulfonic acid (ABTS) [132] and CUPric Reducing Antioxidant Capacity (CUPRAC) [133] values. The grouping allows the identification of the botanical origin:
  - o An extension to multiple dimensions of PCA is the PARAllel FACtor analysis (PARAFAC) [134–136] which can be used on multiway spectral data. It is employed in [134] where fluorescence spectrometry data are first decomposed with PARAFAC in order to identify the representative patterns in honey. An improvement of the traditional PARAFAC specifically for use on chromatography data is the alternating trilinear decomposition algorithm (ATLD) [137].

ATLD can be used to decompose the HPLC data in order to evidence the data related to the phenolic components used as markers; the quantitative data can be subjected to PCA analysis to visualize the clustering potential of the chosen markers in honey [138].

Multiple unsupervised methods are used in some research for exploratory analysis. One example is the research presented in [139], where one way ANOVA, CA and PCA are used on IPC-MS data to explore the relation between concentration in thirty nine elements and the geographical origin of Bracatinga honeydew [139].

4.1.2. Supervised Chemometric Methods

The supervised methods need an initial set of classified or labeled data to adjust the parameters (to train) of a model which is then used as a predictor. The trained predictor can classify new samples in the existing classes. The methods used for food product and in particular for honey characterization can be grouped in:

- DA methods which use the observations of a number of variables for each sample for the separation of samples of the training set in groups and for the allocation of new (test) samples in these groups [140]. DA methods can be grouped after the type of relation used in:
    - Linear Discriminant Analysis (LDA) which builds a discriminator function as a linear combination of the independent variables. It is a common technique used to build predictors for the botanical and geographical origins of honey based on their composition. One recent example is given in [139], that used LDA to develop a predictor for the geographical origin of Bracatinga honeydew honey based on IPC-MS data.

Measurement interferences can make the separation of spectral data harder. In order to facilitate the classification of spectral data, the results of measurements are preprocessed with different mathematical techniques. The techniques mostly employed are Multiplicative Scattering Correction (MSC), normalization, Standard Normal Variate (SNV) transformation, De-Trending (DT) baseline correction, first- and second-order derivatives [141]. Light scattering present in the spectra of food is greatly reduced by MSC and SNV. By applying the first and second derivatives, the baseline and random noise are reduced. The same treatment can be used to preserve the information for quantitative determinations. Individual spectra analysis is enhanced by the use of DT. The baseline shift and curvilinearity variations can be also handled by using DT [142].

Fourier Transform Infrared (FTIR) spectral data are used for the prediction of monofloral honey type and of honey's physico-chemical properties [143]; the preprocessing techniques investigated are MSC, MSC + first derivative, MSC + second derivative, SNV, SNV + DT, SNV + first derivative, SNV + second derivative, first derivative and second derivative. They are compared with no treatment of the spectra. PCA is used for the study of the samples grouping according to their botanical origin. The study shows a distinct grouping for mint, rape, acacia, tilia and sunflower honeys, and mixing for thyme and raspberry honeys. An LDA-based classifier is then developed to reliably identify the link of the sample to one of the seven honey groups. The best results are obtained with the MSC pretreatment. For each of the physico-chemical parameters of interest for the honey quality, i.e., M, pH, electrical conductivity (EC), free acidity, HMF, fructose, glucose and sucrose contents, an individual PLS-R-based predictor is developed. Each predictor uses data from a specific spectral band. SNV combined with the first derivative is the best pretreatment method for predicting pH, electrical conductivity, free acidity, 5-HMF, fructose, glucose and sucrose. The first derivative is the best data pretreatment for the prediction of moisture content [143]:

- Stepwise Linear Discriminant Analysis (SLDA) uses a stepwise inclusion of the independent variables in the model [144].

PCA, MANOVA and SLDA are used to build a correlation between the volatile compounds' composition measured by headspace solid-phase microextraction coupled to gas chromatography/mass spectrometry (HS-SPME/GC-MS) and the geographical provenance of Greek Quercus ilex honey [145]. MANOVA is used to identify the significant VOCs, PCA to construct the groups and SLDA to construct a predictor for the origin of a new sample [145].

- Partial Least Square (PLS) methods are regression-type methods. In opposition to the Ordinary Least Squares (OLS), where all independent variables are used, in PLS a smaller number of uncorrelated components are generated from the independent variables in a similar fashion to PCA [146]. Some examples of using these components for regression in honey analysis are:
    - Simple Partial Least Square Regression (PLSR) [147] is used on the HPLC data decomposed by ATLD to correlate the content in the phenolic compounds used as markers for the honey's antioxidant activity [28];
    - Partial Least Square—Discriminant Analysis (PLS-DA) is a combination between PLS and DA, used when categorical results are needed [148]. The influence of different preprocessing steps (autoscale, variance (std) scaling, min–max scaling, class centroid centering and scaling, smoothing, SNV and Pareto) on the accuracy of a PLS DA predictor for the geographical and botanical origin of honey, is analyzed by [149]. The predictor uses ^{1}H NMR spectra data. A first pretreatment step is the reduction in the data by replacing each six consecutive chemical shifts with their mean. For the geographical origin, identification of the highest accuracy is obtained through autoscale, variance (std) scaling and class centroid centering and scaling. For the botanical origin, the highest accuracy is obtained through the variance (std) scaling data pre-treatment [149];
    - Unfolded PLS-DA UPLS-DA combines unfolded PLS [150] which decompose the sample spectra to extract the relevant information with DA;
    - Multilinear PLS-DA MPLS-DA combines multilinear PLS [151,152] which can use multidimensional data as input with DA.

The suitability of different DA techniques for building adulterant predictors based on fluorescence spectrometry data are compared with PARAFAC [134]. PLS-DA, UPLS-DA and NPLS-DA are used to build classification models. The UPLS-DA model performs the best and the PLS-DA performs the worst [134]:

- Linear discriminant analysis based on partial least-squares (PLS-LDA) in which LDA is performed using PLS as the reduction step [153];
- Orthogonal projections to latent structures discriminant analysis (OPLS-DA) combines Orthogonal projections to latent structures (OPLS), which separates the independent variables into predictive and uncorrelated variables, with DA for a categorical response [154];

- ^{1}H NMR spectra of Chinese honey samples are used to identify adulterated honey. A PCA LDA discriminator and an OPLS-DA one were built, trained, validated and tested. The OPLS-DA has a slightly better accuracy. The OPLS-DA also helped to identify a set of substances with significantly different concentrations in altered and unaltered honey that can be used as a marker for adulteration [155];
- PCA and OPLS-DA on proteomics data obtained with sequential window acquisition of all theoretical fragment ion mass spectra (SWATH-MS) to develop a predictor for honey adulteration, the producing region (Tainan, Changhua, and Taichung), country (Taiwan and Thailand) and botanical sources (longan and litchi) [138]:
    - Orthogonalized partial least squares coupled with linear discriminant analysis (SO-PLS-LDA) is a multi-block discriminant classifier that results from the combination of LDA with the sequential and orthogonalized-partial least squares method (SO-PLS) which is a multi-block regression method [156,157].

A method for detection of honey alteration after heat treatment (4 h at 80 °C) is presented in [158]. The data are obtained through differential pulse voltammetry using three types of Natural Deep Eutectic Solvents (NADES) buffers and a normal buffer with the multiple wells screen-printed carbon electrodes. The data for each type of buffer were first used individually for developing a PLS DA classification models. With the fused data from the four sensors, a multiblock classifier based on SO-PLS-LDA with very good accuracy is developed;

- k-Nearest Neighbors method (kNN) classifies the sample based on the classes of the k-nearest neighbors [159];
- Soft Independent Modeling by Class Analogy Method (SIMCA) uses PCA on the samples of the training set for the construction of the classification models [160].

The correlation between honey sample fluorescence spectra and their botanical (acacia, linden, honeydew, colza, sunflower, chestnut, lavender) and geographical (Romania and France) origin [161] is investigated. The PARAFAC was used to decompose and visualize the fluorescence excitation emission matrices to identify the features that allow classification. On the basis of the obtained PCs, a SIMCA model was developed that predicts the geographical and botanical origin [161].

A comparison between two analytical methods, MIR spectroscopy and Matrix-Assisted Laser Desorption Ionization (MALDI)-Time of Flight (ToF)-based MS (MALDI-ToF-MS) and between multiple chemometric methods PCA-LDA, PCA-kNN and SIMCA for the identification of the botanical origin of the honey is presented [162]. The datasets from each of the analytical methods were used for the development of predictors with each of the enumerated chemometric methods. The SIMCA method has proven better able to identify both monofloral and multifloral honeys due to its "soft" classifier which can identify a sample as an outlier or belonging to multiple groups as opposed to the hard classifier of the other methods which classify a sample strictly in a single existing group. The MIR method due to its higher reproducibility, lower demand for manual labor and laboratory infrastructure is more suitable for implementing an automatic authentication method based on a cloud-located database that can be used worldwide [162].

- Support Vector Machine methods (SVM) use the training set to construct the hyperplane that separates the classes with the largest margin [163]. Support vector machine can be used for regression (SVR) [163] or for classification (SVC) [164]. The Least Squared-Support Vector Machines (LS-SVM) [165] are improved variants.

The performances of four methods for building predictors that use Attenuated Total Reflectance-Fourier Transform Infrared (ATR–FTIR) spectral data are compared to identify adulteration with rice syrup in honey from three different botanical origins (acacia, linden and jujube) [166]. The four used methods are PLS-DA, Der-PLS-DA, LS-SVM and Der-LS-SVM. The methods noted with Der implement a derivative-based pre-processing for removing spectra baseline offsets before modeling.

Hyperspectral imaging combines the spectral information from spectroscopy with spatial information from digital imaging [167]. The data from VIS-NIR hyperspectral imaging are used to develop a machine learning-based classifier for the botanical origin of the honey. The developed classifiers combines two classification methods, SVC and kNN, and attains an accuracy of 91% for close set and 80% for open set cases [168].

A linear classifier based on PLS-DA and a nonlinear classifier based on SVM for the species of honey-producing insects are presented in [169]. The data from the GCMS of honey samples were first investigated with PCA to identify the physico-chemical parameters and volatile compounds relevant for species' determination. The contribution of the parameters to the clustering of the samples were identified using HCA. Both models showed good performances [169].

A new gold nanoparticle sensor array that changes the color in contact with the volatile components in honey was developed [170]. The resulting images from the analysis of honey samples from three different botanical species (acacia, canola, honeydew) are used as training set for three supervised methods LDA, PLS-DA and SVM. The best accuracy in identifying a new sample was obtained with SVM [170];

- Artificial Neural Networks (ANN) are universal approximators that mimic the functioning of biological neurons [171]. Convolutional neural networks (CNN) are ANN in which the connectivity is inspired by the animal visual cortex.

The approach presented in [172] uses data from Raman spectra of pure glucose maltose, fructose and sucrose solution and from commercial honeys to train ANN to identify the honey concentrations in those sugars. Multiple network structures were mentioned. This indicates the possibility of developing a non-invasive, rapid, automatable and cost-effective method for honey-sugar composition analysis [172].

Physico-chemical (moisture, fructose, glucose and sucrose content) and rheological (loss modulus, elastic modulus, complex viscosity, shear storage compliance and shear loss compliance) parameters of the honey samples are used for the identification of their botanical origin. Two predictors were developed, one based on LDA and the other based on ANN, both with good accuracy. A rheological properties' predictor based on the ANN that uses the physico-chemical parameters of honey was also developed [173].

A deep convolutional neural-network was used to classify honey based on the results from a multi-electrode differential pulse voltammetry. The honeys were grouped into 12 classes by PCA and then labeling for the classification was completed on the basis of the type and time of harvesting of the honeys [174].

In [175], the detection of exogenous sugars in honey based on the analysis of the Raman spectra of honey samples was investigated. PLS-DA, PCA-LDA and kNN were used to construct predictors for the quantity and type of multiple adulterants. The study was then extended with an exploratory analysis using PCA and t-distributed stochastic neighbor embedding (tSNE) [176] to correlate the results from Raman spectroscopy with the botanical origin and the possible adulteration. tSNE proved better at separating the adulterations quantitatively. Based on the exploratory analysis, predictors for both the botanical origin and the quantity of adulterants were developed on the basis of CNN, SVR and PLSR. The CNN predictor proved the best in handling the nonlinearities and multimodali-ties of the spectra followed by SVR, and the PLSR was the least accurate. Table 2 presents the chemometrics used in the honey quality analysis.

4.2. Modification of the Quality Parameters Used for Quality Evaluation of Honey during Processing as Analyzed by Chemometrics

The quality of honey is changed during processing and/or storage. Important physico-chemical parameters have been proposed for honey evaluation and its conformity within stipulated legislated limits [77,96], such as: water content, main sugar content (glucose + fructose), disaccharide sucrose, free acidity, diastase activity, electric conductivity, ash content, 5-hydroxymethylfurfural and water-insoluble content. Furthermore, additional attributes, associated with the basic ones above, are evaluated such as total amino acids, color parameters L*, b*, a*, amino acids composition, proline content, minerals, composition in di- and oligosaccharides, phenolic compounds, phenolic content, volatile compounds, viscosity and rheological properties, crystallization, total protein, nitrogen content and sensorial attributes. Major modifications of the physico-chemical parameters and related features by the applied changes regarding the quality of honey changes during processing/storage are discussed below in comparison with those for untreated honeys.

Table 2. Chemometrics used in the honey quality analysis.

Chemometrics	Group Known a Priori	Independent Variables	Dependent Variable	Source Data	Output	Reference
ANOVA MANOVA	no	Categorical	Continuous	IPC-MS	Geographical origin	[139]
				(HS-SPME/GC-MS)	Significant VOC for geographical origin	[145]
PARAFAC		Categorical	Continuous	Fluorescence spectrometry data	Adulterants	[134]
CA	no			DLLME-GC-MS	Geographic grouping of pollutants in honey	[128]
				DLLME-GC-MS	Anthropic grouping of pollutants in honey	[129]
				IPC-MS	Geographical origin	[139]
				GC-MS	Relevant physico-chemical parameters and VOC	[169]
PCA	no	Continuous	Continuous	DLLME-GC-MS	Geographic grouping of pollutants in honey	[128]
				Vitamin B2 and Cu, antioxidant activity	Honey type	[131]
				HPLC data + ATLD	Honey type	[138]
				IPC-MS	Geographical origin	[139]
				(HS-SPME/GC-MS)	Significant groups of geographical origin	[145]
				SWATH-MS	Honey adulteration, geographical and botanical origin	[138]
				GC-MS	Relevant physico-chemical parameters and VOC	[169]
				Raman spectra tSNE	botanical origin and the quantity of adulterants	[177]
LDA	yes	Continuous	Categorical	IPC-MS	Geographical origin	[139]
				FTIR spectra with pre-processing	Botanical origin	[143]
				(HS-SPME/GC-MS)	Geographical origin	[145]
				Fluorescence spectrometry data	Adulterants	[134]
				PCA of MALDI-ToF-MS	Botanical origin	[162]
				PCA of MIR spectra	Botanical origin	[162]
				Gold nanoparticle sensor array	Botanical origin	[170]

Table 2. *Cont.*

Chemometrics	Group Known a Priori	Independent Variables	Dependent Variable	Source Data	Output	Reference
				Physico-chemical and rheological parameters	Botanical origin	[173]
				PCA of Raman spectra	Quantity and type of multiple adulterants	[175]
PLS	yes	Continuous	Continuous	FTIR spectra with pre-processing	pH, electrical conductivity, free acidity, 5-HMF, fructose, glucose and sucrose	[143]
				Raman spectra	Botanical origin and the quantity of adulterants	[177]
PLS-DA	yes	Continuous	Categorical	^{1}H NMR spectra with pre-processing	Geographical and botanical origin	[149]
				^{1}H NMR spectra	Honey adulteration	[155]
				SWATH-MS	Honey adulteration, geographical and botanical origin	[138]
				Differential pulse voltammetry using NADES	Alteration after heat treatment	[158]
				ATR–FTIR spectra with pre-treatment	Adulteration and botanical origin	[166]
				GCMS	Biological origin	[169]
				Gold nanoparticle sensor array	Botanical origin	[170]
				Raman spectra	Quantity and type of multiple adulterants	[175]
kNN	yes	Categorical	Categorical	PCA of MALDI-ToF-MS	Botanical origin	[162]
				PCA of MIR spectra	Botanical origin	[162]
				Raman spectra	Quantity and type of multiple adulterants	[175]
SIMCA	yes	Continous	Categorical	PARAFAC of fluorescence spectra	Geographical and botanical origin	[161]
				MALDI-ToF-MS	Botanical origin	[162]
				MIR spectra	Botanical origin	[162]

Table 2. *Cont.*

Chemometrics	Group Known a Priori	Independent Variables	Dependent Variable	Source Data	Output	Reference
SVM	yes	Continuous	Continuous	ATR–FTIR spectra with pre-treatment	Adulteration and botanical origin	[166]
				GCMS	Biological origin	[169]
				Gold nanoparticle sensor array	Botanical origin	[170]
				PCA and tSNE of Raman spectra	Botanical origin and the quantity of adulterants	[175]
ANN	yes	Continuous	Continuous	Raman spectra	Maltose, fructose and sucrose content	[172]
				Physico-chemical and rheological parameters	Botanical origin	[173]
				Raman spectra	Botanical origin and the quantity of adulterants	[175]

Abbreviations: ANOVA = analysis of variance; MANOVA = multivariate analysis of variance; PARAFAC = parallel factor analysis; HCA = hierarchical cluster analysis; CA = cluster analysis; LDA = linear discriminant analysis; PCA = principal component analysis; PLS = partial least square; PLS-DA = partial least square-discriminant analysis; kNN = k–nearest neighbor; SIMCA = soft independent modeling of class analogy; SVM = support vector machine; ANN = artificial neural network; IPC-MS = inductively coupled plasma-mass spectrometry; HS-SPME/GC-MS = headspace-solid phase microextraction-gas chromatography-mass spectrometry; DLLME-GC-MS = dispersive liquid-liquid microextraction-gas chromatography mass-spectrometry; GCMS = gas chromatography mass spectrometry; HPLC data + ATLD = high-performance liquid chromatography data + alternating trilinear decomposition; SWATH-MS = Sequential Window Acquisition of All Theoretical Mass Spectra; tSNE = t-distributed stochastic neighbor embedding (algorithm); MALDI-ToF-MS = matrix-assisted laser desorption ionization time-of-flight mass spectrometry; MIR = mid-infrared, FTIR = Fourier transform infrared; ^{1}H NMR = proton nuclear magnetic resonance; NADES = natural deep eutectic solvents; ATR–FTIR = attenuated total reflection - Fourier transform infrared.

4.2.1. Free Acidity

Free acidity (FA) originates from organic acids, being in equilibrium with their internal esters lactones and inorganic ions such as phosphates, sulfates, nitrates and chlorides, capable of generating their conjugated acids [36], while its determination constitutes one of the basic parameters ($\leq$50 meq/kg of honey) stipulating limits for assuring the protection of the product's quality from microbial activity [77]. FA is attributed to:

(a) Organic acids. Acidity is mainly derived from the presence of organic acids, up to 0.5% in honeys, contributing to honey flavor, stability against microorganisms, enhancement of chemical reactions and antibacterial and antioxidant activities [178]. The principal organic acid in honey is gluconic acid derived from the activity of the glucose-oxidase enzyme on the glucose substrate, that is in equilibrium with δ-gluconolactone [179–182]. The gluconic acid level, for a specific honey species, is mostly dependent on the time elapsed between the collection of nectar and formation of the final honey by bees for obtaining the final density in the honeycomb cells, while glucose–oxidase activity becomes insignificant when the honey is thickened [183]. Moreover, other organic acids are found in honey such as formic, aspartic, acetic, butyric, citric, fumaric, galacturonic, gluconic, glutamic, butyric, glutaric, 2-hydroxybutyric, glyoxylic, α -hydroxyglutaric, lactic, isocitric, α-ketoglutaric, malic, 2-oxopentanoic, malonic, methylmalonic, propionic, pyruvic, quinic, shikimic, succinic, tartaric, oxalic acid and others [184]; their ratio and abundance are influenced by the honey species enabling discrimination of the honeys [179,185], while some organic acids have exhibited a high discriminant power for the separation of conventional from organic honeys [186];

(b) Lactones. Lactones found in honey are mostly in the form of gluconolactones, constituting part of the organic acids in the intra-esterified form; they contribute a reserved acidity measured when the honey solution becomes alkaline [36]; lactonic acidity is added to FA to yield the total acidity of honey [77]. The pH of honey and its acidity are not parameters directly related to each other because many other components found in honey exert a buffering capacity, therefore, compensating for a part of honey's true acidity [187,188]. Similarly to pH, free and lactonic acidity in the different honeys are dependent on their botanical origin, also influenced by the harvesting season [178,183,187,189–191].

During storage time, the amount of organic acids significantly increases [192]. Several factors contribute to honey FA:

(a) Effect of maturation. During honey maturation, the FA or total acidity is increased while pH is significantly decreased [178]. In a pioneering study covering the introduction of national legislative limits for Talh honey, the free acidity (FA) of Talh honey was determined from Talh tree leaves and flowers (30 $\pm$ 0.99; 34 $\pm$ 0.92 meq/kg) to bee crop (honey stomach) and unripe honey (43 $\pm$ 1.80; 72 $\pm$ 1.56 meq/kg) and finally to ripe honey (77 $\pm$ 1.28 meq/kg), [193], while the highest pH value was recorded in the leaves and kept decreasing as honey production proceeded, obtaining its lowest value in ripe honey (4.91 $\pm$ 0.06);

(b) Effect of storage. Reports have shown a significant effect of storage on honey FA, pH, ($p < 0.05$), with FA increasing and pH decreasing with storage time [181,194,195]. In one kinetic study, exclusively dedicated to the variability of all the three parameters versus 30 months storage for honey stored at room temperature (15–25 °C), lactonic acidity found to increase by storage time ($p < 0.05$), even at a higher degree than FA increased or pH decreased [196], while in some cases lactonic acidity was slightly decreased, and total acidity was increased [181]. Formation of levulinic and formic acids also is derived from 5-HMF transformation, and keep increasing by storage [197]. Evaluation of the variability of FA, pH, lactonic acidity, and total acidity has resulted in estimation of 20 months of storage to be the "best before" period "once opened" [196].

Investigation of the effect of short storage at 35–40 °C for 3 and 6 months with or without the addition of metabisulfites (12 pp) on water content (WC), pH, FA, lactone acidity and total acidity of two honeys, cashew and marmeleiro [198], showed that significant differences were observed for pH, FA, lactone acidity and total acidity compared to the respective parameters for the fresh samples. A reverse correlation between FA and lactone acidity was recorded and attributed to the glucose–oxidase activity that converts glucose to gluconolactone, which is consequently hydrolyzed to gluconic acid. In this study, FA is reduced but lactone acidity is increased with the storage time. The presence of bisulfite acted upon the esterification of gluconic acid to increase the lactone concentration [198];

(c) Effect of dehumidification. Dehumidification of honey in other studies has shown no differences in pH and FA between raw and dehumidified honeys when a group of samples from the stingless bees *H. itama*, *G. thoratica* and *T. apicalis* honeybee species were used. However, samples of *H. itamas* honey had a lower FA and higher pH and ash content values than *G. thoratica* honey samples [199], similarly to honeys of the other bee tribe [200];

(d) Effect of temperature/storage. Storage under different thermal conditions for times up to eight months induced a great increase in the free acidity of Talh honey, a rare type of honey because of its high FA. Talh honey naturally exceeds the permitted level for the FA values (>50 meq/kg) which is attributable to the plant origin. Storage temperature was found to be a factor with the highest significant influencing power on the FA ($p < 0.05$). Although all the values of FA in this study were beyond the standard limit, the results indicated that the stability of the FA of Talh honey was maintained stable at low temperatures (0–25 °C) for up to 6 months without significant effects [194]. In this study, statistical analysis showed the FA to exert a positive correlation with storage period (0.401), storage temperature (0.631), 5-HMF (0.852), color (0.541), moisture (0.440) and EC (0.155). On the other hand, FA was negatively correlated ($p < 0.05$) with glucose (−0.892), pH (−0.851), fructose (−0.821), sucrose (−0.422) and diastase activity (DN) (−0.309). Thus, low pH, DN and sugars are associated with higher FA. The strong positive correlation of FA with the 5-HMF is related to the strong effect of pH on the formation of furfurals generated more by the Amadori Rearrangement Products pathway than the routes of reductones and fission products dominant at pH > 7 [181,201].

4.2.2. Ash Content and Electric Conductivity (EC)

Ash content is indicative of the amount of minerals contained in honey [191]. There is a great dependency of ash content on the type of soil, climatic conditions and environmental pollution. The dissolved salts in the soil are pumped to the flowers used for nectar collected by the bees [31,32], enabling ash content to be a good indicator of the geographic origin of honey [49,202]. Mineral content is strongly correlated with the color and EC, affecting the color and flavor of the honey. Honeydew contains a higher quantity of minerals, therefore, it is commonly used in quality control to distinguish honeydew from floral honey [202,203], while citrus has a lower EC [204]. EC is influenced by the presence of salts, organic acids, minerals, amino acids, proteins, storage time and different sources [187]. The legislative limits require EC $\leq$ 0.8 mS/cm for all honeys, and ash content $\leq$0.6 in general and $\leq$0.1 for honeydew honey and its blends with blossom honey [77,96].

Storage and/or thermal treatment of honey results in an increasing effect of EC on honey [204]. Correlation analysis showed a strong positive correlation ($p < 0.05$) between EC and storage period, HMF and FA, whereas, a negative correlation ($p < 0.05$) was recorded between EC and sucrose, glucose and fructose [192,194,205,206].

4.2.3. Sugars

Carbohydrates are the main constituents of honey [207]. They are produced from nectar sucrose by honey-bees, which is transformed through the catalytic action of several enzymes, mainly α-, and β-glucosidase α- and β-amylase, and β-fructosidase, diastase and invertase, resulting in a composite mixture of monosaccharides, disaccharides and oligosaccharides [183,208]. Glucose and fructose monosaccharides constitute the major honey saccharides, ranging from 65% to 80% of the total soluble solids, followed by disaccharides and trisaccharides, while more than 26 sugars have been identified and quantified in honey [207]. The composition of the carbohydrate fraction of honey is strongly dependent on the plant species from which the nectar is collected by the bees, the bee species, maturation, with geographical and seasonal effect being negligible [183,209]. It has also been reported that bee species have a strong effect on the differentiation of honeys as has been revealed by NMR studies based on the different conformers of glucose and fructose contained in honey [210].

Fructose and glucose are the predominant sugars in honey, with fructose found in higher amounts except rape *(Brassisa napus)*, blue curls *(Trichoderma lanceolatum)* and dandelion *(Taraxacum officinale*) honeys, where glucose is in a higher quantity [98,207]. They are derived from the transformation of nectar's sucrose via the enzymatic action of invertase contained in the salivary glands of bees. However, invertase also owns transglucosylation activity, catalyzing the α-glucosylation of monosaccharides, disaccharides or trisaccharides in honey resulting in the formation of di- and tri-saccharides [203]. More than 30 saccharides have been identified, such as: (a) di-saccharides: sucrose or saccharose (predominant), maltose, turanose, cellobiose, kojibiose, maltulose, trehalose, nigerose, isomaltose, trehalulose, gentiobiose, laminaribiose, palatinose, gentiobiose, (b) tri-saccharides: erlose, theanderose, panose, maltotriose, 1-kestose, isomaltotriose, melezitose, isopanose, gentose, 3-α-isomaltosylglucos, planteose, (c) oligosaccharides: raffinose, isomaltotetraose, isomaltopentaose.

Enzymatic α-glycosylation of monosaccharides to di- and tri- saccharides starts once the mixture of nectar enriched with bees' saliva is placed in combs by the bees. A recent study reports that the acacia honey has ripened when the concentration of turanose is above 1.2 g/100 g honey, which occurs just after the combs are capped by the bees, timing before the 10th day of honey ripening in the honey combs [211]. A strong positive correlation has been recorded between the days of honey maturation in combs and the turanose content ($p < 005$) but a strong negative correlation with the water content as expected to occur during maturation in the beehive [211].

The amount of sucrose is dependent on its botanical source, honey maturity, elevated nectar abundance or artificial bee-feeding [59,183,212]. The steps in honey maturation involve an appreciable decrease in sucrose because of the continuous action of invertase added by the bee. Therefore, the maximum limit of sucrose (<5 g/100 g) is an indicator of freshness or possibly adulteration [134]. The evolution of main honey monosaccharides, glucose, fructose and disaccharides has been investigated by the fractionation of stable carbon isotope ($\delta^{13}C$) towards the examination steps for: (i) flowers, (ii) stamens, (iii) nectar and the (iii) ripened rape honey. It has been reported that $\delta^{13}C$ keeps increasing with the same order with these steps reaching a maximum for the ripened honey, which is significantly different than the $\delta^{13}C$ of the previous steps, a fact attributed to the addition of enzymes for sucrose inversion and the water evaporation caused by the bees' fanning with their wings [213].

During prolonged storage of honey, the amount of fructose, glucose and sucrose decrease. A remarkable decrease of 9% of monosaccharides per year was recorded during prolonged storage [107].

Reported studies for Tahl honey, which by its nature has FA values exceeding international limits, recorded a high decrease in saccharide concentrations especially for samples stored at 35 and 45 °C for different time intervals up to eight months at 0, 25, 35 and 45 °C [194]. Under all the storage temperatures, the sugars (fructose, glucose and sucrose) significantly decreased during the storage period with reducing sugars' amounts to reach levels below legislation limits (Fructose + Glucose >60 g/100 g) [96]. This declining trend for sugars is more pronounced at higher temperatures, 35 and 45 °C. Based on the results, temperatures below 25 °C are suggested for maintaining Talh honey. Statistical analysis showed higher Pearson's coefficients among parameters: positive correlations between sugars and pH and DN, but between sugars and storage period, temperature, color, EC, HMF, FA and moisture negative correlations were recorded.

Very important losses in the monosaccharides of citrus honeys were recorded during storage at different temperatures for 12 months reaching at 13.5, 25 and 25.2% for 10, 20 and 40 °C, respectively, compared to those of fresh honeys [195]. Sucrose is significantly decreased. However, marked increased changes were recorded for other disaccharides, such as nigerose, turanose, maltulose, isomaltose and kojibiose, whereas trisaccharides did not show any trend. In the same line, maltose with an initial amount of 2.5 mg/g changed to 23.2 mg/g after 1 year stored at 40 °C, predominating all the saccharides. Both the presence of invertase, which also acts as transglycosidase on glucose substrates, and the low pH values arising from the elongated storage enhance the conversion of monosaccharides to disaccharides and higher sugars, and count for sucrose decrease and maltose increase in the thermal-treated honeys [194,198,214–216]. High levels of maltose could be used as an indicator of the prolonged storage of honey [195].

Use of thermo-sonication for the honey dehydration processing led to a higher increase in 5-HMF compared to conventional thermal processing. Higher dehydration rates recorded for the thermos-sonication method than bath process count for higher formation rates of 5-HMF as an intermediate product of acid-catalyzed dehydration reaction of hexose and/or by Maillard reaction than conventional thermal treatment [217]. Although the rate of 5-HMF for conventionally heated honeys was the largest for acacia honey, during the processing by microwave the fastest formation of 5-HMF was recorded for lime honey [218]. The conventional and microwave heat processing induced the largest relative increase in the 5-HMF formation in honeydew honey while other studies have reported that during the microwave heating the most rapid increase in 5-HMF was found in lime honey; all these results indicate that the formation of the 5-HMF is dependent on the particular composition of each honey [108,218].

Chemical composition changes for sugars have been studied during thermal treatment and/or storage using FTIR spectroscopy as a tool for mining data [219]. Prominent changes have been recorded for the region 900–1500 cm^{-1} [220], with specific absorptions at 987 and 1040 cm^{-1} where maximum variances were recorded originating from carbohydrates absorption, mainly fructose, glucose, and sucrose [221]. Chemometric discriminating methods applied on FTIR data ranging from 600 to 4000 cm^{-1} discriminated between raw and thermally treated honeys. Furthermore, by using chemometrics the honeys were categorized for (a) 70 °C, for samples treated for 15 min and those for 120 min (overall accuracy 0.947%); (b) 40 °C, for samples treated for 15 min, and those for 120 min (overall accuracy 0.895%), leading to the conclusion that the model is successful for classifying the samples according to different thermal treatments based on the carbohydrate's changes.

4.2.4. HMF

Heating, dehydration, and storage processes play an important role in the formation of 5-HMF. This furanic compound originates from different chemical pathways occurring during honey processing (a) from dehydration of hexoses in acidic conditions [222], and (b) as an intermediate of the Maillard reaction [195,197,223].

The fact that the 5-HMF mostly reaches low concentrations in fresh honeys means that it is commonly employed as a quality parameter for assessing the freshness and/or overheating of honey. Legal regulation bodies have set a maximum 5-HMF content of 40 mg kg^{-1} for honeys in general, and 80 mg kg^{-1} for honeys from tropical climates, including blends [77]. Furthermore, there are safety concerns by consumers for foods of high 5-HMF content, such as DNA mutagenicity and colon carcinogenicity among others [224–226].

Honey freshness based on 5-HMF can be determined with several analytical protocols, however, an easy determination is executed using the ^{1}H NMR spectroscopy, where the integration of signals occurs at 4.65 ppm for the methyl protons or at 9.55 ppm for the proton of aldehyde, enabling its quantification [227].

Storage of cashew honey for up to six months resulted in the excess formation of 5-HMF at levels exceeding the legal limits. Addition of metabisulfite in honey at the start decreased the 5-HMF concentration, a fact attributed to the formation of sulfonic acids of the dehydro-reductone Maillard intermediary interrupting the cyclization prior to the formation of 5-HMF [198]. The 5-HMF can be produced from all hexoses, but actually is selectively derived from keto-hexoses, such as fructose because of (a) enolization of fructose proceeds with a comparatively higher rate than that of glucose because in solution fructose forms less stable ring structures enabling it to spend a larger life-time in the open chain form and (b) fructose becomes involved in an equilibrium reaction by forming di-fructose-di-anhydrides so that the most reactive groups capable for cross-polymerization are intra-disabled favoring an increase in selectivity, whereas glucose forms real oligosaccharides which still have reactive reducing groups, available for cross-polymerization with HMF and reactive intermediates [228–230].

There are many factors influencing 5-HMF levels, such as temperature, storage conditions, water activity, divalent cations concentrated in the media, the flowers' origin, and some chemical properties of honey, including pH, acidity, reducing sugars, and mineral content, of which the pH is the most recorded [195,218,228,231]. For honeys which have an unusual high FA, exceeding the legislative limit, a great increase in FA was followed by a very high increase in 5-HMF, which was more pronounced for honeys stored at 35 °C for twelve months, and for those stored at 45 °C from the very first month. A very strong positive correlation of Pearson's coefficient found between 5-HMF and FA, storage time, and storage temperature [194]. Reports on thermal formation kinetics of 5-HMF fitting to a first- or zero-order reaction, or first-order kinetics of the degradation of amino acid and zero-order kinetics for the formation of product (5-HMF), resulted in activation energies calculated according to the Arrhenius model in various honeydew and floral honeys evidencing that, besides the processing treatment, the composition of honey also has a role on the 5-HMG formation [223,228,232–234].

The effect of thermal processing in the content of 5-HMF in monofloral pine, citrus, thyme, and eucalyptus, blends (Thyme-pine, Erica-pine) and multifloral honeys from Crete, at different temperatures and storage times showed that the 5-HMF changes for Pine honey are the most resistant towards thermal treatment especially for high temperature and prolonged times while thyme was the most vulnerable to 5-HMF increases, followed by citrus honey, a fact attributed to the higher pH value of pine honey [190].

Separation of raw honeys from those heated at different temperature levels was achieved using FTIR data and chemometrics. PCA analysis for eucalyptus honey yields a strong discriminating factor for complete separation of raw samples from those heated at 40 °C/3.5 h and 70 °C/15 min, but not for acacia or orange blossom honeys, based on the spectral features at 990 and 1050 cm^{-1}, where mostly stretching (C-OH) vibrations of carbohydrates occur [221]. Application of DA showed sufficient categorization for (a) raw samples heated to 70 °C for 15 min, and heated samples to 70 °C 120 min (overall accuracy 0.947%) (b) raw heated samples to 40 °C for 15, and samples heated to 40 °C for 120 min samples (overall accuracy 0.895%), thus, enabling classification of the samples according to different thermal conditions, temperature, and duration [221]. The effect of storage has also been investigated with other spectroscopies such as ^{1}H NMR spectroscopy, where mainly sugars and minor components were found to differentiate honeys [29].

During storage or thermal treatment furanic compounds are formed besides 5-HMF, such as 2-HMF and 2-(furan-2-methyl)-4-methoxyfuran-3-(2H)-one (2-Furfural). These are derived by Amadori degradation via: (a) enolization for pH < 7 with 2-HMF formation to prevail (implication of pentose, as xylose is implicated) and (b) enolization at basic pH, 5-HMF (implication of hexose, as glucose) could be produced through reductones and the fission products' route [181,201,235]. In honeys owning mostly an acidic pH, 5-HMF prevails whereas in wood and chestnut honeys, 2-Furfural is predominant because of the higher pH values [236], however, different conclusions have been reported as well [197]. The Maillard reaction is commenced by nitrogen-containing compounds, and carbonyl compounds such as amino acids in foods resulting in Schiff base intermediates and rearrangement to Amadori or Heyns products. The α-Dicarbonyl compounds (α-DCs), which are highly reactive critical intermediates in the Maillard reaction [237,238], are affected by pH, nectar composition, and storage. Artificially matured honeys significantly exceed in α-DCs those that are naturally matured [237], proving α-DCs to be a more sensitive indicator for heat treatment than HMF [237,239].

4.2.5. Components in Crystallization

Crystallization and viscosity affect sensorial properties of honeys, in consequence, the consumers' acceptability. Heating pretreatment delays the crystallization of honeys while maintaining low viscosity values, with the most common methods of crystal prevention including the pasteurization method (treatment at high temperatures), storage at low temperature, microwave or ultra sound pre-treatment, storage at very low temperature and filtration using sieves (pores < 80 μm) [34,38]. Heating procedure at 70 °C/15 min manages to extend the non-crystallization period, but it negatively affects the quality of the honey and its shelf life [240].

The ratio of fructose to glucose (F/G) and their relationship with water content govern the rate of honey crystallization, and as a consequence, have a predominant role for controlling the rheological properties of honeys [241,242]. An effect of the botanical source on the honey crystallization and stability has also been recorded [243]. Molecular dynamics revealed that the crystallization of honey was different from that of pure glucose regarding the morphology and conformational stability, with glucose/fructose at 2.5:1 to result in crystallization of the same stability as the crystals found in honey described of the same glucose/fructose ratio [244].

The parameter Fructose/Water (F/G) can predict crystallization because glucose is less soluble than fructose. When F/G < 1.4, fast crystallization of honey occurs but when ratio is > 1.54 honey exhibits no tendency to crystallize [245,246]. Strong relationships among sugars (fructose, glucose, sucrose, melezitose, and maltose), palynological characteristics, sugar ratios (F + G, F/G, G/W), and moisture content were revealed when statistics applied towards the establishment of a predictive base for the crystallization tendency of monofloral honeys [98]. PCA discriminated rape and sunflower honey that exhibited the highest F + G mean values (>75%) which are significantly different compared to honeydew, eucalyptus, bramble, heather, chestnut honeys ($p < 0.05$). Honeydew exhibited the lowest F + G concentration (< 60%) than the others ($p < 0.05$). This parameter is a marker for distinguishing honeydew honey from blossom [182]. Rape and sunflower, being blossom honeys, are known from the literature to easily crystallize during storage, therefore, they are expected to have the highest F/G ratio in the PCA components plot [98]. Water content is another parameter that affects crystallization. The ratio G/W < 1.7 is indicative of slow or no crystallization but ratio > 2 for complete or fast crystallization [107]. The higher the glucose content and the lower the water content, the higher the crystallization rate proceeds. Many sugars crystallize if their concentration is above the saturated level [205]. Glucose in high concentration leads honey to crystallization because it is less soluble and crystallizes faster than fructose.

The effect of the phase of honey during thermal treatment conditions on the moisture, 5-HMF content, lightness, and yellowness has been recorded, where moisture loss of crystallized and Bi-phase honeys are greatly affected by the thermal treatment at a low temperature of 39 °C/30 min and at a high temperature of 55 °C/24 h, while liquid samples were not affected [247]. Chemometrics were applied on FTIR spectroscopic data, which express structural modifications induced by the heating process originating from the dissolution of glucose crystals, acquired at 1470 (O-H stretching vibrations), 1935 (H_2O-OH bonds of water molecules), 2100 (carbohydrates), and 1690 cm^{-1} for fructose, for highlighting the ratio on F/G, showing (a) a high predictive ability (0.78 MMC) to discriminate honeys treated at room temperature from those at high temperature. Conventional and chemometric approaches showed that changes in physico-chemical parameters and NIR spectroscopic characteristics were larger for crystallized and Bi-phase honeys than for those in liquid phase, therefore, they are phase related [247].

Prediction for the type of liquefaction treatment can be achieved for honeys liquefied at 40 °C or 72 °C first and afterwards stored for 12 months at different temperatures, based on changes in the physico-chemical parameters associated with the crystallization occurrence in honeys versus time [240]. A PCA statistical model showed separation for each group of samples with each one subjected to different liquefaction treatment, while HCA categorized samples into three groups, one for heated honeys at low temperatures (0–18 °C), a second for unheated (fresh + heated at a low temperature), and a third group containing the samples heated at a high temperature level [240]. Samples stored at −18 °C showed no difference in the physico-chemical parameters from the fresh ones; it may affect honey viscosity [220].

Dehydration of honeys at 40 °C goes through several steps: at the two first steps, water loss occurs, the third step is characterized by volatiles' loss, whereas, at the last dehydration step both volatiles and water are removed from samples [248]. Kinetics of honey heating at 40 °C for time intervals varying up to 90 h were performed with Synchronous 2D correlation spectroscopy for monitoring changes at 995 nm for rape and 990 nm for chaste honeys versus increasing dehydration times. NIR spectra showed differentiation at 900–990 nm for absorption assigned to the second overtone stretching vibrations of NH of aromatic amines and OH of CHOH groups originated from sugars, enzymes, fragrant aroma, water, and other components of honey [249].

The presence of oligosaccharides can modify the honey's tendency for crystallization [98] Honeydew honeys containing a higher content of melezitose do not easily crystallize [203,207,250]. In this line, crystallization prevention has been reported based on the addition of trehalose to the honey samples, a fact attributed to trehalose's high hydrophilic character that prevents the formation of internal hydrogen bonds but enhances the tendency to form hydrogen bonds with surrounding macromolecules resulting in preventing glucose crystallization and the expulsion of water molecules from its supersaturated solution [251]. Although the addition of trehalose leaves the G/W ratio the same, trehalose interferes with the dynamics of water molecules by hindering glucose from crystallizing in the supersaturated honey solution [252].

The crystallization process is directly associated with the chemical composition of honey [98]. The ease with which glucose monohydrate forms crystals in honey is due to its ability to adopt miscellaneous geometric forms leading to a change not only in the consistency but also in the water binding during the crystallization process [85,246]. A recent approach, developed for detecting honey liquefaction based on (i) the water impact on pH and the monosaccharides and (ii) disaccharides' contents of honey, predicts if honey has previously undergone liquefaction above 30 °C based on the modification of EC behavior and pH changes, both induced by the irreversible changes of the molecular structure [107,205].

4.2.6. Amino Acids/Proteins

Amino acids are contained in honeys accounting for 1% (*w*/*w*) [13]. They are originated mostly from pollen, and in a lesser degree from animal and vegetable sources, among them fluids and the nectar secretions of salivary glands and pharynx of bees [253]. Several amino acids are contained in honeys such as arginine, aspartic, serine, glutamic, asparagine, glutamine, threonine, proline, phenylalanine, histidine, β-alanine, glycine, α-alanine, γ-aminobutyric, tyrosine, α-aminobutyric, tryptophan, methionine, valine, isoleucine, leucine, ornithine, and lysine [253]. The composition of amino acids in honey is dependent on the pollen species from where the nectar originates, therefore, it can serve as an index for botanical differentiation [56,254]. Proline, originated from the salivary secretion of bees acting on honey collection and maturation, has been found in higher amounts than the other amino acids in honeys, as the major contributor with 50–85% of the total amino acids [208,255,256], while in some studies, phenylalanine has been reported to be the predominant one [253,257]. In general, honeydew contains higher amounts of amino acids than floral honeys [203], and can be differentiated from raw floral honey by amino acids composition in all amino acids except for phenylalanine, tyrosine, and proline [255]. Chemometrics on the amino acid profile of honeydew *Mimosa scabrella Bentham* species resulted in differentiation based on geographical origin [57].

Proteins such as diastases (α-, β- amylases) for hydrolysis of a-D-(1,4) glycosidic bonds, invertase for sucrose inversion to equimolar amount of glucose and fructose, and glucose-oxidase for oxidation of glucose to δ-gluconolactone to final hydrolysis to gluconic acid, are the most common enzymes in honey [180,198,258].

The concentration of amino acids is reduced during the thermal treatment of honey or prolonged storage [203,233]. Several compounds are formed from the reaction of the carbonyl group of a reducing sugar with the free amino group originated from amino acid or protein (Maillard reaction) resulting in the formation of dark compounds followed the initial step of Amadori compounds, darkening the honey's color. Amino acids found to easily trigger the Maillard reaction are lysine, proline, γ-aminobutyric acid, and arginine [192].

The effect of storage at room temperature on the composition of amino acids of honeydew, floral, and blends of honey for different time intervals up to 24 months showed the total amino acid content to decrease with storage, with losses ranging from 45 to 52% for glutamic acid, arginine, and glutamine in honeydew honey [258], while in Brazilian honeydew, the "Bracatinga" glutamic acid, the most sensitive to prolonged storage, decreased by 84% compared with fresh honeys [192]. In general, honeydew honeys have a higher content of individual amino acids [203,259] than floral honeys except for histidine, tyrosine, phenylalanine, leucin, and proline. Although the amount of each amino acid decreases over a storage period, their concentration and ratio in 24 months-stored honeys can still discriminate the honey type (88.7% correct assignment) [255]. The fact that the order for honeydew or floral amino acid concentrations, found in raw honeys as glutamine < asparagine < aspartic acid < glutamic acid << proline, is different than the one in stored honeys is evidence for a selective decrease in amino acids because of the different sensitivity of each of them to trigger the Maillard reaction [192]. Although it is expected that the proportion of each amino acid in processed honeys varies a lot since its reactivity is different, the loss of amino acids occurs in a way associated with the type of honey (honeydew, floral, blends) [255]. Results reported for jujube and chaste honeys showed a strong correlation of the decrease in the concentration of each amino acid for different heating treatments with the increase in 5-HMF formed during storage time. Low pH favors the Maillard reaction: it was found that chaste honey, with lower pH and higher total amino acid content, undergoes reactions of high rates for both the disappearance of the dominant amino acids and the formation of 5-HMF which were found to be several times faster than the respective reactions recorded for the jujube honey [257].

Heating or long-term storage of honey has been associated with the formation of furosine (2-furoylmethyl-lysine) generated by acid catalysis of Amadori compound fructosyl-lysine, which has been proposed as a useful indicator of the extent of the degradation of foods because of their exposure to processing or extended storage [260,261]. In addition, 2-furoylmethyl adducts of arginine, gamma-Aminobutyric acid (GABA), and proline have been detected. Furosine content is higher in honeydew honey than in other commercial/fresh honey as derived from the higher content of honeydew in amino acids, and its increase is associated with the increase in 5-HMF [260]. Samples may be considered properly heat-treated, when the furosine concentration is low, while the absence of 2-furoylmethyl-amino acids could be an indicator for honey freshness or a short period of honey storage under proper conditions [260].

Derivatives of 5-HMF, N-(1-deoxy—1-fructosyl) phenylalanine, (Fru-Phe), formed through the Amadori pathway towards thermal processing and storage of honeys, can serve as indicators of thermal process and storage. In this line, in Artificial Mature Acacia

Honeys Fru-Phe is rapidly increasing, reaching 2.4 times the initial concentration while for Natural Mature Acacia Honey the increase was only 1.2 times, with respect to the initial concentrations at the start of the 24 months' storage period [262]. Chemometrics were performed to screen for differential compounds that distinguish the two groups resulted in several compounds, however, only Fru-Phe was the one chosen because it fulfills the criteria: (a) to be present in all Artificial Mature Acacia Honeys and be in low concentrations or absent in Natural Mature Acacia Honey samples, and (b) to be stable and easily determined [237].

Reports have pointed out that D-amino acids are formed during heating of aqueous solutions of L-amino acids in the time course of the non-enzymic browning through formation of stable Amadori compounds, upon which racemization occurs [263], with the racemization to be dependent, particularly, on steric hindrance and electronic properties of the amino acid side chains [236].

4.2.7. Diastase Activity

Heating processes for crystallization prevention and improvement of rheological properties of honey and elongated storage time result in denaturation of the contained enzymes as directly calculated by enzyme activities [107,195,264] or by the loss of ability to inhibit biofilm formation of certain bacteria species during thermal liquefaction of crystallized honeys [107].

Diastase is the most thermal tolerant among honey enzymes, therefore, its presence is the criterion of freshness. According to the legislation limits, diastase activity (DN) in honey in general must be ≥ 8 (Schade scale), except for honeys with low natural enzyme content (e.g., citrus), where diastase ≥ 3 which is accompanied by a limit for 5-HMF not exceeding 15 mg/kg in order to ensure that the low DN value does not originate from heating or elongated storage of honey [77]. Diastase hydrolyzes starch and dextrin to oligosaccharides, and exhibits activity significantly positively correlated with invertase activity [265]. It originates from pollen digestion by bees [208] and from nectar (blossom and honeydew), and therefore is dependent on the botanical origin [264,266]. It is also dependent on the age of the bees, the nectar harvesting season, the physiological period of the colony, the large quantity of nectar flow, its sugar content since a high flow of concentrated nectar results in a decrease in enzyme concentration and pollen consumption [228], and the bee tribe [267] while low diastase activity occurs in artificial honeybees feeding [59]. Consequently, DN alone is not an indicative parameter for honey overheating treatment [77].

High heating temperatures decrease diastase content [194,198,268], with up to 40% reported for the storage of Greek honeys for one year [59]. Chemometrics showed a negative correlation ($p < 0.05$) for diastase activity with storage time, storage temperature, FA, and HMF [194], while the diastase content decreased concomitantly with heating at higher temperatures [264].

Studies on the effect of heating temperature on the denaturation of diastase have shown a different behavior of diastase activation towards isothermal or transient patterns of heating; specifically, after isothermal treatment, diastase activity remained constant, even increased, for heating periods 600 and 1200 s for each specific temperature [190]. This peculiar behavior has been explained on the basis of the Eyring theory regarding the enzymatic recovery after the enzyme returns to the native-like state for not succeeding in overcoming the energy barrier opposed by the isothermal pattern heating [228,269]. Pine honeydew exhibits higher change-resistance heating followed by citrus, multifloral, thyme, and cotton honeys. Storage for an elongated time decreases diastase activity in honeys. Kinetics regarding storage time showed that diastase activity of pre-heated honeys from different origins were significantly different ($p < 0.05$) at both the initial and ending storage time, while the correlation of honey origin with heating exposure period was significant ($p < 0.05$) [270].

4.2.8. Water Content

The water content of honey is dependent on different factors such as the botanical and geographical origin of nectar [182,202,203,249,271], harvesting season [178], intensity of nectar flux [36], soil and climatic conditions, bee tribe [199], degree of honey maturation in the beehive, and practices followed in harvest and extraction [183]. Some properties of honey such as viscosity, density, crystallization, color, and flavor are influenced by the water content [246].

However, it is not the water (or moisture M) content but the water activity that is responsible for the quality and process attributes of honey, since water activity represents the fraction of water not tightly bound to solids (mostly sugars) contained in honey, which becomes available for yeasts' and bacterial growth [203,271]. There is a linear relationship between water activity and water content [203,271], therefore, it is reliable to use the term water content /moisture instead. The water content of honey must be kept low, $\leq$20%, except for heather $\leq$23%, as stipulated by the legislation limits [77]. In this line, water content is deliberately reduced by dehumidification to lower than a 20% level to prevent fermentation, especially in honeys produced by stingless bees which produce honeys of a water content higher than the above limits [199].

During the ripening process of honey, the moisture keeps decreasing with the storage beehive cells to be totally capped with beeswax when the percentage of water in honey is appropriate [183].

The effect of honeycomb material on water evaporation and the maturation of honey inside the beehive has been investigated by the replacement of beeswax by a composite material consisted by 90% paraffin and 10% beeswax, and used along with the pure beeswax honeycombs within the same beehive [272]. Results showed that the water content of honeys ripening in the paraffin-based honeycombs was significantly higher than those in the beeswax honeycombs, negatively influencing the honey ripening. This was attributed to the hydrophilic character of the function groups of esters and unbound aliphatic alcohols and acids, and the presence of lipolytic enzymes incorporated into the wax by bees during comb construction enabling cells that allow for moisture transfer. ^{1}H NMR spectroscopic data showed that honeys of elevated water content in paraffin combs were accompanied with a higher concentration of acetic and citric acids [272].

Water content is important for the phase stability of honey, with the ratio G/W to be proposed for the prediction of honey crystallization [98,245]. Honeydew honeys exerting a G/W ratio > 1.7 are prone to crystallization compared to blossom honeys possessing a ratio between 1.17 to 1.27 that make them vulnerable to granulate [271].

Water activity is increased during the crystallization of stored honey, because reduction in the water molecules bound to glucose occurs, from five in liquid to one in the crystalline phase, releasing water and causing the liquid phase to become less concentrated, in consequence, leading to an increase in the water activity [107,205,243,244]. On the contrary, thermal liquefaction of honey results in a decrease in water activity of honey [107]. Moisture exerted high discriminant power for the differentiation of honeys of different origin based on the storage period and botanical origin after they were first liquefied at 40 °C/60 min, and afterwards when stored at 14 °C for 60 and 180 days [243]. Significant differences ($p < 0.001$) have been reported for the water content of honeys of three different phases to follow the order: crystallized > bi-phase > liquid, at each different storage temperature [247].

Results derived from this bibliographic search shows that experimental data give a wealth of information containing hidden trends and correlations among variables that cannot been realized with other ways than chemometrics [273]. Spectroscopic methods, most commonly NMR and FTIR, and chromatography, providing the chemical fingerprint of honey are ideal to be utilized for honey chemometrics [273]. Although NMR spectroscopy has not been extensively used for honey processing and storage, the FTIR spectroscopy is currently the most common data platform for application chemometrics for quality control as well for exploring the maturation, the aging, and the degradation of honey versus processing methods and storage conditions towards quality control the preservation and/or the improvement of honey quality. In this way, it is possible that the useful information is extracted and separated from the non-useful one, also solving the problem of spectral noise. As described, honey is a complicated chemical system, with a number of parameters to influence its quality and safety, even more so since these variables are strongly correlated to each other through chemical interacting pathways and system in equilibrium states; therefore, chemometrics, as a method for multivariate data analysis, can be used for fingerprinting analysis and chemical profiling for honey affected by processing and storage.

4.3. Chemometrics Used in Recent Studies Related to Honey Quality Analysis during Storage and Processing

In this part of the review paper, the combination of chemometrics with analytical techniques mentioning the quality characteristics in honey during processing and storage are discussed. Specific emphasis is given to the chemometric methods used in each study and the outcome after their application. A brief presentation of very recent studies available in literature is shown in Table 3.

Segato et al. (2019) studied how heating treatments may change various physico-chemical characteristics and color of honey samples in three phases (liquid, bi-phasal, and crystallized). NIR measurements were taken, and interpretation took place by applying PCA. Only the first two principal components (PCs) were important. PLS-DA and SVM have been applied on the NIR data. They worked with a training set and all the remaining samples formed the blind test. Cross-validation was conducted using a leave-one-out procedure. The three most important absorbance bands were found to be around 1420, 1905, and 2130 nm. The PCA results showed that neither mild heating nor overheating resulted in changes of the NIR data of the liquid samples; thus, they were not sensitive to temperature changes. However, overheating at 55 °C for 24 h affected the native conformation of glucose crystals in bi-phasal samples, and reduction in moisture and an increase in the HMF content, as well as a strong color change (intense browning) generated. The highest temperature strongly affected the NIR data of the crystallized and bi-phase honeys. The importance of chemometrics was underlined again, this time by SVM which showed that mild heating (39 °C for 30 min) did not affect the NIR data making this combination suitable for pre-treatment analysis if needed [240].

Oroian et al. (2017) studied five different honey species, all from Romania, overall 50 samples of honey. They measured important properties such as pH (3.88–6.39), aw (0.476–0.603), free acidity (3.40–37.10), MC (14.44–19.80%), EC (109.9–1276.8 μS/cm), ash content (0.05–0.63%), fructose content (33.64–47.31%), glucose (22.06–38.25%), sucrose (0–2.71%), as well as fructose and glucose contents sum (66.62–79.94%), etc. Based on LDA, 94% of the samples were correctly classified. Only the first two principal components (PCs) were important. The samples were grouped in five clusters based on their species, thus by botanical origin. However, in the meantime, the physico-chemical characteristics that may change during storage were also discussed. Regarding PCA, conductivity and ash content were found to be very important factors for the clustering of samples, however, free acidity and hue angle were not, due to similar species' origins. The authors supported that pH is a factor influencing the extraction and storage of honey, as it affects stability, texture, and shelf life of honey. In addition, more than 20% moisture can speed up the fermentation reactions during storage [172].

Table 3. Research studies related to quality characteristics in honey during processing and storage.

Quality Characteristics Analyzed	Analytical Method(s)	Chemometric Method(s)	Reference
Water content, EC, AC, pH and FA, HMF, IM, proline, sugar profile (fructose, glucose, maltose and sucrose)	Digital refractometer, EC meter, electrical furnace, pH meter, spectrophotometer, HPLC-RID and ATR-FTIR	PCA	[199]
Free amino acids, color and 5-HMF	HPLC-DAD and colorimeter	PCA, HCA and OPLS-DA	[257]
Water content, pH, sugar content (glucose, fructose, and sucrose) and HMF	Refractometry, pH meter, GC-FID, UHPLC-PAD and ATR-FTIR	PCA and LDA	[221]
EC, water content, pH and HMF	pH meter, EC meter and ^{1}H-NMR	PCA, PLS-DA, OPLS-DA and HCA	[29]
Glucose, fructose and sucrose content, pH, water content, a_w, refraction index, Brix concentration, FA, ash content, EC and color parameters	Chromatography, pH meter, refractometer, conductivity meter and colorimeter	PCA and LDA	[173]
Water content, electric conductivity, pH, free acidity and lactones, diastase index, UV/Vis spectrophotometer, color and NIR measurements	Refractometer, conductivity meter, pH meter, UV/Vis spectrophotometer, UHPLC-RID, visible spectrophotometer and NIR spectrometer	PCA, PLS-DA and SVM	[247]
Reduction of water content and volatile components by evaluating H-bonds forming or collapsing in the vibrations of H-bonded groups due to thermal hydration	NIR	Synchronous 2D correlation analysis	[248]
N-(1-deoxy-1-fructosyl) phenylalanine (Fru-Phe), an Amadori compound which is produced in the first stages of the Maillard reaction due to thermal hydration	UHPLC-Q-TOF-MS, HR-MS and NMR	PCA	[262]
Spectral regions related to age, temperature, and syrup adulteration of honey	NIR	ASCA	[219]
HMF, diastase activity and phenolic content	UV–visible spectrometry and chromatography	PCA and HCA	[240]

Table 3. *Cont.*

Quality Characteristics Analyzed	Analytical Method(s)	Chemometric Method(s)	Reference
Physico-chemical properties (liquefaction time, diastase number, color and viscosity and HMF formation)	Rheometer, spectrophotometer, HPLC	PCA	[112]
Water content, acidity, water activity, glucose, fructose, sucrose, glucose/water ratio, glucose/fructose ratio, textural parameters (hardness, springiness, cohesiveness, adhesiveness, viscosity, chewiness and gumminess), microbial number and content of crystals	Refractometer, HPLC, UV-VIS spectrophotometer, colorimeter, texture analyzer, stereomicroscope	PCA	[115]

Abbreviations: ASCA = ANOVA-simultaneous component analysis, a_w = water activity, AC = ash content, ATR-FTIR = Attenuated total reflection Fourier–transform infrared, DAD = diode array detector, EC = electrical conductivity, FA = free acidity, GC-FID = gas chromatography with flame ionization detector, HCA = hierarchical cluster analysis, 5-HMF = 5-hydroximethylfurfural, HPLC = high-performance liquid chromatography, HPLC-DAD = high-performance liquid chromatography with diode-array detection, HPLC-RID = high-performance liquid chromatography with a refractive index detector, HR-MS = high resolution-mass spectrometry, IM = insoluble matter, LDA = linear discriminant analysis, M = moisture content, NIR = near infrared, NMR = nuclear magnetic resonance, PCA = principal component analysis, PLS = partial least square, PLS-DA = partial least square—discriminant analysis, OPLS-DA = orthogonal projections to latent structures discriminant analysis, SVM = support vector machine, UHPLC-PAD = ultra-high-performance liquid chromatography equipped with a photodiode-array detector, UHPLC-Q-TOF-MS = ultra-high-performance liquid chromatography equipped with Quadrupole-time of flight and mass spectrometer, UHPLC-RID = ultra-high-performance liquid chromatography equipped with refractive Index Detector.

Olawode et al. (2018) analyzed 10 honey samples based on their pH (3.75–4.38), EC (99–659 μS/cm), and moisture (14.2–17.7%) and their measurements agreed with quality limits. In addition, ^{1}H-NMR profiling was measured, and after that PCA and PLS-DA were applied. The aim of the study was to classify the samples based on botanical or geographical origins; however, they also discussed processing and storage. Concerning ^{1}H-NMR profiling, the peaks at 8.46 ppm (formic acid) and 9.49 ppm (HMF) were important in all honey samples. All the honeys contained HMF. The authors confirmed that the HMF in honey increases due to storage and thermal treatment of honey during processing [29]. D-glucose, D-fructose, maltose, and sucrose profiles were similar for all the honey samples and did not seem to be important in chemometric analysis. Normalization of the data using Log 2 transformation and Pareto scaling were used to make the metabolites' concentration reasonably normal and more comparable. OPLS-DA supervised chemometric method was also applied, as it is a powerful tool for data reduction and identification of the most important spectral points for discrimination of samples. The authors stated that the OPLS-DA models produced were less complex and more meaningful than PLS-DA. OPLS-DA may replace PLS-DA, due to its ability to discriminate between variations in the normalized data that are important for predicting grouping. HCA took place using the centroid option for the observations, and Euclidean distances were calculated. The HCA dendrogram and the PLS-DA score plot were both in agreement [29].

Zhao et al. (2018) managed to discriminate overheated honey (industrial treatment) based on the analysis of amino acid and 5-HMF contents as well as color values after different thermal treatments. Two categories of honey were used, such as a light-colored honey (chaste honey) and a dark-colored honey (jujube honey). The authors found out that the concentrations of most amino acids in honey decreased after heat treatment, and also 5-HMF and proline in jujube honey, as well as 5-HMF and phenylalanine in chaste honey [257]. HCA, PCA, and OPLS-DA chemometric methods were applied to study the similarities and differences of the samples. More particularly, HCA and PCA, so the non-supervised methods were applied to the data with data mean-centered, UV-scaled, and log-transformed. A cluster tree was produced after application of HCA by using the group distance method, thus the distance represents the degree of similarity, based on heating time and heating temperature of honey samples. OPLS-DA was performed to discriminate the samples, and mean-centered, Pareto scaled, and log-transformed data were used. The seven-fold internal cross-validation was used to validate the OPLS-DA models, as well as permutation tests (20 times). Furthermore, their study may be used for identification of overheated (ultra-high temperature) processed honey (65 °C for 10 h or 80 °C for 8 h) versus moderate thermal conditions [257].

Ismail et al. (2021) used a variety of physico-chemical properties as well as characterization based on ATR-FTIR measurements to distinguish Malaysian honey samples from different species, dehumidification process, and geographical origins by applying chemometrics. Dehumidified honey samples were treated by dehydration using a dehumidifier at a temperature between 35 °C and 38 °C. Their study was innovative since no study has investigated the effect of dehumidification on the proline level in honey. The mean concentration of proline in dehumidified honey samples (14.97 mg kg^{-1}) was significantly higher than in raw honey samples (5.52 mg kg^{-1}). The conclusion was that physico-chemical properties, ATR-FTIR, and chemometrics are capable of differentiating honey samples according to the dehumidification process and geographical origin but not by species [199].

Regarding chemometric analysis, PCA was used and successfully grouped raw and dehumidified honey using both physico-chemical properties and FTIR data. The data were centered to 0 and scaled using a unit variance. For the visualization of results various techniques were applied, such as 2-D and 3-D score plots, biplot (overlay between 2-D score plots and loading plots), and eigenvalues plots. The statistical difference was tested using the unpaired *t*-test with Welch's correction for normally distributed data, Mann–Whitney

U test, and Spearman's r test for parameters with non-normal distribution. *p*-values of less than 0.05 and 0.01 were considered significant [199].

Some honeys found in the local Malaysian market were dehumidified, and this trend is expected to increase in the future, as the authors stated; their goal was to observe the differences between raw and dehumidified honey. Twenty-five samples of dehumidified honey were compared with 49 samples of raw honey of the same species. The dehumidified group was found to have significantly lower water content (WC), fructose, and sucrose, however the group possessed significantly higher electrical conductivity (EC), insoluble matter (IM) content, and proline, as presented in Figure 2A. The parameters pH, free acidity (FA), glucose, maltose, HMF, and ash content (AC) were similar between raw and dehumidified samples. A clear separation was observed in the PCA biplot with the raw samples at the left side of the quadrant while dehumidified samples at the right side, as shown in Figure 2B.

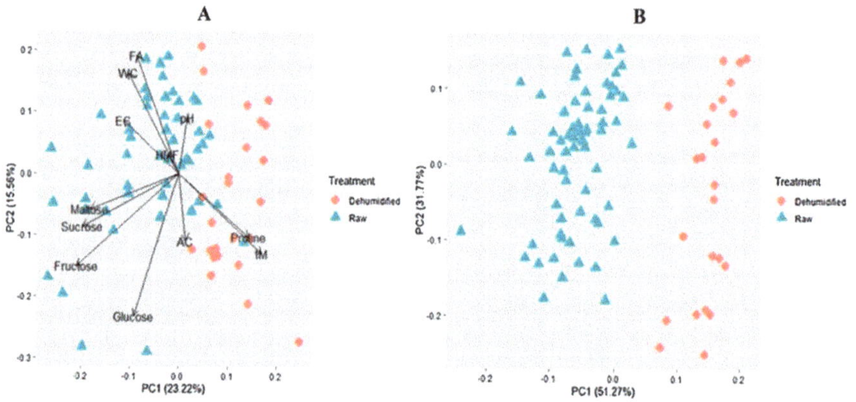

Figure 2. (**A**) PCA loading plot and (**B**) PCA score plot. Adopted from [196].

Chemometric interpretation of the ATR-FTIR measurements showed that there were different wavenumbers in the raw honey when compared with the dehumidified stingless bee honey, as shown in Figure 3. The wavenumbers at 3242 cm^{-1} due to OH stretching, 2934 cm^{-1} is related to -CH stretching of carboxylic acids, 1657 cm^{-1} because of OH deformation, 1256 and 1040 cm^{-1} corresponding to the C-O stretch in the COH group and the C-C stretch in the carbohydrate structure were significantly present in raw samples. On the other hand, the wavenumbers at 700–978 cm^{-1} were prominent in dehumidified samples representing the out-of-plane OH deformation, C=O in-plane deformation, and vibrations of the CH_2 group of L-proline.

Antonova et al. (2021) worked with honey samples to determine adulterations or changes caused due to thermal treatment. They stated that chemometrics had a significant role for the ATR-FTIR data interpretation. Three species of raw honey before and after thermal treatment, for various exposure periods and different temperatures have been tested. Calibration and validation models produced by chemometric analysis showed that the most useful region was 800–1500 cm^{-1} which contained characteristic bands of sugar transformations. Cross-validation took place based on the train-test-split approach, which randomly splits the data into a training set and a test set containing 75 and 25% of the data, respectively [215]. PCA successfully distinguished the samples between manually thermally treated and raw honey. It was used to initially decrease the huge load of data, so as to produce a new, smaller set of variables. PCA is not suitable for quantitative studies, but it is very useful for a general overview of the samples. The authors found that among the three species, eucalyptus honey changed the most after thermal treatment. As shown in Figure 4B, thermal processing caused significant changes in the ratio of the intensities of the bands at 990 cm^{-1} and 1050 cm^{-1}, which represent fructose and glucose, respectively, and the reason could be either the Maillard reaction or sugar changes. As seen in Figure 4A, thermal treatment for 15 and 120 min at 70 °C cannot be separated very well by PCA [215].

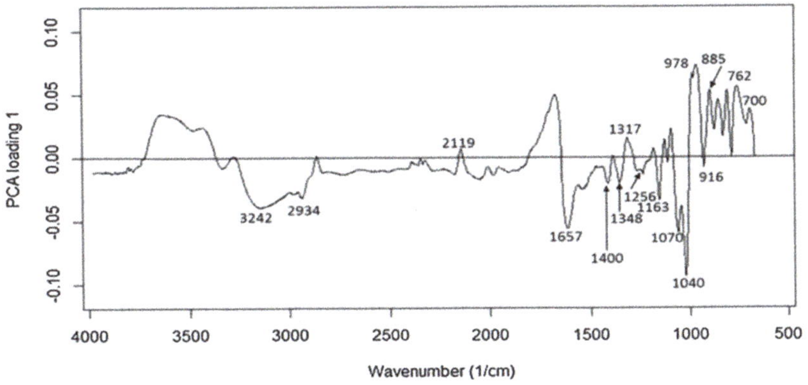

Figure 3. ATR–FTIR spectrum. Adopted from [193].

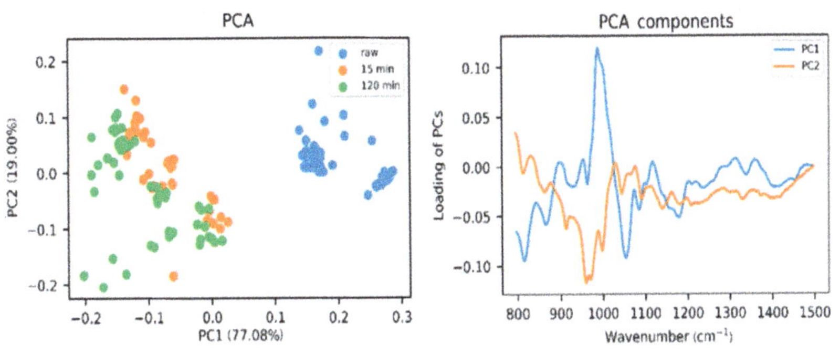

(A) **(B)**

Figure 4. Based on ATR-FTIR data (**A**) PCA score plot and (**B**) loading plot for raw and thermally treated eucalyptus honey samples. Adopted from [215].

On the other hand, LDA quantitatively discriminated against different conditions of thermal treatment at 70 °C, as presented in Figure 5 left. A confusion matrix was calculated for validation. In addition, heating at 40 °C at different periods is seen in Figure 5 right. Of course, the lower the heating temperature, the lower the classification that was possible, but raw honey was still clearly discriminated from heated honey. LDA also proved that the most important spectral wavenumber for differentiation is approximately at 990 cm^{-1}. Antonova et al. (2021) stated that ATR-FTIR spectroscopy with chemometric methods proves a powerful technique as they can detect if heating was due to transport or storage or the intended adulteration of raw honey [221].

Chen et al. (2014) have studied manually dehydrated raw honey by using synchronous two-dimensional (2D) NIR correlation spectroscopy. Honey samples were taken from six different dehydration stages using a drum wind-drying method with the temperature monitored at 40 °C. The second overtone of O–H and N–H groups vibration upon their H-bonds forming or collapsing due to the interactions between water and solute. Problems such as baseline shift and overlapping peaks of raw spectra were solved by performing preprocessing using a chemometric software. The 25-point Savitzky–Golay (SG) quadratic polynomial smoothing was used, as well as 25-point quadratic polynomial first derivative to reduce baseline shift and enhance the spectral features. The synchronous 2D correlation contour maps between 900–1080 nm were obtained from the first derivative spectra of chaste honey at different drying stages. The authors concluded that absorption in the NIR short wave region was much weaker than that in the middle and long wave region (1100–2500 nm), and it was also overshaded by water absorption. Only the use of advanced chemometrics can handle that problem as the application enlarges the signal by minimizing the noise [248].

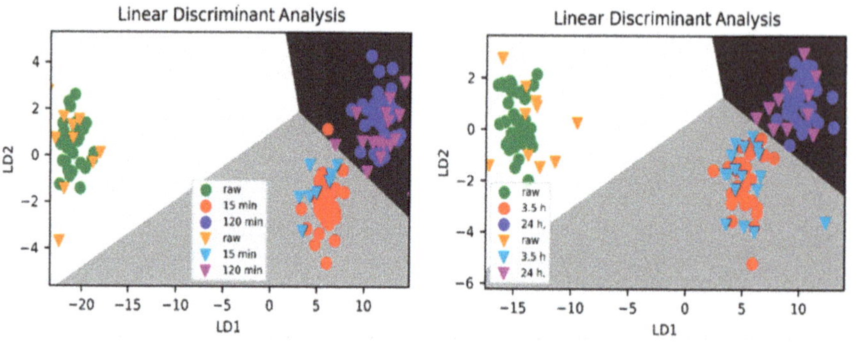

Figure 5. LDA plots based on ATR-FTIR data of the region 800–1500 cm^{-1}; left—Samples heated at 70 °C, right—Samples heated at 40 °C. Dots: training set, triangles: test set. Adopted from [215].

Yan et al. (2022) investigated a unique marker (327.1321 Da) that is produced after thermal heating (dehydration) by using UHPLC-Q-TOF-MS, HRMS, and NMR. N-(1-deoxy-1-fructosyl) phenylalanine (Fru-Phe) was identified, and it was an Amadori compound. The concentration of Fru-Phe was almost stable in naturally heated samples for more than 2 years of storage, but it increased in manually heated samples, and this marker can indicate honey fraud. The PCA method was used to visualize the groups of samples and to check that the groups were well-separated [262].

Rust et al. (2021) used NIR spectroscopy combined with chemometric analysis for determining syrup-adulterated honey. The effects of age, storage temperature, syrup adulteration (10 and 20% w/w), and irradiation treatment were captured, and ANOVA-simultaneous component analysis (ASCA) was used to treat the data which is a method combining ANOVA and PCA. Pre-processing by standard normal variate (SNV) was performed to eliminate unwanted multiplicative effects from the spectra. The proposed method was successful to detect syrup-adulterated honey [219].

Pasias et al. (2022) tried to determine the optimum conditions for honey storage. To obtain a non-crystallized product that lasts over a year, samples must be treated by heating at 72 °C or stored at −18 °C in order to maintain the same quality (low HMF content, high diastase activity, and high phenolic content) as with the fresh non-heated samples. PCA and HCA chemometric methods were applied to cluster the samples, and to distinguish and confirm the outcomes of the study [240]. Önür et al. (2018) assessed three methods of processing in honey, such as thermal, ultrasound (US), and high hydrostatic pressure (HHP) processing on the liquefaction of honey. US generated rapid dissolution of crystals. HHP gave shorter liquefaction times, as well as relatively lower HMF formation observed with HHP treatments. Chemometric analysis took place via the PCA method, and the contribution was in explaining the importance and the relation among various factors especially in the case of ultrasonication variables which were interrelated [112]. Scripcă and Amariei (2021) also used PCA to study the correlations between the honey types and the sugars, color parameters, and texture parameters. The PCA method was also important to see the variation and to finalize the main principle components [115].

5. Conclusions

Dewatering/dehumidification and thermal treatments, together with storage, are the most important stages of honey processing. They influence honey composition such as the concentration of sugars, organic acids, amino acids, and the content of HMF and phenolics, physico-chemical characteristics such as the free acidity and diastase activity, water content or processes such as the honey crystallization and melting. Honey composition and the physico-chemical parameters undergo significant changes through chemical and enzymatic reactions that take place and subsequently the changes in chemical and physical structures may affect its quality.

Chemometric analysis is a useful and necessary approach, as well as an alternative way of handling experimental data. Validation increases the robustness of chemometric models to underline the quality of chemometric studies. Without using chemometrics, the results of the studies would not be possible to be drawn, thus the use of chemometrics seems crucial. Between 2014 and 2022, the growing number of research papers which use chemometrics in analyzing honey samples shows its current importance and effectiveness in the honey industry. The main reasons justifying the increasing demand for chemometrics are initially to interpret the huge quantities of measurements obtained by the analytical methods, and after this, the visualization of samples in groups for the unsupervised methods. Moreover, chemometrics are used to clarify the reason for grouping by identifying a marker responsible for grouping the samples. In addition, it enables prediction for a possible classification of unknown samples for the supervised methods. The research potential of chemometrics seems to be very positive. The application of chemometrics on honey composition/physico-chemical parameters alternations has become a valuable tool for revealing trends and patterns hidden in data based on which processing/storage affect honey quality. Chemometrics enable the explanation and understanding of the meaning of the numbers, in order to optimize quality control and safety protection leading to a better service to society, both to the industry performance and to consumers by ensuring high quality products.

Author Contributions: Conceptualization, M.T., M.M., C.D. and I.D.M.; investigation, M.T., M.M., C.D. and I.D.M.; resources, M.T., M.M., C.D. and I.D.M.; writing—original draft preparation, M.T., M.M., C.D. and I.D.M.; writing—review and editing, M.T., M.M., C.D., I.D.M. and S.A.; supervision, M.T. and M.M.; project administration, M.T. All authors have read and agreed to the published version of the manuscript.

Funding: M.M. acknowledges the support of Lucian Blaga University of Sibiu in the Project financed by Lucian Blaga University of Sibiu through the research grant LBUS-IRG-2022-08.

Data Availability Statement: Not applicable.

Conflicts of Interest: The authors declare no conflict of interest.

References

1. Eyer, M.; Neumann, P.; Dietemann, V. A Look into the Cell: Honey Storage in Honey Bees, Apis mellifera. *PLoS ONE* **2016**, *11*, e0161059. [CrossRef]
2. Srivastava, P.K. Status Report on Bee Keeping and Honey Processing. 2020. Available online: https://msmedikanpur.gov.in/cmdatahien/reports/diffIndustries/Status%20Report%20on%20Bee%20keeping%20&%20Honey%20Processing%202019-202020.pdf (accessed on 21 January 2022).
3. Long, L. *A Global History*; Reaction Books Ltd.: London, UK, 2017; pp. 8–15.
4. Dunne, J.; Höhn, A.; Franke, G.; Neumann, K.; Breunig, P.; Gillard, T.; Walton-Doyle, C.; Evershed, R.P. Honey-collecting in prehistoric West Africa from 3500 years ago. *Nat. Commun.* **2021**, *12*, 2227. [CrossRef]
5. Dranca, F.; Ropciuc, S.; Pauliuc, D.; Oroian, M. Honey adulteration detection based on composition and differential scanning calorimetry (DSC) parameters. *LWT* **2022**, *168*, 113910. [CrossRef]
6. Marlowe, F.W.; Berbesque, J.C.; Wood, B.; Crittenden, A.; Porter, C.; Mabulla, A. Honey, Hadza, hunter-gatherers, and human evolution. *J. Hum. Evol.* **2014**, *71*, 119–128. [CrossRef]
7. Chirsanova, A.; Capcanari, T.; Boiştean, A.; Khanchel, I. Bee honey: History, characteristics, properties, benefits and adulteration in the beekeeping sector. *J. Soc. Sci.* **2021**, *4*, 98–114. [CrossRef]
8. Gündoğdu, E.; Cakmakci, S.; Şat, İ. An Overview of Honey: Its Composition, Nutritional and Functional Properties. *J. Food Sci. Eng.* **2019**, *9*, 10–14. [CrossRef]
9. Khan, R.U.; Naz, S.; Abudabos, A.M. Towards a better understanding of the therapeutic applications and corresponding mechanisms of action of honey. *Environ. Sci. Pollut. Res. Int.* **2017**, *24*, 27755–27766. [CrossRef]

10. Louppis, A.P.; Karabagias, I.K.; Kontakos, S.; Kontominas, M.G.; Papastephanou, C. Botanical discrimination of Greek unifloral honeys based on mineral content in combination with physicochemical parameter analysis, using a validated chemometric approach. *Microchem. J.* **2017**, *135*, 180–189. [CrossRef]
11. Notardonato, I.; Passarella, S.; Ianiri, G.; Di Fiore, C.; Russo, M.V.; Avino, P. Analytical Method Development and Chemometric Approach for Evidencing Presence of Plasticizer Residues in Nectar Honey Samples. *Int. J. Environ. Res. Public Health* **2020**, *17*, 1692. [CrossRef]
12. Nayik, G.A.; Nanda, V. A chemometric approach to evaluate the phenolic compounds, antioxidant activity and mineral content of different unifloral honey types from Kashmir, India. *LWT* **2016**, *74*, 504–513. [CrossRef]
13. da Silva, P.M.; Gauche, C.; Gonzaga, L.V.; Costa, A.C.O.; Fett, R. Honey: Chemical composition, stability and authenticity. *Food Chem.* **2016**, *196*, 309–323. [CrossRef]
14. Ajibola, A.; Chamunorwa, J.P.; Erlwanger, K.H. Nutraceutical values of natural honey and its contribution to human health and wealth. *Nutr. Metab.* **2012**, *9*, 61. [CrossRef]
15. Liyanage, D.; Bauddhaloka, M. Health benefits and traditional uses of honey: A review. *J. Apith.* **2017**, *2*, 9–14.
16. Mironescu, M.; Fratila, L.; Hupert, A.; Mironescu, I.D. Obtaining and Characterisation of Starch-Based Edible Films Incorporating Honey, Propolis and Bee Bread. *Acta Univ. Cibiniensis. Ser. E Food Technol.* **2019**, *23*, 193–198. [CrossRef]
17. Camacho-Bernal, G.I.; Nelly del Socorro, C.-C.; Ramírez-Moreno, E.; Delgado-Olivares, L.; Zafra-Rojas, Q.Y.; Castañeda-Ovando, A.; Suárez-Jacobo, Á. Addition of Bee Products in Diverse Food Sources: Functional and Physicochemical Properties. *Appl. Sci.* **2021**, *11*, 8156. [CrossRef]
18. Aurongzeb, M.; Azim, M.K. Antimicrobial properties of natural honey: A review of literature. *Pak. J. Biochem. Mol. Biol.* **2011**, *44*, 118–124.
19. Almasaudi, S. The antibacterial activities of honey. *Saudi J. Biol. Sci.* **2021**, *28*, 2188–2196. [CrossRef]
20. Khalil, I.; Moniruzzaman, M.; Boukraâ, L.; Benhanifia, M.; Islam, A.; Islam, N.; Sulaiman, S.A.; Gan, S.H. Physicochemical and antioxidant properties of Algerian honey. *Molecules* **2012**, *17*, 11199–11215. [CrossRef]
21. Kek, S.P.; Chin, N.L.; Yusof, Y.A.; Tan, S.W.; Chua, L.S. Classification of entomological origin of honey based on its physicochemical and antioxidant properties. *Int. J. Food Prop.* **2017**, *20*, S2723–S2738. [CrossRef]
22. Erejuwa, O.O.; Sulaiman, S.A.; Ab Wahab, M.S. Honey: A novel antioxidant. *Molecules* **2012**, *17*, 4400–4423. [CrossRef]
23. Ali, A.M.; Kunugi, H. Propolis, Bee Honey, and Their Components Protect against Coronavirus Disease 2019 (COVID-19): A Review of In Silico, In Vitro, and Clinical Studies. *Molecules* **2021**, *26*, 1232. [CrossRef]
24. Ranneh, Y.; Akim, A.M.; Hamid, H.A.; Khazaai, H.; Fadel, A.; Zakaria, Z.A.; Albujja, M.; Bakar, M.F.A. Honey and its nutritional and anti-inflammatory value. *BMC Complement. Med. Ther.* **2021**, *21*, 30. [CrossRef]
25. Premratanachai, P.; Chanchao, C. Review of the anticancer activities of bee products. *Asian Pac. J. Trop. Biomed.* **2014**, *4*, 337–344. [CrossRef]
26. Directive 2014/63/EU. Directive 2014/63/EU of the European Parliament and of the Council Amending Council Directive 2001/110/EC Relating to Honey. 2014. Available online: https://eur-lex.europa.eu/legal-content/EN/TXT/HTML/?uri=CELEX%3A32014L0063 (accessed on 21 January 2022).
27. Karabagias, I.K.; Karabournioti, S. Discrimination of Clover and Citrus Honeys from Egypt According to Floral Type Using Easily Assessable Physicochemical Parameters and Discriminant Analysis: An External Validation of the Chemometric Approach. *Foods* **2018**, *7*, 70. [CrossRef]
28. Zhang, X.-H.; Qing, X.-D.; Mu, S.-T.; Wang, D.; Zheng, J.-J.; Zhou, S.-J.; Kang, C.; Liu, Z. Authentication of honey of different nectar sources and antioxidant property evaluation by phenolic composition analysis with chemometrics. *Food Control* **2021**, *124*, 107900. [CrossRef]
29. Olawode, E.O.; Tandlich, R.; Cambray, G. 1H-NMR Profiling and Chemometric Analysis of Selected Honeys from South Africa, Zambia, and Slovakia. *Molecules* **2018**, *23*, 578. [CrossRef]
30. Abdulkhaliq, A.; Swaileh, K.M. Physico-chemical properties of multi-floral honey from the West Bank, Palestine. *Int. J. Food Prop.* **2017**, *20*, 447–454. [CrossRef]
31. Tomczyk, M.; Zaguła, G.; Puchalski, C.; Dżugan, M. Transfer of Some Toxic Metals from Soil to Honey Depending on Bee Habitat Conditions. *Acta Univ. Cibiniensis. Ser. E Food Technol.* **2020**, *24*, 49–59. [CrossRef]
32. Lazor, P.; Tomáš, J.; Tóth, T.; Tóth, J.; Čéryová, S. Monitoring of air pollution and atmospheric deposition of heavy metals by analysis of honey. *J. Microbiol. Biotechnol. Food Sci.* **2021**, *4*, 522–533.
33. Shamsudin, S.; Selamat, J.; Sanny, M.; A, R.S.; Jambari, N.N.; Khatib, A. A Comparative Characterization of Physicochemical and Antioxidants Properties of Processed Heterotrigona itama Honey from Different Origins and Classification by Chemometrics Analysis. *Molecules* **2019**, *24*, 3898. [CrossRef]
34. Krishnan, R.; Mohammed, T.; Kumar, G.S.; Arunima, S. Honey crystallization: Mechanism, evaluation and application. *J. Pharm. Innov.* **2021**, *10*, 222–231. [CrossRef]
35. Grégrová, A.; Kružík, V.; Vrácovská, E.; Rajchl, A.; Čížková, H. Evaluation of factors affecting crystallization of disparate set of multi-flower honey samples. *Agron. Res.* **2015**, *13*, 1215–1226.
36. Machado De-Melo, A.A.; Almeida-Muradian, L.B.D.; Sancho, M.T.; Pascual-Maté, A. Composition and properties of Apis mellifera honey: A review. *J. Apic. Res.* **2018**, *57*, 5–37. [CrossRef]

37. Sanova, P.; Jitka, S.; Hrubcová, B.; Šeráková, P. Segmentation of Honey Buyers' Behaviour by Conjoint Analysis. *Sci. Agric. Bohem.* **2017**, *48*, 55–62. [CrossRef]
38. Luo, X.; Dong, Y.; Gu, C.; Zhang, X.; Ma, H. Processing Technologies for Bee Products: An Overview of Recent Developments and Perspectives. *Front. Nutr.* **2021**, *8*, 727181. [CrossRef] [PubMed]
39. Thrasyvoulou, A.; Tananaki, C.; Goras, G.; Karazafiris, E.; Dimou, M.; Liolios, V.; Kanelis, D.; Gounari, S. Legislation of honey criteria and standards. *J. Apic. Res.* **2018**, *57*, 88–96. [CrossRef]
40. Israili, Z.H. Antimicrobial properties of honey. *Am. J. Ther.* **2014**, *21*, 304–323. [CrossRef]
41. Piotr, S. Antimicrobial Activity of Honey. In *Honey Analysis*; Vagner de Alencar Arnaut de, T., Ed.; IntechOpen: Rijeka, Croatia, 2017; p. 10.
42. Martinello, M.; Mutinelli, F. Antioxidant Activity in Bee Products: A Review. *Antioxidants* **2021**, *10*, 71. [CrossRef]
43. Santos, M.C.; Nascimento, P.A.M.; Guedes, W.N.; Pereira-Filho, E.R.; Filletti, É.R.; Pereira, F.M.V. Chemometrics in analytical chemistry—An overview of applications from 2014 to 2018. *Eclet. Quim.* **2019**, *44*, 11–25. [CrossRef]
44. Kuś, P.M.; Rola, R. LC-QqQ-MS/MS methodology for determination of purine and pyrimidine derivatives in unifloral honeys and application of chemometrics for their classification. *Food Chem.* **2021**, *348*, 129076. [CrossRef]
45. Liu, T.; Ming, K.; Wang, W.; Qiao, N.; Qiu, S.; Yi, S.; Huang, X.; Luo, L. Discrimination of honey and syrup-based adulteration by mineral element chemometrics profiling. *Food Chem.* **2021**, *343*, 128455. [CrossRef] [PubMed]
46. Drivelos, S.A.; Danezis, G.P.; Halagarda, M.; Popek, S.; Georgiou, C.A. Geographical origin and botanical type honey authentication through elemental metabolomics via chemometrics. *Food Chem.* **2021**, *338*, 127936. [CrossRef] [PubMed]
47. Zhang, J.; Chen, H.; Fan, C.; Gao, S.; Zhang, Z.; Bo, L. Classification of the botanical and geographical origins of Chinese honey based on 1H NMR profile with chemometrics. *Food Res. Int.* **2020**, *137*, 109714. [CrossRef] [PubMed]
48. Ghanavati Nasab, S.; Javaheran Yazd, M.; Marini, F.; Nescatelli, R.; Biancolillo, A. Classification of honey applying high performance liquid chromatography, near-infrared spectroscopy and chemometrics. *Chemom. Intell. Lab. Syst.* **2020**, *202*, 104037. [CrossRef]
49. Revenga-Parra, M.; Robledo, S.N.; Martínez-Periñán, E.; González-Quirós, M.M.; Colina, A.; Heras, A.; Pariente, F.; Lorenzo, E. Direct determination of monosaccharides in honey by coupling a sensitive new Schiff base Ni complex electrochemical sensor and chemometric tools. *Sens. Actuators B Chem.* **2020**, *312*, 127848. [CrossRef]
50. Rodríguez-Flores, M.S.; Escuredo, O.; Míguez, M.; Seijo, M.C. Differentiation of oak honeydew and chestnut honeys from the same geographical origin using chemometric methods. *Food Chem.* **2019**, *297*, 124979. [CrossRef] [PubMed]
51. Stöbener, A.; Naefken, U.; Kleber, J.; Liese, A. Determination of trace amounts with ATR FTIR spectroscopy and chemometrics: 5-(hydroxymethyl)furfural in honey. *Talanta* **2019**, *204*, 1–5. [CrossRef]
52. Wang, X.; Yang, S.; He, J.; Chen, L.; Zhang, J.; Jin, Y.; Zhou, J.; Zhang, Y. A green triple-locked strategy based on volatile-compound imaging, chemometrics, and markers to discriminate winter honey and sapium honey using headspace gas chromatography-ion mobility spectrometry. *Food Res. Int.* **2019**, *119*, 960–967. [CrossRef]
53. Ciulu, M.; Serra, R.; Caredda, M.; Salis, S.; Floris, I.; Pilo, M.I.; Spano, N.; Panzanelli, A.; Sanna, G. Chemometric treatment of simple physical and chemical data for the discrimination of unifloral honeys. *Talanta* **2018**, *190*, 382–390. [CrossRef]
54. Ferreiro-González, M.; Espada-Bellido, E.; Guillén-Cueto, L.; Palma, M.; Barroso, C.G.; Barbero, G.F. Rapid quantification of honey adulteration by visible-near infrared spectroscopy combined with chemometrics. *Talanta* **2018**, *188*, 288–292. [CrossRef]
55. Devi, A.; Jangir, J.; Anu-Appaiah, K.A. Chemical characterization complemented with chemometrics for the botanical origin identification of unifloral and multifloral honeys from India. *Food Res. Int.* **2018**, *107*, 216–226. [CrossRef] [PubMed]
56. Sun, Z.; Zhao, L.; Cheng, N.; Xue, X.; Wu, L.; Zheng, J.; Cao, W. Identification of botanical origin of Chinese unifloral honeys by free amino acid profiles and chemometric methods. *J. Pharm. Anal.* **2017**, *7*, 317–323. [CrossRef] [PubMed]
57. Azevedo, M.S.; Seraglio, S.K.T.; Rocha, G.; Balderas, C.B.; Piovezan, M.; Gonzaga, L.V.; Falkenberg, D.D.B.; Fett, R.; de Oliveira, M.A.L.; Costa, A.C.O. Free amino acid determination by GC-MS combined with a chemometric approach for geographical classification of bracatinga honeydew honey (*Mimosa scabrella* Bentham). *Food Control* **2017**, *78*, 383–392. [CrossRef]
58. Spiteri, M.; Rogers, K.M.; Jamin, E.; Thomas, F.; Guyader, S.; Lees, M.; Rutledge, D.N. Combination of 1H NMR and chemometrics to discriminate manuka honey from other floral honey types from Oceania. *Food Chem.* **2017**, *217*, 766–772. [CrossRef]
59. Pasias, I.N.; Kiriakou, I.K.; Proestos, C. HMF and diastase activity in honeys: A fully validated approach and a chemometric analysis for identification of honey freshness and adulteration. *Food Chem.* **2017**, *229*, 425–431. [CrossRef]
60. Popek, S.; Halagarda, M.; Kursa, K. A new model to identify botanical origin of Polish honeys based on the physicochemical parameters and chemometric analysis. *LWT* **2017**, *77*, 482–487. [CrossRef]
61. Karabagias, I.K.; Louppis, A.P.; Karabournioti, S.; Kontakos, S.; Papastephanou, C.; Kontominas, M.G. Characterization and geographical discrimination of commercial *Citrus* spp. honeys produced in different Mediterranean countries based on minerals, volatile compounds and physicochemical parameters, using chemometrics. *Food Chem.* **2017**, *217*, 445–455. [CrossRef]
62. Li, S.; Zhang, X.; Shan, Y.; Su, D.; Ma, Q.; Wen, R.; Li, J. Qualitative and quantitative detection of honey adulterated with high-fructose corn syrup and maltose syrup by using near-infrared spectroscopy. *Food Chem.* **2017**, *218*, 231–236. [CrossRef]
63. Arroyo-Manzanares, N.; García-Nicolás, M.; Castell, A.; Campillo, N.; Viñas, P.; López-García, I.; Hernández-Córdoba, M. Untargeted headspace gas chromatography—Ion mobility spectrometry analysis for detection of adulterated honey. *Talanta* **2019**, *205*, 120123. [CrossRef]

64. Rodríguez-Flores, M.S.; Falcão, S.I.; Escuredo, O.; Seijo, M.C.; Vilas-Boas, M. Description of the volatile fraction of Erica honey from the northwest of the Iberian Peninsula. *Food Chem.* **2021**, *336*, 127758. [CrossRef]
65. Khansaritoreh, E.; Salmaki, Y.; Akbari Azirani, T.; Henareh, F.; Alizadeh, K.; Ramezani, E.; Zarre, S.; Beckh, G.; Behling, H. The sources and quality of Iranian honey. *Heliyon* **2021**, *7*, e06651. [CrossRef] [PubMed]
66. Zawawi, N.; Chong, P.J.; Mohd Tom, N.N.; Saiful Anuar, N.S.; Mohammad, S.M.; Ismail, N.; Jusoh, A.Z. Establishing Relationship between Vitamins, Total Phenolic and Total Flavonoid Content and Antioxidant Activities in Various Honey Types. *Molecules* **2021**, *26*, 4399. [CrossRef] [PubMed]
67. Saklani, S.; Kuma, N. Quality Honey Production, Processing, and Various Mechanisms for Detection of Adulteration. In *A Miraculous Product of Nature*; Kumar, B., Agrawal, O.P., Hajam, Y.A., Eds.; CRC Press: Boca Raton, FL, USA, 2022.
68. Habryka, C.; Socha, R.; Juszczak, L. The Effect of Enriching Honey with Propolis on the Antioxidant Activity, Sensory Characteristics, and Quality Parameters. *Molecules* **2020**, *25*, 1176. [CrossRef] [PubMed]
69. Kowalski, S.; Makarewicz, M. Functional properties of honey supplemented with bee bread and propolis. *Nat. Prod. Res.* **2017**, *31*, 2680–2683. [CrossRef] [PubMed]
70. Sowa, P.; Tarapatskyy, M.; Puchalski, C.; Jarecki, W.; Dżugan, M. A novel honey-based product enriched with coumarin from Melilotus flowers. *J. Food Meas. Charact.* **2019**, *13*, 1748–1754. [CrossRef]
71. Guldas, M.; Gurbuz, O.; Cakmak, I.; Yildiz, E.; Sen, H. Effects of honey enrichment with Spirulina platensis on phenolics, bioaccessibility, antioxidant capacity and fatty acids. *LWT* **2022**, *153*, 112461. [CrossRef]
72. Tomczyk, M.; Miłek, M.; Sidor, E.; Kapusta, I.; Litwińczuk, W.; Puchalski, C.; Dżugan, M. The Effect of Adding the Leaves and Fruits of Morus alba to Rape Honey on Its Antioxidant Properties, Polyphenolic Profile, and Amylase Activity. *Molecules* **2020**, *25*, 84. [CrossRef]
73. Štajner, D.; Popović, B.M.; Čanadanović-Brunet, J.; Đilas, S.; Ćetković, G. Nutritive composition and free radical scavenger activity of honey enriched with of *Rosa* spp. *LWT-Food Sci. Technol.* **2014**, *55*, 408–413. [CrossRef]
74. Miłek, M.; Grabek-Lejko, D.; Stępień, K.; Sidor, E.; Mołoń, M.; Dżugan, M. The enrichment of honey with Aronia melanocarpa fruits enhances its in vitro and in vivo antioxidant potential and intensifies its antibacterial and antiviral properties. *Food Funct.* **2021**, *12*, 8920–8931. [CrossRef]
75. Šedík, P.; Horská, E.; Ivanišová, E.; Kačániová, M.; Krasnodębski, A. Consumer behaviour of young generation in slovakia towards cocoa-enriched honey. *Potravinárstvo/Slovak J. Food Sci.* **2019**, *13*, 18–24. [CrossRef]
76. Šedík, P.; Cristina Bianca, P.; Ivanišová, E. Interdisciplinary Approach Towards Consumer Acceptability of Flavoured Honey: Case of Young Generation in Slovakia. *Bull. UASVM Food Sci. Technol.* **2020**, *77*, 2020. [CrossRef] [PubMed]
77. Directive 2001/110/EC. European Commission Council Directive 2001/110/EC of 20 December 2001 Relating to Honey. *Off. J. Eur. Communities* **2002**, *10*, 47–52.
78. Chin, N.L.; Sowndhararajan, K. A Review on Analytical Methods for Honey Classification, Identification and Authentication. In *Honey Analysis—New Advances and Challenges*; De Toledo, V.D.A.A., Dechechi Chambó, E., Eds.; IntechOpen Limited: London, UK, 2020.
79. Oroian, M.; Amariei, S.; Escriche, I.; Leahu, A.; Damian, C.; Gutt, G. Chemical Composition and Temperature Influence on the Rheological Behaviour of Honeys. *Int. J. Food Prop.* **2014**, *17*, 2228–2240. [CrossRef]
80. Witczak, M.; Juszczak, L.; Gałkowska, D. Non-Newtonian behaviour of heather honey. *J. Food Eng.* **2011**, *104*, 532–537. [CrossRef]
81. Tafere, D.A. Chemical composition and uses of Honey: A Review. *J. Food Nutr. Res.* **2021**, *4*, 194–201.
82. Prica, N.; Živkov Baloš, M.; Jakšić, S.; Mihaljev, Ž.; Kartalović, B.; Babić, J.; Savić, S. Moisture and Acidity as Indicators of the Quality of Honey Originating from Vojvodina Region. *Arch. Vet. Sci.* **2015**, *7*, 99–109. [CrossRef]
83. Kędzierska-Matysek, M.; Florek, M.; Wolanciuk, A.; Skałecki, P. Effect of freezing and room temperatures storage for 18 months on quality of raw rapeseed honey (Brassica napus). *J. Food Sci. Technol.* **2016**, *53*, 3349–3355. [CrossRef]
84. Aljuhaimi, F.; Özcan, M.M.; Ghafoor, K.; Babiker, E.E. Determination of physicochemical properties of multifloral honeys stored in different containers. *J. Food Process. Preserv.* **2018**, *42*, e13379. [CrossRef]
85. Cimo', G.; Conte, P. Conformational Redistribution of Honey Components following Different Storage Conditions. *Int. J. Spectrosc.* **2015**, *2015*, 354327. [CrossRef]
86. Mouhoubi-Tafinine, Z.; Ouchemoukh, S.; Bey, M.; Louaileche, H.; Tamendjari, A. Effect of storage on hydroxymethylfurfural (HMF) and color of some Algerian honey. *Int. Food Res. J.* **2018**, *25*, 1044–1050.
87. Juliano, C.; Magrini, G.A. Methylglyoxal, the Major Antibacterial Factor in Manuka Honey: An Alternative to Preserve Natural Cosmetics? *Cosmetics* **2019**, *6*, 1. [CrossRef]
88. Bucekova, M.; Valachova, I.; Kohutova, L.; Prochazka, E.; Klaudiny, J.; Majtan, J. Honeybee glucose oxidase—Its expression in honeybee workers and comparative analyses of its content and H2O2-mediated antibacterial activity in natural honeys. *Naturwissenschaften* **2014**, *101*, 661–670. [CrossRef] [PubMed]
89. Chen, C.; Campbell, L.T.; Blair, S.E.; Carter, D.A. The effect of standard heat and filtration processing procedures on antimicrobial activity and hydrogen peroxide levels in honey. *Front. Microbiol.* **2012**, *3*, 265. [CrossRef] [PubMed]
90. Bogdanov, S. Honey as Nutrient and Functional Food. *Proteins* **2016**, *15*, 1400–2700.
91. Sowa, P.; Grabek-Lejko, D.; Wesołowska, M.; Swacha, S.; Dżugan, M. Hydrogen peroxide-dependent antibacterial action of Melilotus albus honey. *Lett. Appl. Microbiol.* **2017**, *65*, 82–89. [CrossRef] [PubMed]

92. Guttentag, A.; Krishnakumar, K.; Cokcetin, N.; Harry, E.; Carter, D. Factors affecting the production and measurement of hydrogen peroxide in honey samples. *Access Microbiol.* **2021**, *3*, 000198. [CrossRef]
93. Bianchin, J.N.; Nardini, G.; Merib, J.; Dias, A.N.; Martendal, E.; Carasek, E. Screening of volatile compounds in honey using a new sampling strategy combining multiple extraction temperatures in a single assay by HS-SPME-GC-MS. *Food Chem.* **2014**, *145*, 1061–1065. [CrossRef]
94. Castro-Vázquez, L.; Elena Alañon, M.; Gonzalez-Viñas, M.A.; Soledad Pérez-Coello, M. Changes in the volatile fractions and sensory properties of heather honey during storage under different temperatures. *Eur. Food Res. Technol.* **2012**, *235*, 185–193. [CrossRef]
95. Visquert, M.; Vargas, M.; Escriche, I. Effect of postharvest storage conditions on the colour and freshness parameters of raw honey. *Int. J. Food Sci. Technol.* **2014**, *49*, 181–187. [CrossRef]
96. Codex Alimentarious, C. Standars for Honey. 2019, pp. 1–9. Available online: https://extension://bfdogplmndidlpjfhoijckpakkdjkkil/pdf/viewer.html?file=https%3A%2F%2Fwww.fao.org%2Ffao-who-codexalimentarius%2Fsh-proxy%2Fen%2F%3Flnk%3D1%26url%3Dhttps%25253A%25252F%25252Fworkspace.fao.org%25252Fsites%25252Fcodex%25252FStandards%25252FCXS%252B12-1981%25252Fcxs_012e.pdf (accessed on 10 November 2022).
97. Lastriyanto, A.; Wibowo, S.A.; Erwan, E.J.F.; Batoro, J.; Masyithoh, D.; Lamerkabel, J. Moisture Reduction of Honey in Dehumidification and Evaporation Processes. *J. Mech. Sci. Technol.* **2020**, *4*, 153–163. [CrossRef]
98. Escuredo, O.; Dobre, I.; Fernández-González, M.; Seijo, M.C. Contribution of botanical origin and sugar composition of honeys on the crystallization phenomenon. *Food Chem.* **2014**, *149*, 84–90. [CrossRef]
99. Singh, S.; Gill, R.S.; Singh, P.P. Desiccant honey dehydrator. *Int. J. Ambient. Energy* **2011**, *32*, 62–69. [CrossRef]
100. Gill, R.S.; Hans, V.S.; Singh, S.; Pal Singh, P.; Dhaliwal, S.S. A small scale honey dehydrator. *J. Food Sci. Technol.* **2015**, *52*, 6695–6702. [CrossRef] [PubMed]
101. Yap, S.K.; Chin, N.L.; Yusof, Y.A.; Chong, K.Y. Quality characteristics of dehydrated raw Kelulut honey. *Int. J. Food Prop.* **2019**, *22*, 556–571. [CrossRef]
102. Abid, A.; Yunus, M.A.M.; Sahlan, S.; Ramli, M.M.; Amin, M.R.A.; Lotpi, Z.F.M.; Anuar, N.H.K. An Optimized Honey Dehydration System with Drying Air Temperature and Relative Humidity Control. In Proceedings of the 2019 IEEE International Conference on Automatic Control and Intelligent Systems (I2CACIS), Selangor, Malaysia, 29 June 2019; pp. 305–310.
103. Morawski, M.; Malec, M.; Niezgoda-Żelasko, B. Effective Condensing Dehumidification in a Rotary-Spray Honey Dehydrator. *Energies* **2022**, *15*, 100. [CrossRef]
104. Ramli, A.S.; Basrawi, F.; Daing Idris, D.M.N.; bin Yusof, M.H.; Khalil Ibrahim, T.; Mustafa, Z.; Sulaiman, S.A. A new dewatering technique for stingless bees honey. *MATEC Web Conf.* **2017**, *131*, 03014. [CrossRef]
105. Jiang, M.; Bai, X.; Sun, J.; Zhu, W. Implication of ultrasonic power and frequency for the ultrasonic vacuum drying of honey. *Dry. Technol.* **2021**, *39*, 1389–1400. [CrossRef]
106. Aydoğan Coşkun, B.; Coklar, H.; Akbulut, M. Effect of heat treatment for liquefaction and pasteurization on antioxidant activity and phenolic compounds of Astragalus and sunflower-cornflower honeys. *Food Sci. Technol.* **2019**, *40*, 629–634. [CrossRef]
107. Villacrés-Granda, I.; Proaño, A.; Coello, D.; Debut, A.; Vizuete, K.; Ballesteros, I.; Granda-Albuja, G.; Rosero-Mayanquer, H.; Battino, M.; Giampieri, F.; et al. Effect of thermal liquefaction on quality, chemical composition and antibiofilm activity against multiresistant human pathogens of crystallized eucalyptus honey. *Food Chem.* **2021**, *365*, 130519. [CrossRef]
108. Kowalski, S. Changes of antioxidant activity and formation of 5-hydroxymethylfurfural in honey during thermal and microwave processing. *Food Chem.* **2013**, *141*, 1378–1382. [CrossRef]
109. Kabbani, D.; Sepulcre, F.; Wedekind, J. Ultrasound-assisted liquefaction of rosemary honey: Influence on rheology and crystal content. *J. Food Eng.* **2011**, *107*, 173–178. [CrossRef]
110. Akhmazillah, M.F.N.; Farid, M.M.; Silva, F.V.M. High pressure processing (HPP) of honey for the improvement of nutritional value. *Innov. Food Sci. Emerg. Technol.* **2013**, *20*, 59–63. [CrossRef]
111. Quintero-Lira, A.; Ángeles Santos, A.; Aguirre-Álvarez, G.; Reyes-Munguía, A.; Almaraz-Buendía, I.; Campos-Montiel, R.G. Effects of liquefying crystallized honey by ultrasound on crystal size, 5-hydroxymethylfurfural, colour, phenolic compounds and antioxidant activity. *Eur. Food Res. Technol.* **2017**, *243*, 619–626. [CrossRef]
112. Önür, İ.; Misra, N.N.; Barba, F.J.; Putnik, P.; Lorenzo, J.M.; Gökmen, V.; Alpas, H. Effects of ultrasound and high pressure on physicochemical properties and HMF formation in Turkish honey types. *J. Food Eng.* **2018**, *219*, 129–136. [CrossRef]
113. Kuś, P.M.; Jerković, I. New Sample Preparation Method for Honey Volatiles Fingerprinting Based on Dehydration Homogeneous Liquid–Liquid Extraction (DHLLE). *Molecules* **2018**, *23*, 1769. [CrossRef]
114. Janghu, S.; Bera, M.B.; Nanda, V.; Rawson, A. Study on Power Ultrasound Optimization and Its Comparison with Conventional Thermal Processing for Treatment of Raw Honey. *Food Technol. Biotechnol.* **2017**, *55*, 570–579. [CrossRef]
115. Scripcă, L.A.; Amariei, S. The Use of Ultrasound for Preventing Honey Crystallization. *Foods* **2021**, *10*, 773. [CrossRef]
116. Peláez-Acero, A.; Cobos-Velasco, J.E.; González-Lemus, U.; Espino-Manzano, S.O.; Aguirre-Álvarez, G.; González-Montiel, L.; Figueira, A.C.; Campos-Montiel, R.G. Bioactive compounds and antibacterial activities in crystallized honey liquefied with ultrasound. *Ultrason. Sonochem.* **2021**, *76*, 105619. [CrossRef]
117. Dżugan, M.; Grabek-Lejko, D.; Sidor, E.; Tomczyk, M. The impact of ultrasound decrystallization on enzymatic, antioxidant and antibacterial properties of honey. *Innov. Food Sci. Emerg. Technol.* **2021**, *71*, 102709. [CrossRef]

118. Chiozzi, V.; Agriopoulou, S.; Varzakas, T. Advances, Applications, and Comparison of Thermal (Pasteurization, Sterilization, and Aseptic Packaging) against Non-Thermal (Ultrasounds, UV Radiation, Ozonation, High Hydrostatic Pressure) Technologies in Food Processing. *Appl. Sci.* **2022**, *12*, 2202. [CrossRef]
119. Al-Habsi, N.A.; Niranjan, K. Effect of high hydrostatic pressure on antimicrobial activity and quality of Manuka honey. *Food Chem.* **2012**, *135*, 1448–1454. [CrossRef]
120. Singh, I.; Singh, S. Honey moisture reduction and its quality. *J. Food Sci. Technol.* **2018**, *55*, 3861–3871. [CrossRef] [PubMed]
121. Stojković, M.; Cvetković, D.; Savić, A.; Topalić-Trivunović, L.; Velemir, A.; Papuga, S.; Žabić, M. Changes in the physicochemical, antioxidant and antibacterial properties of honeydew honey subjected to heat and ultrasound pretreatments. *J. Food Sci. Technol.* **2021**, *58*, 2555–2566. [CrossRef] [PubMed]
122. Saxena, S.; Gautam, S.; Sharma, A. Microbial Decontamination of Honey of Indian Origin Using Gamma Radiation and Its Biochemical and Organoleptic Properties. *J. Food Sci.* **2010**, *75*, M19–M27. [CrossRef]
123. Sircar, A.; Yadav, K. Application of Geothermal water for Honey Processing. In Proceedings of the PROCEEDINGS, 43rd Workshop on Geothermal Reservoir Engineering, Stanford, CA, USA, 12–14 February 2018.
124. Yadav, K.; Sircar, A. Application of low enthalpy geothermal fluid for space heating and cooling, honey processing and milk pasteurization. *Case Stud. Therm. Eng.* **2019**, *14*, 100499. [CrossRef]
125. Bystrzanowska, M.; Tobiszewski, M. Chemometrics for Selection, Prediction, and Classification of Sustainable Solutions for Green Chemistry—A Review. *Symmetry* **2020**, *12*, 2055. [CrossRef]
126. Alkarkhi, A.; Alqaraghuli, W.A.A. Multivariate Analysis of Variance. In *Applied Statistics for Environmental Science with R*; Alkarkhi, A., Alqaraghuli, W.A.A., Eds.; Elsevier: Amsterdam, The Netherlands, 2020; pp. 87–112.
127. Nielsen, F. *Introduction to HPC with MPI for Data Science*; Springer: Cham, Switzerland, 2016.
128. Passarella, S.; Guerriero, E.; Quici, L.; Ianiri, G.; Cerasa, M.; Notardonato, I.; Protano, C.; Vitali, M.; Russo, M.V.; De Cristofaro, A.; et al. Dataset of PAHs determined in home-made honey samples collected in Central Italy by means of DLLME-GC-MS and cluster analysis for studying the source apportionment. *Data Br.* **2022**, *42*, 108136. [CrossRef]
129. Passarella, S.; Guerriero, E.; Quici, L.; Ianiri, G.; Cerasa, M.; Notardonato, I.; Protano, C.; Vitali, M.; Russo, M.V.; De Cristofaro, A.; et al. PAHs presence and source apportionment in honey samples: Fingerprint identification of rural and urban contamination by means of chemometric approach. *Food Chem.* **2022**, *382*, 132361. [CrossRef]
130. Bro, R.; Smilde, A.K. Principal component analysis. *Anal. Methods* **2014**, *6*, 2812–2831. [CrossRef]
131. Kaygusuz, H.; Tezcan, F.; Bedia Erim, F.; Yildiz, O.; Sahin, H.; Can, Z.; Kolayli, S. Characterization of Anatolian honeys based on minerals, bioactive components and principal component analysis. *LWT* **2016**, *68*, 273–279. [CrossRef]
132. Ilyasov, I.R.; Beloborodov, V.L.; Selivanova, I.A.; Terekhov, R.P. ABTS/PP Decolorization Assay of Antioxidant Capacity Reaction Pathways. *Int. J. Mol. Sci.* **2020**, *21*, 1131. [CrossRef]
133. Özyürek, M.; Güçlü, K.; Apak, R. The main and modified CUPRAC methods of antioxidant measurement. *TrAC Trends Anal. Chem.* **2011**, *30*, 652–664. [CrossRef]
134. Antônio, D.C.; de Assis, D.C.S.; Botelho, B.G.; Sena, M.M. Detection of adulterations in a valuable Brazilian honey by using spectrofluorimetry and multiway classification. *Food Chem.* **2022**, *370*, 131064. [CrossRef] [PubMed]
135. Andersen, C.M.; Bro, R. Practical aspects of PARAFAC modeling of fluorescence excitation-emission data. *J. Chemom.* **2003**, *17*, 200–215. [CrossRef]
136. Bro, R. PARAFAC. Tutorial and applications. *Chemom. Intell. Lab. Syst.* **1997**, *38*, 149–171. [CrossRef]
137. Wu, H.-L.; Nie, J.-F.; Yu, Y.-J.; Yu, R.-Q. Multi-way chemometric methodologies and applications: A central summary of our research work. *Anal. Chim. Acta* **2009**, *650*, 131–142. [CrossRef] [PubMed]
138. Zheng, Y.-F.; Wu, M.-C.; Chien, H.-J.; Wang, W.-C.; Kuo, C.-Y.; Lai, C.-C. Honey proteomic signatures for the identification of honey adulterated with syrup, producing country, and nectar source using SWATH-MS approach. *Food Chem.* **2021**, *354*, 129590. [CrossRef]
139. Silva, B.; Gonzaga, L.V.; Maltez, H.F.; Samochvalov, K.B.; Fett, R.; Costa, A.C.O. Elemental profiling by ICP-MS as a tool for geographical discrimination: The case of bracatinga honeydew honey. *J. Food Compos. Anal.* **2021**, *96*, 103727. [CrossRef]
140. Alkarkhi, A.F.M.; Alqaraghuli, W.A.A. *Easy Statistics for Food Science with R*; Academic Press: London, UK, 2018.
141. Rinnan, Å.; Berg, F.V.D.; Engelsen, S.B. Review of the most common pre-processing techniques for near-infrared spectra. *TrAC Trends Anal. Chem.* **2009**, *28*, 1201–1222. [CrossRef]
142. Moros, J.; Garrigues, S.; Guardia, M.D.L. Vibrational spectroscopy provides a green tool for multi-component analysis. *TrAC Trends Anal. Chem.* **2010**, *29*, 578–591. [CrossRef]
143. Pauliuc, D.; Ciursă, P.; Ropciuc, S.; Dranca, F.; Oroian, M. Physicochemical parameters prediction and authentication of different monofloral honeys based on FTIR spectra. *J. Food Compos. Anal.* **2021**, *102*, 104021. [CrossRef]
144. Kubus, M. Discriminant Stepwise Procedure. *Folia Oeconomica* **2014**, *3*, 302.
145. Karabagias, I.K.; Karabagias, V.K.; Nayik, G.A.; Gatzias, I.; Badeka, A.V. A targeted chemometric evaluation of the volatile compounds of Quercus ilex honey in relation to its provenance. *LWT* **2022**, *154*, 112588. [CrossRef]
146. Dijkstra, T.K. Latent Variables and Indices: Herman Wold's Basic Design and Partial Least Squares. In *Handbook of Partial Least Squares: Concepts, Methods and Applications*; Esposito Vinzi, V., Chin, W.W., Henseler, J., Wang, H., Eds.; Springer: Berlin/Heidelberg, Germany, 2010; pp. 23–46.

147. Wold, S.; Eriksson, L.; Kettaneh, N. PLS in Data Mining and Data Integration. In *Handbook of Partial Least Squares: Concepts, Methods and Applications*; Esposito Vinzi, V., Chin, W.W., Henseler, J., Wang, H., Eds.; Springer: Berlin/Heidelberg, Germany, 2010; pp. 327–357.
148. Lee, L.C.; Liong, C.Y.; Jemain, A.A. Partial least squares-discriminant analysis (PLS-DA) for classification of high-dimensional (HD) data: A review of contemporary practice strategies and knowledge gaps. *Analyst* **2018**, *143*, 3526–3539. [CrossRef] [PubMed]
149. Hategan, A.R.; Guyon, F.; Magdas, D.A. The improvement of honey recognition models built on 1H NMR fingerprint through a new proposed approach for feature selection. *J. Food Compos. Anal.* **2022**, *114*, 104786. [CrossRef]
150. Olivieri, A.C.; Escandar, G.M.; Goicoechea, H.C.; de la Peña, A.M. Chapter 7—Unfolded and Multiway Partial Least-Squares with Residual Multilinearization: Fundamentals. In *Data Handling in Science and Technology*; de la Peña, A.M., Goicoechea, H.C., Escandar, G.M., Olivieri, A.C., Eds.; Elsevier: Amsterdam, The Netherlands, 2015; Volume 29, pp. 347–363.
151. Bro, R. Multiway calibration. Multilinear PLS. *J. Chemom.* **1996**, *10*, 47–61. [CrossRef]
152. Liland, K.H.; Indahl, U.G.; Skogholt, J.; Mishra, P. The canonical partial least squares approach to analysing multiway datasets—N-CPLS. *J. Chemom.* **2022**, *36*, e3432. [CrossRef]
153. Tang, L.; Peng, S.; Bi, Y.; Shan, P.; Hu, X. A new method combining LDA and PLS for dimension reduction. *PLoS ONE* **2014**, *9*, e96944. [CrossRef]
154. Bylesjö, M.; Rantalainen, M.; Cloarec, O.; Nicholson, J.K.; Holmes, E.; Trygg, J. OPLS discriminant analysis: Combining the strengths of PLS-DA and SIMCA classification. *J. Chemom.* **2006**, *20*, 341–351. [CrossRef]
155. He, C.; Liu, Y.; Liu, H.; Zheng, X.; Shen, G.; Feng, J. Compositional identification and authentication of Chinese honeys by 1H NMR combined with multivariate analysis. *Food Res. Int.* **2020**, *130*, 108936. [CrossRef]
156. Biancolillo, A.; Næs, T. Chapter 6—The Sequential and Orthogonalized PLS Regression for Multiblock Regression: Theory, Examples, and Extensions. In *Data Handling in Science and Technology*; Cocchi, M., Ed.; Elsevier: Amsterdam, The Netherlands, 2019; Volume 31, pp. 157–177.
157. Næs, T.; Måge, I.; Segtnan, V.H. Incorporating interactions in multi-block sequential and orthogonalised partial least squares regression. *J. Chemom.* **2011**, *25*, 601–609. [CrossRef]
158. Azcarate, S.M.; Elia Dazat, R.; Camiña, J.M.; Silva, M.F.; Gomez, F.J.V. NADES-modified voltammetric sensors and information fusion for detection of honey heat alteration. *Food Control* **2022**, *140*, 109144. [CrossRef]
159. Cunningham, P.; Delany, S.J. K-Nearest Neighbour Classifiers-A Tutorial. *ACM Comput. Surv.* **2021**, *54*, 1–25. [CrossRef]
160. Brereton, R.G. One Class Classifiers. In *Chemometrics for Pattern Recognition*; Sons, J.W., Ed.; John Wiley & Sons: Hoboken, NJ, USA, 2009; pp. 233–287.
161. Suciu, R.C.; Guyon, F.; Magdas, D.A. Application of emission—Excitation matrices in parallel with factor analysis with other chemometric techniques for honey classification. *J. Food Compos. Anal.* **2022**, *107*, 104401. [CrossRef]
162. Brendel, R.; Schwolow, S.; Gerhardt, N.; Schwab, J.; Rau, P.; Oest, M.; Rohn, S.; Weller, P. MIR spectroscopy versus MALDI-ToF-MS for authenticity control of honeys from different botanical origins based on soft independent modelling by class analogy (SIMCA)—A clash of techniques? *Spectrochim. Acta A Mol. Biomol. Spectrosc.* **2021**, *263*, 120225. [CrossRef] [PubMed]
163. Awad, M.; Khanna, R. Support Vector Machines for Classification. In *Efficient Learning Machines*; Awad, M., Khanna, R., Eds.; Apress: Berkeley, CA, USA, 2015; pp. 39–66.
164. Awad, M.; Khanna, R. Support Vector regression. In *Efficient Learning Machines*; Awad, M., Khanna, R., Eds.; Apress: Berkeley, CA, USA, 2015; pp. 67–80.
165. Suykens, J.A.K.; Van, G.T.; De Brabanter, J.; De Moor, B.; Vandewalle, J. *Least Squares Support Vector Machines*; World Scientific Pub. Co.: Singapore, 2002.
166. Li, Q.; Zeng, J.; Lin, L.; Zhang, J.; Zhu, J.; Yao, L.; Wang, S.; Yao, Z.; Wu, Z. Low risk of category misdiagnosis of rice syrup adulteration in three botanical origin honey by ATR-FTIR and general model. *Food Chem.* **2020**, *332*, 127356. [CrossRef] [PubMed]
167. ElMasry, G.; Sun, D.-W. CHAPTER 1—Principles of Hyperspectral Imaging Technology. In *Hyperspectral Imaging for Food Quality Analysis and Control*; Sun, D.-W., Ed.; Academic Press: San Diego, CA, USA, 2010; pp. 3–43.
168. Noviyanto, A.; Abdulla, W.H. Honey botanical origin classification using hyperspectral imaging and machine learning. *J. Food Eng.* **2020**, *265*, 109684. [CrossRef]
169. Sharin, S.N.; Sani, M.S.A.; Jaafar, M.A.; Yuswan, M.H.; Kassim, N.K.; Manaf, Y.N.; Wasoh, H.; Zaki, N.N.M.; Hashim, A.M. Discrimination of Malaysian stingless bee honey from different entomological origins based on physicochemical properties and volatile compound profiles using chemometrics and machine learning. *Food Chem.* **2021**, *346*, 128654. [CrossRef]
170. Chaharlangi, M.; Tashkhourian, J.; Bordbar, M.M.; Brendel, R.; Weller, P.; Hemmateenejad, B. A paper-based colorimetric sensor array for discrimination of monofloral European honeys based on gold nanoparticles and chemometrics data analysis. *Spectrochim. Acta A Mol. Biomol. Spectrosc.* **2021**, *247*, 119076. [CrossRef]
171. Awad, M.; Khanna, R. Deep Neural Networks. In *Efficient Learning Machines: Theories, Concepts, and Applications for Engineers and System Designers*; Awad, M., Khanna, R., Eds.; Apress: Berkeley, CA, USA, 2015; pp. 127–147.
172. Özbalci, B.; Boyaci, İ.H.; Topcu, A.; Kadılar, C.; Tamer, U. Rapid analysis of sugars in honey by processing Raman spectrum using chemometric methods and artificial neural networks. *Food Chem.* **2013**, *136*, 1444–1452. [CrossRef]
173. Oroian, M.; Ropciuc, S.; Buculei, A. Romanian honey authentication based on physico-chemical parameters and chemometrics. *J. Food Meas. Charact.* **2017**, *11*, 719–725. [CrossRef]

174. Wójcik, S.; Ciepiela, F.; Baś, B.; Jakubowska, M. Deep learning assisted distinguishing of honey seasonal changes using quadruple voltammetric electrodes. *Talanta* **2022**, *241*, 123213. [CrossRef] [PubMed]
175. Wu, X.; Xu, B.; Ma, R.; Niu, Y.; Gao, S.; Liu, H.; Zhang, Y. Identification and quantification of adulterated honey by Raman spectroscopy combined with convolutional neural network and chemometrics. *Spectrochim. Acta A Mol. Biomol. Spectrosc.* **2022**, *274*, 121133. [CrossRef] [PubMed]
176. Schubert, E.; Gertz, M. Intrinsic t-Stochastic Neighbor Embedding for Visualization and Outlier Detection. In Proceedings of the Similarity Search and Applications, Munich, Germany, 4–6 October 2017; pp. 188–203.
177. Wu, X.; Xu, B.; Ma, R.; Gao, S.; Niu, Y.; Zhang, X.; Du, Z.; Liu, H.; Zhang, Y. Botanical origin identification and adulteration quantification of honey based on Raman spectroscopy combined with convolutional neural network. *Vib. Spectrosc.* **2022**, *123*, 103439. [CrossRef]
178. Wu, M.-C.; Wu, C.-Y.; Klaithin, K.; Tiong, K.K.; Peng, C.-C. Effect of harvest time span on physicochemical properties, antioxidant, antimicrobial, and anti-inflammatory activities of Meliponinae honey. *J. Sci. Food Agric.* **2022**, *102*, 5750–5758. [CrossRef] [PubMed]
179. An, N.; Cai, W.-J.; Zhu, Q.-F.; Wang, W.; Hussain, D.; Feng, Y.-Q. Metabolic profiling of organic acids in honey by stable isotope labeling assisted liquid chromatography-mass spectrometry. *J. Food Compos. Anal.* **2020**, *87*, 103423. [CrossRef]
180. Kawashima, H.; Suto, M.; Suto, N. Stable carbon isotope ratios for organic acids in commercial honey samples. *Food Chem.* **2019**, *289*, 49–55. [CrossRef]
181. Apriceno, A.; Girelli, A.M.; Scuto, F.R.; Tarola, A.M. Determination of furanic compounds and acidity for Italian honey quality. *Flavour Fragr. J.* **2018**, *33*, 411–419. [CrossRef]
182. Karabagias, I.K.; Badeka, A.V.; Kontakos, S.; Karabournioti, S.; Kontominas, M.G. Botanical discrimination of Greek unifloral honeys with physico-chemical and chemometric analyses. *Food Chem.* **2014**, *165*, 181–190. [CrossRef]
183. Zhang, G.-Z.; Tian, J.; Zhang, Y.-Z.; Li, S.-S.; Zheng, H.-Q.; Hu, F.-L. Investigation of the Maturity Evaluation Indicator of Honey in Natural Ripening Process: The Case of Rape Honey. *Foods* **2021**, *10*, 2882. [CrossRef]
184. Costa dos Santos, A.; Carina Biluca, F.; Brugnerotto, P.; Valdemiro Gonzaga, L.; Carolina Oliveira Costa, A.; Fett, R. Brazilian stingless bee honey: Physicochemical properties and aliphatic organic acids content. *Food Res. Int.* **2022**, *158*, 111516. [CrossRef]
185. Seraglio, S.K.T.; Bergamo, G.; Brugnerotto, P.; Gonzaga, L.V.; Fett, R.; Costa, A.C.O. Aliphatic organic acids as promising authenticity markers of bracatinga honeydew honey. *Food Chem.* **2021**, *343*, 128449. [CrossRef] [PubMed]
186. Consonni, R.; Bernareggi, F.; Cagliani, L.R. NMR-based metabolomic approach to differentiate organic and conventional Italian honey. *Food Control* **2019**, *98*, 133–140. [CrossRef]
187. Machado, A.M.; Tomás, A.; Russo-Almeida, P.; Duarte, A.; Antunes, M.; Vilas-Boas, M.; Graça Miguel, M.; Cristina Figueiredo, A. Quality assessment of Portuguese monofloral honeys. Physicochemical parameters as tools in botanical source differentiation. *Food Res. Int.* **2022**, *157*, 111362. [CrossRef] [PubMed]
188. Belay, A.; Haki, G.D.; Birringer, M.; Borck, H.; Lee, Y.-C.; Cho, C.-W.; Kim, K.-T.; Bayissa, B.; Baye, K.; Melaku, S. Sugar profile and physicochemical properties of Ethiopian monofloral honey. *Int. J. Food Prop.* **2017**, *20*, 2855–2866. [CrossRef]
189. Karabagias, I.K.; Louppis, A.P.; Kontakos, S.; Drouza, C.; Papastephanou, C. Characterization and Botanical Differentiation of Monofloral and Multifloral Honeys Produced in Cyprus, Greece, and Egypt Using Physicochemical Parameter Analysis and Mineral Content in Conjunction with Supervised Statistical Techniques. *J. Anal. Methods Chem.* **2018**, *2018*, 7698251. [CrossRef]
190. Blidi, S.; Gotsiou, P.; Loupassaki, S.; Grigorakis, S.; Calokerinos, A.C. Effect of Thermal Treatment on the Quality of Honey Samples from Crete. *Adv. Food Sci. Eng.* **2017**, *1*, 1–8. [CrossRef]
191. Majewska, E.; Drużyńska, B.; Wołosiak, R. Determination of the botanical origin of honeybee honeys based on the analysis of their selected physicochemical parameters coupled with chemometric assays. *Food Sci. Biotechnol.* **2019**, *28*, 1307–1314. [CrossRef]
192. Seraglio, S.K.T.; Bergamo, G.; Molognoni, L.; Daguer, H.; Silva, B.; Gonzaga, L.V.; Fett, R.; Costa, A.C.O. Quality changes during long-term storage of a peculiar Brazilian honeydew honey: "Bracatinga". *J. Food Compos. Anal.* **2021**, *97*, 103769. [CrossRef]
193. Raweh, H.S.A.; Ahmed, A.Y.B.H.; Iqbal, J.; Alqarni, A.S. Monitoring and evaluation of free acidity levels in Talh honey originated from Talh tree Acacia gerrardii Benth. *J. King Saud Univ. Sci.* **2022**, *34*, 101678. [CrossRef]
194. Raweh, H.S.A.; Badjah-Hadj-Ahmed, A.Y.; Iqbal, J.; Alqarni, A.S. Impact of Different Storage Regimes on the Levels of Physicochemical Characteristics, Especially Free Acidity in Talh (*Acacia gerrardii* Benth.) Honey. *Molecules* **2022**, *27*, 5959. [CrossRef]
195. Castro-Vázquez, L.; Díaz-Maroto, M.C.; González-Viñas, M.A.; de la Fuente, E.; Pérez-Coello, M.S. Influence of Storage Conditions on Chemical Composition and Sensory Properties of Citrus Honey. *J. Agric. Food Chem.* **2008**, *56*, 1999–2006. [CrossRef] [PubMed]
196. Mar Cavia, M.D.; Fernández-Muiño, M.A.; Francisco Huidobro, J.; Teresa Sancho, M. 'Best before period once opened' for honey samples from oceanic climates on the basis of their acidity types. *Int. J. Food Sci. Technol.* **2008**, *43*, 1929–1934. [CrossRef]
197. Yang, W.; Zhang, C.; Li, C.; Huang, Z.Y.; Miao, X. Pathway of 5-hydroxymethyl-2-furaldehyde formation in honey. *J. Food Sci. Technol.* **2019**, *56*, 2417–2425. [CrossRef]
198. Moreira, R.F.A.; De Maria, C.A.B.; Pietroluongo, M.; Trugo, L.C. Chemical changes in the non-volatile fraction of Brazilian honeys during storage under tropical conditions. *Food Chem.* **2007**, *104*, 1236–1241. [CrossRef]
199. Ismail, N.F.; Maulidiani, M.; Omar, S.; Zulkifli, M.F.; Mohd Radzi, M.N.F.; Ismail, N.; Jusoh, A.Z.; Roowi, S.; Yew, W.M.; Rudiyanto, R.; et al. Classification of stingless bee honey based on species, dehumidification process and geographical origins using physicochemical and ATR-FTIR chemometric approach. *J. Food Compos. Anal.* **2021**, *104*, 104126. [CrossRef]

200. Vázquez, L.; Verdú, A.; Miquel, A.; Burló, F.; Carbonell-Barrachina, A.A. Changes in physico-chemical properties, hydroxymethylfurfural and volatile compounds during concentration of honey and sugars in Alicante and Jijona turrón. *Eur. Food Res. Technol.* **2007**, *225*, 757–767. [CrossRef]
201. Ames, J.M. Control of the Maillard reaction in food systems. *Trends Food Sci. Technol.* **1990**, *1*, 150–154. [CrossRef]
202. Karabagias, I.K.; Badeka, A.; Kontakos, S.; Karabournioti, S.; Kontominas, M.G. Characterisation and classification of Greek pine honeys according to their geographical origin based on volatiles, physicochemical parameters and chemometrics. *Food Chem.* **2014**, *146*, 548–557. [CrossRef]
203. Bentabol Manzanares, A.; García, Z.H.; Galdón, B.R.; Rodríguez, E.R.; Romero, C.D. Differentiation of blossom and honeydew honeys using multivariate analysis on the physicochemical parameters and sugar composition. *Food Chem.* **2011**, *126*, 664–672. [CrossRef]
204. Escriche, I.; Kadar, M.; Juan-Borrás, M.; Domenech, E. Suitability of antioxidant capacity, flavonoids and phenolic acids for floral authentication of honey. Impact of industrial thermal treatment. *Food Chem.* **2014**, *142*, 135–143. [CrossRef]
205. Płowaś-Korus, I.; Masewicz, Ł.; Szwengiel, A.; Rachocki, A.; Baranowska, H.M.; Medycki, W. A novel method of recognizing liquefied honey. *Food Chem.* **2018**, *245*, 885–889. [CrossRef] [PubMed]
206. Karabagias, V.K.; Karabagias, I.K.; Gatzias, I. The impact of different heating temperatures on physicochemical, color attributes, and antioxidant activity parameters of Greek honeys. *J. Food Process Eng.* **2018**, *41*, e12668. [CrossRef]
207. de la Fuente, E.; Ruiz-Matute, A.I.; Valencia-Barrera, R.M.; Sanz, J.; Martínez Castro, I. Carbohydrate composition of Spanish unifloral honeys. *Food Chem.* **2011**, *129*, 1483–1489. [CrossRef]
208. Sahin, H.; Kolayli, S.; Beykaya, M. Investigation of Variations of Invertase and Glucose Oxidase Degrees against Heating and Timing Options in Raw Honeys. *J. Chem.* **2020**, *2020*, 5398062. [CrossRef]
209. Beliz, H.-D.; Grosch, W.; Schieberle, P. Sugars, Sugar Alcohols and Honey. In *Food Chemistry*, 4th ed.; Springer: Berlin/Heidelberg, Germany, 2009; pp. 883–889.
210. Razali, M.T.; Zainal, Z.A.; Maulidiani, M.; Shaari, K.; Zamri, Z.; Mohd Idrus, M.Z.; Khatib, A.; Abas, F.; Ling, Y.S.; Rui, L.L.; et al. Classification of Raw Stingless Bee Honeys by Bee Species Origins Using the NMR- and LC-MS-Based Metabolomics Approach. *Molecules* **2018**, *23*, 2160. [CrossRef]
211. Yan, S.; Wang, W.; Zhao, W.; Tian, W.; Wang, X.; Wu, L.; Xue, X. Identification of the maturity of acacia honey by an endogenous oligosaccharide: A preliminary study. *Food Chem.* **2023**, *399*, 134005. [CrossRef]
212. Karabagias, I.K.; Vlasiou, M.; Kontakos, S.; Drouza, C.; Kontominas, M.G.; Keramidas, A.D. Geographical discrimination of pine and fir honeys using multivariate analyses of major and minor honey components identified by 1H NMR and HPLC along with physicochemical data. *Eur. Food Res. Technol.* **2018**, *244*, 1249–1259. [CrossRef]
213. Li, H.; Wu, M.; She, S.; Lin, G.; Zhou, J.; Chen, L. Study on stable carbon isotope fractionation of rape honey from rape flowers (*Brassica napus* L.) to its unifloral ripe honey. *Food Chem.* **2022**, *386*, 132754. [CrossRef]
214. Cavia, M.M.; Fernández-Muiño, M.A.; Alonso-Torre, S.R.; Huidobro, J.F.; Sancho, M.T. Evolution of acidity of honeys from continental climates: Influence of induced granulation. *Food Chem.* **2007**, *100*, 1728–1733. [CrossRef]
215. White, J.W., Jr.; Riethof, M.; Kushnir, I. Composition of Honey. VI. The Effect of Storage on Carbohydrates, Acidity and Diastase Content. *J. Food Sci.* **1961**, *26*, 63–71. [CrossRef]
216. Echigo, T.; Takenaka, T. Changes in Erlose Contents by Honeybee Invertase. *Agric. Biol. Chem.* **1973**, *47*, 177–183. [CrossRef]
217. Chong, K.Y.; Chin, N.L.; Yusof, Y.A. Thermosonication and optimization of stingless bee honey processing. *Food Sci. Technol. Int.* **2017**, *23*, 608–622. [CrossRef] [PubMed]
218. Biluca, F.C.; Della Betta, F.; de Oliveira, G.P.; Pereira, L.M.; Gonzaga, L.V.; Costa, A.C.O.; Fett, R. 5-HMF and carbohydrates content in stingless bee honey by CE before and after thermal treatment. *Food Chem.* **2014**, *159*, 244–249. [CrossRef] [PubMed]
219. Rust, A.; Marini, F.; Allsopp, M.; Williams, P.J.; Manley, M. Application of ANOVA-simultaneous component analysis to quantify and characterise effects of age, temperature, syrup adulteration and irradiation on near-infrared (NIR) spectral data of honey. *Spectrochim. Acta A Mol. Biomol. Spectrosc.* **2021**, *253*, 119546. [CrossRef]
220. Kędzierska-Matysek, M.; Matwijczuk, A.; Florek, M.; Barłowska, J.; Wolanciuk, A.; Matwijczuk, A.; Chruściel, E.; Walkowiak, R.; Karcz, D.; Gładyszewska, B. Application of FTIR spectroscopy for analysis of the quality of honey. *BIO Web Conf.* **2018**, *10*, 02008. [CrossRef]
221. Antonova, O.; Calvo, J.; Seifert, A. Rapid Detection of Thermal Treatment of Honey by Chemometrics-Assisted FTIR Spectroscopy. *Foods* **2021**, *10*, 2892. [CrossRef]
222. Grainger, M.N.C.; Manley-Harris, M.; Lane, J.R.; Field, R.J. Kinetics of conversion of dihydroxyacetone to methylglyoxal in New Zealand mānuka honey: Part I—Honey systems. *Food Chem.* **2016**, *202*, 484–491. [CrossRef]
223. Turhan, I.; Tetik, N.; Karhan, M.; Gurel, F.; Reyhan Tavukcuoglu, H. Quality of honeys influenced by thermal treatment. *LWT-Food Sci. Technol.* **2008**, *41*, 1396–1399. [CrossRef]
224. Lee, Y.C.; Shlyankevich, M.; Jeong, H.K.; Douglas, J.S.; Surh, Y.J. Bioactivation of 5-Hydroxymethyl-2-Furaldehyde to an Electrophilic and Mutagenic Allylic Sulfuric Acid Ester. *Biochem. Biophys. Res. Commun.* **1995**, *209*, 996–1002. [CrossRef]
225. Bruce, W.R.; Archer, M.C.; Corpet, D.E.; Medline, A.; Minkin, S.; Stamp, D.; Yin, Y.; Zhang, X.-M. Diet, aberrant crypt foci and colorectal cancer. *Mutat. Res./Fundam. Mol. Mech. Mutagen.* **1993**, *290*, 111–118. [CrossRef]

226. Bakhiya, N.; Monien, B.; Frank, H.; Seidel, A.; Glatt, H. Renal organic anion transporters OAT1 and OAT3 mediate the cellular accumulation of 5-sulfooxymethylfurfural, a reactive, nephrotoxic metabolite of the Maillard product 5-hydroxymethylfurfural. *Biochem. Pharmacol.* **2009**, *78*, 414–419. [CrossRef] [PubMed]
227. Lemus Ringele, G.B.; Beteinakis, S.; Papachristodoulou, A.; Axiotis, E.; Mikros, E.; Halabalaki, M. NMR Metabolite Profiling in the Quality and Authentication Assessment of Greek Honey—Exploitation of STOCSY for Markers Identification. *Foods* **2022**, *11*, 2853. [CrossRef] [PubMed]
228. Khan, Z.S.; Nanda, V.; Bhat, S.; Khan, A. Kinetic Studies of HMF Formation and Diastase Activity in Two Different Honeys of Kashmir. *Int. J. Curr. Microbiol. Appl. Sci.* **2015**, *4*, 97–107.
229. Román-Leshkov, Y.; Chheda, J.N.; Dumesic, J.A. Phase Modifiers Promote Efficient Production of Hydroxymethylfurfural from Fructose. *Science* **2006**, *312*, 1933–1937. [CrossRef] [PubMed]
230. Semchyshyn, H.M. Fructation in vivo: Detrimental and protective effects of fructose. *Biomed. Res. Int.* **2013**, *2013*, 343914. [CrossRef]
231. Bath, P.K.; Singh, N. A Research Note Chemical Changes in Helianthus Annuus and Eucalyptus Lanceolatus Honey during Storage. *J. Food Qual.* **2000**, *23*, 443–451. [CrossRef]
232. Fallico, B.; Zappalà, M.; Arena, E.; Verzera, A. Effects of conditioning on HMF content in unifloral honeys. *Food Chem.* **2004**, *85*, 305–313. [CrossRef]
233. Boonchiangma, S.; Chanthai, S.; Srijaranai, S. Chemical Compositions and Non-Enzymatic Browning Compounds of Thai Honey: A Kinetic Study. *J. Food Process Eng.* **2011**, *34*, 1584–1596. [CrossRef]
234. Bulut, L.; Kilic, M. Kinetics of Hydroxymethylfurfural Accumulation and Color Change in Honey during Storage in Relation to Moisture Content. *J. Food Process. Preserv.* **2009**, *33*, 22–32. [CrossRef]
235. Kasiotis, K.M.; Machera, K. Improved liquid chromatographic photo diode array mass spectrometric determination of 5-hydroxymethylfurfural and related furfurals in honey. *Toxicol. Environ. Chem.* **2017**, *99*, 578–589. [CrossRef]
236. Kim, J.S.; Lee, Y.S. The effect of pH on the formation of furfural compounds in the glucose and fructose with amino acid enantiomers in maillard reaction. *J. Food Sci. Nutr.* **2008**, *13*, 54–59. [CrossRef]
237. Yan, S.; Sun, M.; Zhao, L.; Wang, K.; Fang, X.; Wu, L.; Xue, X. Comparison of Differences of α-Dicarbonyl Compounds between Naturally Matured and Artificially Heated Acacia Honey: Their Application to Determine Honey Quality. *J. Agric. Food Chem.* **2019**, *67*, 12885–12894. [CrossRef] [PubMed]
238. Arena, E.; Ballistreri, G.; Tomaselli, F.; Fallico, B. Survey of 1,2-Dicarbonyl Compounds in Commercial Honey of Different Floral Origin. *J. Food Sci.* **2011**, *76*, C1203–C1210. [CrossRef] [PubMed]
239. Pita-Calvo, C.; Guerra-Rodríguez, M.E.; Vázquez, M. Analytical Methods Used in the Quality Control of Honey. *J. Agric. Food Chem.* **2017**, *65*, 690–703. [CrossRef] [PubMed]
240. Pasias, I.N.; Raptopoulou, K.G.; Makrigennis, G.; Ntakoulas, D.D.; Lembessis, D.; Dimakis, V.; Katsinas, R.; Proestos, C. Finding the optimum treatment procedure to delay honey crystallization without reducing its quality. *Food Chem.* **2022**, *381*, 132301. [CrossRef]
241. Lazaridou, A.; Biliaderis, C.G.; Bacandritsos, N.; Sabatini, A.G. Composition, thermal and rheological behaviour of selected Greek honeys. *J. Food Eng.* **2004**, *64*, 9–21. [CrossRef]
242. Oroian, M.; Amariei, S.; Escriche, I.; Gutt, G. A Viscoelastic Model for Honeys Using the Time–Temperature Superposition Principle (TTSP). *Food Bioprocess Technol.* **2013**, *6*, 2251–2260. [CrossRef]
243. Schiassi, M.C.E.V.; Souza, V.R.D.; Alves, N.A.; Lago, A.M.T.; Silva, S.H.; Carvalho, G.R.; de Resende, J.V.; Queiroz, F. Effect of botanical origin on stability and crystallization of honey during storage. *Br. Food J.* **2022**, *124*, 2689–2704. [CrossRef]
244. Ma, Y.; Zhang, B.; Li, H.; Li, Y.; Hu, J.; Li, J.; Wang, H.; Deng, Z. Chemical and molecular dynamics analysis of crystallization properties of honey. *Int. J. Food Prop.* **2017**, *20*, 725–733. [CrossRef]
245. Dobre, I.; Georgescu, L.A.; Alexe, P.; Escuredo, O.; Seijo, M.C. Rheological behavior of different honey types from Romania. *Food Res. Int.* **2012**, *49*, 126–132. [CrossRef]
246. Dettori, A.; Tappi, S.; Piana, L.; Dalla Rosa, M.; Rocculi, P. Kinetic of induced honey crystallization and related evolution of structural and physical properties. *LWT* **2018**, *95*, 333–338. [CrossRef]
247. Segato, S.; Merlanti, R.; Bisutti, V.; Montanucci, L.; Serva, L.; Lucatello, L.; Mirisola, M.; Contiero, B.; Conficoni, D.; Balzan, S.; et al. Multivariate and machine learning models to assess the heat effects on honey physicochemical, colour and NIR data. *Eur. Food Res. Technol.* **2019**, *245*, 2269–2278. [CrossRef]
248. Chen, G.; Sun, X.; Huang, Y.; Chen, K. Tracking the dehydration process of raw honey by synchronous two-dimensional near infrared correlation spectroscopy. *J. Mol. Struct.* **2014**, *1076*, 42–48. [CrossRef]
249. Xagoraris, M.; Lazarou, E.; Kaparakou, E.H.; Alissandrakis, E.; Tarantilis, P.A.; Pappas, C.S. Botanical origin discrimination of Greek honeys: Physicochemical parameters versus Raman spectroscopy. *J. Sci. Food Agric.* **2021**, *101*, 3319–3327. [CrossRef]
250. Escuredo, O.; Míguez, M.; Fernández-González, M.; Carmen Seijo, M. Nutritional value and antioxidant activity of honeys produced in a European Atlantic area. *Food Chem.* **2013**, *138*, 851–856. [CrossRef] [PubMed]
251. Kosar, F.; Akram, N.A.; Sadiq, M.; Al-Qurainy, F.; Ashraf, M. Trehalose: A Key Organic Osmolyte Effectively Involved in Plant Abiotic Stress Tolerance. *J. Plant Growth Regul.* **2019**, *38*, 606–618. [CrossRef]
252. Amariei, S.; Norocel, L.; Scripcă, L.A. An innovative method for preventing honey crystallization. *Innov. Food Sci. Emerg. Technol.* **2020**, *66*, 102481. [CrossRef]

253. Kivrak, İ. Free Amino Acid Profiles of 17 Turkish Unifloral Honeys. *J. Liq. Chromatogr. Relat. Technol.* **2015**, *38*, 855–862. [CrossRef]
254. Silici, S.; Karaman, K. Chemometric Approaches for the Characterization of Turkish Rhododendron and Honeydew Honeys Depending on Amino Acid Composition. *J. Liq. Chromatogr. Relat. Technol.* **2014**, *37*, 864–877. [CrossRef]
255. Iglesias, M.T.; de Lorenzo, C.; Polo, M.D.C.; Martín-Álvarez, P.J.; Pueyo, E. Usefulness of Amino Acid Composition To Discriminate between Honeydew and Floral Honeys. Application to Honeys from a Small Geographic Area. *J. Agric. Food Chem.* **2004**, *52*, 84–89. [CrossRef]
256. Dimins, F.; Cinkmanis, I.; Radenkovs, V.; Augspole, I.; Valdovska, A. Analysis of 18 Free Amino Acids in Honeybee and Bumblebee Honey from Eastern and Northern Europe and Central Asia Using HPLC-ESI-TQ-MS/MS Approach Bypassing Derivatization Step. *Foods* **2022**, *11*, 2744. [CrossRef] [PubMed]
257. Zhao, H.; Cheng, N.; Zhang, Y.; Sun, Z.; Zhou, W.; Wang, Y.; Cao, W. The effects of different thermal treatments on amino acid contents and chemometric-based identification of overheated honey. *LWT* **2018**, *96*, 133–139. [CrossRef]
258. Iglesias, M.T.; Martín-Álvarez, P.J.; Polo, M.C.; de Lorenzo, C.; González, M.; Pueyo, E. Changes in the Free Amino Acid Contents of Honeys During Storage at Ambient Temperature. *J. Agric. Food Chem.* **2006**, *54*, 9099–9104. [CrossRef] [PubMed]
259. Pérez, R.A.; Iglesias, M.T.; Pueyo, E.; González, M.; de Lorenzo, C. Amino Acid Composition and Antioxidant Capacity of Spanish Honeys. *J. Agric. Food Chem.* **2007**, *55*, 360–365. [CrossRef] [PubMed]
260. Sanz, M.L.; del Castillo, M.D.; Corzo, N.; Olano, A. 2-Furoylmethyl Amino Acids and Hydroxymethylfurfural As Indicators of Honey Quality. *J. Agric. Food Chem.* **2003**, *51*, 4278–4283. [CrossRef] [PubMed]
261. Nie, S.; Huang, J.; Hu, J.; Zhang, Y.; Wang, S.; Li, C.; Marcone, M.; Xie, M. Effect of pH, temperature and heating time on the formation of furan in sugar–glycine model systems. *Food Sci. Hum. Wellness* **2013**, *2*, 87–92. [CrossRef]
262. Yan, S.; Wang, X.; Wu, Y.; Wang, K.; Shan, J.; Xue, X. A metabolomics approach revealed an Amadori compound distinguishes artificially heated and naturally matured acacia honey. *Food Chem.* **2022**, *385*, 132631. [CrossRef]
263. Quintas, P.Y.; Keunchkarian, S.; Romero, L.; Canizo, B.V.; Wuilloud, R.G.; Castells, C.B. Determination of amino acid content and its enantiomeric composition in honey samples from Mendoza, Argentina. *J. Food Process. Preserv.* **2021**, *45*, e15966. [CrossRef]
264. Huang, Z.; Liu, L.; Li, G.; Li, H.; Ye, D.; Li, X. Nondestructive Determination of Diastase Activity of Honey Based on Visible and Near-Infrared Spectroscopy. *Molecules* **2019**, *24*, 1244. [CrossRef]
265. Serrano, S.; Espejo, R.; Villarejo, M.; Jodral, M.L. Diastase and invertase activities in Andalusian honeys. *Int. J. Food Sci. Technol.* **2007**, *42*, 76–79. [CrossRef]
266. Juan-Borrás, M.; Domenech, E.; Hellebrandova, M.; Escriche, I. Effect of country origin on physicochemical, sugar and volatile composition of acacia, sunflower and tilia honeys. *Food Res. Int.* **2014**, *60*, 86–94. [CrossRef]
267. Vit, P.; Pulcini, P. Diastase and invertase activities in Meliponini and Trigonini honeys from Venezuela. *J. Apic. Res.* **1996**, *35*, 57–62. [CrossRef]
268. Czipa, N.; Phillips, C.J.C.; Kovács, B. Composition of acacia honeys following processing, storage and adulteration. *J. Food Sci. Technol.* **2019**, *56*, 1245–1255. [CrossRef]
269. Tosi, E.; Martinet, R.; Ortega, M.; Lucero, H.; Ré, E. Honey diastase activity modified by heating. *Food Chem.* **2008**, *106*, 883–887. [CrossRef]
270. Sahinler, N. Effects of heating and storage on hydroxy methylfurfural and diastase activity of different Turkish honeys. *J. Apic. Res.* **2007**, *46*, 34–39. [CrossRef]
271. Serin, S.; Turhan, K.N.; Turhan, M. Correlation between water activity and moisture content of Turkish flower and pine honeys. *Food Sci. Technol.* **2018**, *38*, 238–243. [CrossRef]
272. Svečnjak, L.; Jović, O.; Prđun, S.; Rogina, J.; Marijanović, Z.; Car, J.; Matošević, M.; Jerković, I. Influence of beeswax adulteration with paraffin on the composition and quality of honey determined by physico-chemical analyses, 1H NMR, FTIR-ATR and HS-SPME/GC–MS. *Food Chem.* **2019**, *291*, 187–198. [CrossRef] [PubMed]
273. Aleixandre-Tudo, J.L.; Castello-Cogollos, L.; Aleixandre, J.L.; Aleixandre-Benavent, R. Chemometrics in food science and technology: A bibliometric study. *Chemom. Intell. Lab. Syst.* **2022**, *222*, 104514. [CrossRef]

foods

Review

How Chemometrics Can Fight Milk Adulteration

Silvia Grassi [1], Maria Tarapoulouzi [2], Alessandro D'Alessandro [3], Sofia Agriopoulou [4], Lorenzo Strani [3,*] and Theodoros Varzakas [4,*]

[1] Department of Food, Environmental and Nutritional Sciences (DeFENS), Università degli Studi di Milano, Via Celoria, 2, 20133 Milano, Italy
[2] Department of Chemistry, Faculty of Pure and Applied Science, University of Cyprus, P.O. Box 20537, Nicosia CY-1678, Cyprus
[3] Department of Chemical and Geological Sciences, University of Modena and Reggio Emilia, Via Campi 103, 41125 Modena, Italy
[4] Department of Food Science and Technology, University of the Peloponnese, Antikalamos, 24100 Kalamata, Greece
* Correspondence: lostrani@unimore.it (L.S.); t.varzakas@uop.gr (T.V.)

Abstract: Adulteration and fraud are amongst the wrong practices followed nowadays due to the attitude of some people to gain more money or their tendency to mislead consumers. Obviously, the industry follows stringent controls and methodologies in order to protect consumers as well as the origin of the food products, and investment in these technologies is highly critical. In this context, chemometric techniques proved to be very efficient in detecting and even quantifying the number of substances used as adulterants. The extraction of relevant information from different kinds of data is a crucial feature to achieve this aim. However, these techniques are not always used properly. In fact, training is important along with investment in these technologies in order to cope effectively and not only reduce fraud but also advertise the geographical origin of the various food and drink products. The aim of this paper is to present an overview of the different chemometric techniques (from clustering to classification and regression applied to several analytical data) along with spectroscopy, chromatography, electrochemical sensors, and other on-site detection devices in the battle against milk adulteration. Moreover, the steps which should be followed to develop a chemometric model to face adulteration issues are carefully presented with the required critical discussion.

Keywords: fraud; authentication; dairy; clustering; classification; regression; validation

Citation: Grassi, S.; Tarapoulouzi, M.; D'Alessandro, A.; Agriopoulou, S.; Strani, L.; Varzakas, T. How Chemometrics Can Fight Milk Adulteration. *Foods* **2023**, *12*, 139. https://doi.org/10.3390/foods12010139

Academic Editor: Gianfranco Picone

Received: 26 November 2022
Revised: 10 December 2022
Accepted: 22 December 2022
Published: 27 December 2022

1. Introduction

Milk and milk products provide the human body with valuable nutritional components such as proteins, carbohydrates, vitamins, minerals, organic acids, and fat [1,2]. Milk's high protein content has attracted many consumers, making it a popular nutritional commodity [3]. The increasing consumption of milk and dairy products leads to many cases of adulteration [4,5]. A range of possible milk adulterants is described by Nascimento et al. [4].

The prices of milk differ primarily depending on the type of animal from which they come, whereas its availability is significantly affected by the season. These two factors are enough to cause problems in its market, as practices of replacing it with cheaper milk are common [6]. Goat's milk shows a nutritional profile superior to that of cows, as a result of which it is a priority for consumers not only in traditional dairy products such as cheese and yogurt, but also in liquid form. Its low production combined with its beneficial nutritional content makes this category of milk an attractive target for adulteration. Goat's milk is easily mixed with water, whey as well as cow's milk which is much cheaper. The latest fraud is increasingly worrying people because of their sensitivity to lactose and the allergic disorders that can be caused by cow's milk proteins [7]. An equally important adulteration is related to the substitution of goat's milk with sheep's milk. In this case, the lower price of goat's milk compared to sheep's milk pushes the producers to this adulteration [6].

Fraud in milk production is carried out by admixture or substitution of inferior substances and sometimes dangerous products. The economically motivated adulteration (EMA) is the most important, aiming to gain profit by the addition of extraneous water, glucose or other sugars, non-dairy proteins such as soybean and pea protein isolates [8], various substances such as melamine, urea, maltodextrin, cheese whey (a byproduct of cheese production) [9], hypochlorite, dichromate, salicylic acid [10], and reconstituted milk powders to correct protein and/or density values [11]. A famous case of adulteration was recorded in China in 2013 when the substance melamine was detected in milk powder in infant milk products, which was added to increase the apparent protein content, with dramatic consequences for public health [12].

The deliberate addition of formaldehyde to raw milk is also illegal and considered a major adulteration, which aims to increase the shelf life of milk at room temperature. High moisture content is responsible for the rapid spoilage of milk. Therefore, formaldehyde provides preservative and antiseptic properties, and the ability to improve the appearance including the smell of milk. Furthermore, formaldehyde is toxic at low concentrations and is classified as a human carcinogen by the International Agency for Research on Cancer (IARC) [12,13].

Another form of adulteration is the replacement of milk fat with vegetable fats of lower economic value [14]. Among others, soybean oil has been mentioned in the adulteration of milk [15]. In addition, the recent EU regulations for foods designated as PDO (protected designation of origin), PGI (protected geographical indication), and TSG (traditional specialty guaranteed) require the inclusion on the label of the geographical origin of food. In the case of dairy products such as cheeses produced in a defined area with specific physicochemical and sensorial features, their geographical origin is put forward as an important indication [16].

Chemometrics plays a dominant role in the field of food adulteration as it relates a multitude of chemical analytical characteristics to the qualitative and quantitative analysis of food [17]. Deriving a fingerprint of each sample and reflecting its complex chemical composition could be a way to solve such difficult analytical tasks. Then, chemometric techniques can be used to develop classification models to classify samples into authentic/adulterated ones, or regression models aiming at quantifying a specific adulterant [8,18–21]. In this direction, both specific and non-specific fingerprinting can be implemented. Specific chemical analysis is based on the detection of organic species, mainly achieved by chromatographic techniques.

The non-specific fingerprinting approach relies on the implementation of instrumental methods to obtain a multivariate description of the chemical composition of the sample. These non-specific fingerprints can be obtained by different methodologies such as Fourier transform infrared spectroscopy (FT-IR), mid-infrared spectroscopy (MIR), Raman spectrometry, nuclear magnetic resonance (NMR), or mass spectrometry [22]. All these methodologies have been used in studies, which are relevant to authenticity and chemometrics in milk and dairy products [23–25]. In addition, near-infrared (NIR) spectroscopy has been used by several researchers to detect various forms of adulteration in both cow's milk and cow's milk products [26–28].

Vibrational spectroscopic techniques are rapid, low-cost, and non-destructive tests that require only limited training for processing. Results are evaluated using chemometric models to extract meaningful information that distinguishes different and significant groups by removing redundant data [29].

Data processing can be completed by principal component analysis (PCA) since it is amongst the most fundamental methods for multivariate data exploration [18]. PCA has been used along with other methodologies to help to differentiate fresh milk and reconstituted skim milk powder samples [11].

kNN (k-nearest neighbor), PLS-DA (partial least squares-discriminant analysis), and SIMCA (soft independent modeling of class analogy) are the most popular classification methods [30]. kNN and PLS-DA have been used for the detection of various types of adulteration, such as water, urea, cow's whey, and cow's milk in goat's milk samples [31]. SIMCA could also be employed to model the class of fresh types of milk. When addressing a specific adulterant quantification, the goal could be achieved by means of partial least squares (PLS) regression analysis, as demonstrated for the prediction of fresh milk adulteration with reconstituted skim milk powders [11].

Finally, in order to validate a chemometric approach, a sampling strategy should be followed taking into account the size and the representativeness of the sample along with intrinsic variability [32]. Sampling is closely associated with robustness and reliability. Other key parameters of authenticity and fraud not to be ignored are the heterogeneity of a food matrix and the presence of an undeclared substance to the geographical origin discrimination.

In this framework, the aim of this work is to give an overview of the recent application of different chemometric techniques—from clustering to classification and regression applied—to several analytical data—encompassing spectroscopy, chromatography, and electrochemical sensors—to fight milk adulteration. Further, a critical discussion is presented to schematize the steps which should be followed to develop a chemometric model to face adulteration issues.

2. Chemometric Approaches

2.1. Clustering

The definition of "cluster analysis" or "clustering" encompasses the techniques which split a set of samples (observations) into several groups or clusters. The outcome is usually represented as a vector of data, or a point (scatter) in a multidimensional space [33]. Clustering falls in the general category of unsupervised pattern recognition and numerical and mathematical taxonomy [33,34]. Natural grouping of data takes place based on some inherent similarity, as clustering is performed without any group labels, and this justifies the unsupervised pattern recognition [33,35]. Furthermore, it takes place based on similarities of the samples within the same group and others in different groups. Therefore, homogeneity is dominant within the same groups [34]. In practice, the most common approach to define similarity is the distance among the patterns; by lowering the distance (e.g., Euclidean distance which is a well-used dissimilarity measure) between the two objects, higher similarity and vice versa will be obtained [35,36].

Clustering is a valuable component of data analysis or machine learning-based applications such as regression, prediction, data mining, etc. [35]. Saxena et al. (2017) [35] stated that there are various ways to categorize clustering methods because it is difficult to define a cluster. In their paper, they suggested division into two different groups such as hierarchical and partitioning techniques, or in three categories based on application, density-based methods, model-based methods, and grid-based methods.

Hierarchical methods initially group the objects into small clusters of some samples, and these are next grouped into larger clusters, thus a dendrogram is produced, which is a tree-based depiction of each observation [36]. Optimization- partitioning methods split the samples into a few groups to optimize a particular feature e.g., total within-group distances. In this category, algorithms like k-*means clustering*, *Fuzzy* c-*means clustering*, etc., are included [33–35]. Density-based clustering is focused on the probability that data objects are drawn from a specific probability distribution and the overall distribution of the data is assumed to be a mixture of several distributions. Data points can be derived from different types of density functions (e.g., multivariate Gaussian or *t*-distribution), or from the same families but with different parameters. Model-based clustering works by detecting feature details for each cluster, where each cluster represents a concept or class. Decision trees and neural networks are the two most frequently used methods in this category. Grid-based clustering divides the space into a finite number of cells that make a grid structure on which all the operations for clustering are performed [35].

Recently, many evaluation criteria have been developed, and these are internal and external. Internal quality parameters include the sum of squared error, scatter criteria, Condorcet's criterion, the C-criterion, category utility metrics, and edge cut metrics. External quality criteria are related to the mutual information-based measure, Rand index, F-measure, Jaccard index, Fowlkes–Mallows index, and confusion matrix [35].

Clustering is applied to perform data reduction or compression for handling huge loads of data. It helps in compressing data information by grouping them into different sets of clusters. This helps us to choose what is useful or not by saving time from data processing along with data reduction [35]. Other uses contain data mining, document retrieval, image segmentation, and pattern classification [33].

In order to explore the use and development of clustering methods recently, Table 1 has been prepared to summarize the studies related to milk adulteration and authenticity.

Table 1. Recent studies (2015–2021) related to milk adulteration and authenticity in combination with clustering analysis.

Type of Milk	Target	Analytical Method(s)	Clustering Method	Approach	Reference
		Milk adulteration			
Cow's, sheep's, and water buffalo's origin milk	Adulteration from different species' origin milk	FTIR	HCA	method	[37]
Bovine milk	Adulteration with urea	EIS	HCA	Euclidean distance	[36]
UTH milk samples (skimmed and semi-skimmed) and raw milk	Adulteration with cheese whey, based on quantification of caseinomacropeptide	FTIR-ATR	HCA	Euclidean distance and Ward's method	[38]
Cow milk	Adulteration with melamine and urea	Electrochemical biosensor	HCA	Ward's method	[39]
Bovine milk	Adulteration with formaldehyde, based on aldehydes and ketones	Colorimetric sensor array	HCA	-	[40]
UHT whole bovine milk and UHT goat milk	Adulteration with soymilk in bovine and goat milk, as well as bovine milk in goat milk.	NMR	CA	The minimum distance method	[41]
Raw cow milk	Adulteration with Sodium Salicylate, Dextrose, Hydrogen Peroxide, Ammonium Sulphate	Sensor system	k-means clustering algorithm	-	[42]
		Milk authentication			
Powder and liquid milk	Type of milk based on metal profiles	ICP-OES	HCA	Euclidean distance and Ward's method	[43]
Organic and conventional milk	Type of milk (organic vs. conventional) based on organic status and trace element content	ICP-MS	HCA	Euclidean distance and Ward's method	[44]
Malaysian vs. milk from other countries	Geographical origin, based on metal content	ICP-MS	HCA	Ward's method	[45]

Table 1. *Cont.*

Type of Milk	Target	Analytical Method(s)	Clustering Method	Approach	Reference
-	Geographical origin, isotope ratios, metals, and fatty acids	CF-IRMS ($\delta^{18}O$), EA-IRMS ($\delta^{13}C$ and $\delta^{15}N$), GC (fatty acids), ICP-OES (Na, K, Mn, P, Zn, Ca, Fe, and Mg), and ICP-MS (other metals)	HCA	-	[46]
Cow milk	Geographical origin, based on stable isotope ratios	IRMS and CRDS	HCA	-	[47]
Raw milk	Geographical origin, based on stable isotope ratios and metal content	IRMS and ICP-MS	HCA and k-means clustering algorithm	HCA: Euclidean distance and Ward's method K means: 200 iterations and 25 random starting points	[48]
Cow, goat, camel, donkey, and yak milk	Species recognition based on sn-2 and sn-1,3 fatty acid composition and sterols	GC, GC-MS	HCA	-	[49]
Fresh buffalo, bovine, and donkey milk as well as processed milk samples (pasteurized and dried skimmed powder)	Species recognition based on amino acids, non-amino acids, and citric acid cycle metabolites	GC-MS	HCA	Euclidean distance and Ward's method	[50]
Reconstituted milk vs. UHT milk	Different content of peptides, lipids, and nucleic acids	UPLC–Q-TOF-MS combined with UPLC–MS/MS	HCA	-	[51]
Cow milk	Fat globule characteristics (diameter, membrane surface, and yield), fat, protein, fatty acids, calcium content	IR (fat, protein, and lactose contents), GC (fatty acids composition), atomic absorption spectrophotometry (calcium content)	HCA	Euclidean distance	[52]
Cow, goat, buffalo, and camel milk	Different seasons of milk collection, based on sterols in milk fat of different species' origin of milk	GC–MS-SIM	HCA	Euclidean distance	[53]

Abbreviations: CA = cluster analysis, CF-IRMS = continuous flow-isotope ratio mass spectrometer, CRDS = cavity ring-down spectroscopy, EA-IRMS = element analysis-isotope mass spectrometry, EIS = electrochemical impedance spectroscopy, FCM = fuzzy c-means, FTIR-ATR = Fourier transform infrared-attenuated total reflection, FTIR = Fourier transform infrared spectroscopy, GC = gas chromatography, GC-MS = gas chromatography-mass spectrometry, GC-MS-SIM = gas chromatography-mass spectrometry-single ion monitoring mode, HCA = hierarchical cluster analysis, ICP-MS = inductively coupled plasma mass spectrometry, ICP-OES = inductively coupled plasma emission spectroscopy, IR = infrared, IRMS = isotopic ratio mass spectrometry, UHT = ultra-high temperature, UPLC–MS/MS = UPLC–tandem mass spectrometry, UPLC–Q-TOF-MS = ultra-high performance liquid chromatography-quadrupole time-of-flight mass spectrometry.

Regarding milk adulteration studies, Cirak et al. [37] focused on determining milk species adulteration by using FTIR. HCA was conducted based on Ward's algorithm after having calculated the initial derivate by using a standard method. The produced 2D-dendrogram indicated that the types of origins (sheep, cow, and water buffalo origin, and adulterated samples in binary mixtures) were clustered correctly. Minetto et al. [36] applied HCA to detect urea in raw bovine milk samples, and the Euclidean distance was used to build the dendrogram. HCA helped them to find the more appropriate number of clusters which was used later in the classification of the samples. Vinciguerra et al. [38] used HCA as an exploratory treatment on the pre-processed measurements obtained by FTIR-ATR. By using both the Euclidean distance and Ward's method, a dendrogram was generated, however no pattern related to the caseinomacropeptide concentration was observed in the dendrogram, and multivariate regression was followed. Qualitatively, the adulterated groups with caseinomacropeptide were separated correctly in 3 groups: raw milk, skimmed milk, and semi-skimmed milk. Adulteration with melamine and urea in cow's milk was also studied by Ezhilan et al. [39], who developed an electrochemical biosensor to detect the two adulterants simultaneously. HCA application was useful to study the interrelationship of the factors affecting the model for measurements taken by using various combinations of concentrations of the adulterants. Mostafapour et al. (2021) [40] used a colorimetric array device. The authors commented that even if there are differences in the colorimetric schemes of the analytes, it is not a proper manner to group the samples after visual examination, thus chemometrics is used to perform the clustering. The HCA dendrograms showed highly accurate clustering of the studied carbonyl compounds, particularly eight different aldehydes and ketones. In addition, HCA showed that one sample from formaldehyde and one sample from acetophenone has been misclassified. Li et al. [41] used NMR to detect the metabolites as markers of different milk types. Clustering analysis (CA) was very useful as it provided similarities for the same species of milk as well as variations in different milk species by applying the minimum distance method. CA also separated the three milk types and showed that NMR and metabolites can differentiate these milk products. Sowmya et al. [42] during the pre-processing steps applied cluster analysis, i.e., the k-means clustering algorithm. The algorithm proceeded by calculating the centroid point of the dataset and the groups' mean points to build the new groups required. The aim was to see the grouping of samples, to identify the similarities in the same categories, and to check if the adulterants can be clustered by using raw spectra. Intraclass variation was performed.

Regarding milk authenticity, Souza et al. [43] studied the metal profile of powder and liquid milk samples to differentiate them based on the type of milk. HCA successfully confirmed the initial outcome of PCA, and it allows the visualization of a sample's trend to form two groups. Whole cow powder milk, whole goat powder milk, skimmed cow powder milk, and milk compounds powder fell in the first group due to their similar composition. A sample from the last group clustered at a longer distance from its group due to the high content of Zn. The second group consisted of whole and skimmed cow liquid milk and some yogurts. Rodriguez-Bermudez et al. [44] by applying HCA revealed a correct clustering based on the type of milk, organic vs. conventional. It was obvious that the variables (metal content) in both the organic and conventional sets were distinct. To determine the geographical origin, Zain et al. [45] measured the metal content of milk samples and due to different environmental conditions, and the samples clustered successfully by HCA. Ca, Na, Fe, Zn, Mn, K, Ba, and Mg are the metals that were significant for the samples' grouping regarding geographical origin. Xu et al. [46] worked also in terms of geographical origin by measuring isotope ratios, metals, and fatty acids and then by applying HCA. $\delta^{18}O$ measurements were taken by having the milk in the fluid state, but for $\delta^{13}C$, $\delta^{15}N$, and elemental and fatty acid measurements lyophilization took place. HCA aided to picture the correlation between the sample and each variable as HCA heatmaps were created. In addition, geographical origin was the target of Amenzou et al. [47], who studied the $^{13}C/^{12}C$, $^{15}N/^{14}N$, $^{18}O/^{16}O$. The application of HCA was very important to visualize the samples in 3 important clusters. The stable isotope ratios analysis in combination with chemometrics showed a very good capability to indicate the geographical origin of milk. In a similar study, Podkolzin and Solovev [48] used HCA and the k-mean clustering algorithm and both methods showed an equal number of clusters with almost the same content. Karrar et al. [49] used HCA to evaluate the similarity in terms of sn-2 and sn-3 fatty acids in different milk-origin samples. HCA heatmaps were produced to present the content of sn-2 and sn-3 fatty acids in the samples. Bhumireddy et al. [50] applied HCA to group the samples based on intrinsic similarities in their GC-MS measurements. HCA heatmaps were produced using the log-transformed and normalized values of the relative abundance of 17 amino acids, and their high and low expressions in each sample were presented with different colors. Tan et al. [51] employed HCA to proceed to the clustering of the different biomarkers (peptides, lipids, and nucleic acids) and to demonstrate the chemical properties of the important metabolites. It must be also noted that the results indicated that the processing that takes place to produce milk powders influences the nutritional loss of peptides and lipids. HCA heatmaps showed that nutritional components were found to be in lower concentrations in reconstituted milk compared to ultra-high-temperature milk. Couvreur and Hurtaud [52] studied the parameters of fat globule characteristics (diameter, membrane surface, and yield), fat, protein, fatty acids, and calcium content in milk concerning diet composition, milking frequency, breed, stage of lactation, parity and residual/cisternal milk. Based on the principal components of PCA, HCA was performed which indicated 4 independent clusters of milk. A minor relationship was observed between fat content and fat globule diameter in milk, especially for the Normandy breed at the very end of the lactation. Dhankhar et al. [53] proposed a method to study the influence of season on the variability of sterols in different species' origins. Buffalo milk has a very different sterol profile compared to other animal species. In addition, seasonal variation affected especially cholesterol content compared to other minor sterols, and winter milk had a lower level of cholesterol compared to other seasons. The authors commented that the variation based on season was not able to be satisfactorily explained by PCA. However, HCA correctly grouped the 4 species of animals into 4 clusters by the sterol content. Squared Euclidean distance between objects was applied in HCA, to give the natural grouping of samples. The HCA dendrogram allowed the visualization of the similarity or dissimilarity of the measurements in 2D.

As can be observed, HCA is the main representative of the clustering methods. It is also important to note that after CA, most of the studies presented above proceeded to classification and/or regression analysis, which are presented in the next sections of this paper. Overall, in the aforementioned-studies, CA was used as a step to visualize the samples in clusters and to understand the interrelationships of the samples' datasets, before proceeding to supervised methods.

2.2. Classification

The capability to assign an object to a class on the basis of its characteristics belongs to the pattern recognition field. There are many methods to classify objects and one of the applications of chemometrics is the classification of objects in groups depending on their characteristics expressed as results of a set of measurements [54]. Classification methods could be distinguished into "discriminant" and "class-modeling" techniques (Table 2).

In the first case, the technique tries to discriminate among the object's groups dividing the model hyperspace into several regions equal to the number of classes and assigning each object to a specific region of the hyperspace on the base of its characteristics. In this way, each sample may belong to just one class. In the case of class modeling instead, the technique tries to model the analogies between objects of a class rather than observe the differences. So, each group of objects is modeled separately, and, at the end, an object could be assigned to one or more classes, or rejected as non-included in none of the classes (Figure 1).

Table 2. Main classification methods cited.

Classification Method	Extended Name	Abbreviation
Discriminant	Partial least squares-discriminant analysis	PLS-DA
	Orthogonal partial least squares-discriminant analysis	OPLS-DA
	One class-partial least squares	OC-PLS
	Quadratic discriminant analysis	QDA
	Random forest	RF
	Support vector machine	SVM
	Linear discriminant analysis	LDA
	k-nearest neighbors	kNN
	Extreme learning machine	ELM
	Ensemble of extreme learning machine	EELM
Class-modeling	Soft independent modeling of class analogy	SIMCA
	Data-driven soft independent modeling of class analogy	DD-SIMCA
	Unequal class models	UNEQ

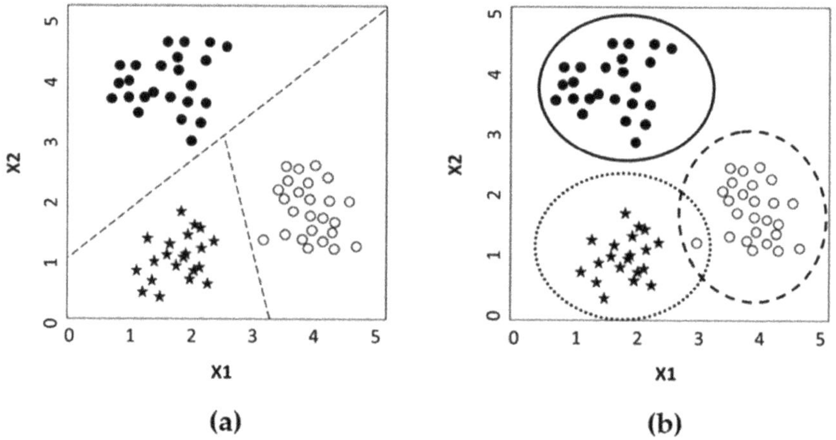

Figure 1. Example of difference between discriminant (**a**) modeling and (**b**) classification methods. In (a) the hyperspace is divided into regions equal to the category number.

In the discriminant classification, some methods may be counted: kNN, PLS-DA, LDA, and QDA. Instead, class-modeling techniques may be included: SIMCA, DD-SIMCA, and UNEQ [55].

Describing the details of all classification methods is out of the scope of this work, and here we will consider only the most used techniques (discriminant or class-modeling) applied to the milk and dairy product classification in milk adulteration in the last years.

A basic distinction between supervised and unsupervised classification techniques will be maintained. Supervised classification methods require some knowledge "a priori" of the classes and the method to assign or not assign samples to a certain class; in contrast, the unsupervised methods just classify samples on the base of their characteristics [56].

In recent years, the number of studies that use chemometrics to properly elaborate and interpret analytical results is largely increasing. The power of the chemometric technique is evident in all the cases where the output of an instrumental analytical technique is a spectrum, like in visible and/or infrared spectroscopy (VIS, VIS-NIRS, NIRS), nuclear magnetic resonance (NMR), or spectrometry (CG-MS, LC-MS).

Regarding classification used in milk adulteration, in the last five years, there have been several examples that used chemometrics and in Table 3 some relevant examples have been reported.

The use of chemometrics on instrumental data requires some preliminary steps, like data pre-processing or data dimension reduction. A short illustration of these steps has been reported below. In general, the application of a specific classification technique in place of another one depends on the data structure. In some cases, using one method rather than another one leads to the same results; in others, the application of a specific method could improve classification efficiency.

The classification statistical techniques most used in the last years for milk applications were PLS-DA as a pure classification technique and SIMCA as a class-modeling approach. Kamboj [57], for example, used PLS-DA to detect water adulteration in milk from NIRS spectra. Chung [58], working on isotope ratio data, used OPLS-DA to perform classification. The paper did not extensively explain the reason for this choice. Jin [59] used the least squares support vector machine (LS-SVM) for qualitative analysis of adulterated milk identification using 2D autocorrelation spectroscopic data. Karunathilaka [60] used Raman spectroscopy data from two different instruments and SIMCA for not-target classification to detect milk powder adulteration. Galvan [61], on data coming from low-cost spectroscopic devices (NIR and energy dispersive X-ray fluorescence—EDXRF), used more than one technique: PLS-DA for the EDXRF data and C-support vector classification (C-SVC) for NIR data. In the end, they concluded that DD-SIMCA was more useful to classify the samples with good accuracy (98.9%). Other two interesting uses of PLS-DA applied to NIR data were conducted by Ejeahalaka et al. [62] on cow's milk and by Di Donato et al. [63] on donkey's milk. DD-SIMCA is a one-class classification algorithm proposed in 2017 by Zontov [64]. The algorithm in the first phase is similar to the SIMCA algorithm, with a preliminary PCA. Then the PCA results were used to calculate the orthogonal distance and score distance for each object. These distances were then used to individuate a threshold limit value of the classification area. New samples were then classified in the orthogonal vs. score plot and assigned to the class when under the acceptance area defined for a given alpha value. Wang [65] evaluated four different classification methods (RF, LDA, SVM, and kNN) when dealing with milk authentication by infrared spectroscopy. To evaluate the best algorithm, the means of precision, accuracy, recall (true positive divided by the sum of true positive and false negative), and another parameter F1 (that together evaluate precision and recall) were calculated for each performance evaluation of all classes and for every classifier. The results indicate that RF had the best performance. In a work about image analysis [66] applied to recognize goat's milk (as a target class) from other milk species adulterants, two methods were tested: OC-PLS and DD-SIMCA. In this case, OC-PLS was not recommended and DD-SIMCA was preferred. Chen [67] used ELM and extreme ELM (EELM) to classify six types of milk of different brands analyzed by NIRS. ELM is a regression and classification algorithm. It is simple and efficient and extremely fast. Vargas [56] applied PLS on the voltammetric characterization of fresh cow's milk and from milk powder, using as Y the percentage of adulteration with reconstituted milk. Potocnik [68] in his paper used DA and OPLS-DA to elaborate data from isotopic ratios on types of milk to verify their geographical origins. Similarly, Xie [69] performed similar work on geographical discrimination of milk from Mongolia using isotope ratio, elements, and amino acids composition. In this paper, the chemometric analysis was performed with OPLS-DA. Tommasini [70], again using NMR, in this case, to classify the breed of cow, used PLS-DA analysis to distinguish between milk from different cow breeds, Friesian vs. autochthonous. PLS-DA and OPLS-DA, together with HCA and RF, were also cited by Sundelkide [71] to elaborate on the NMR spectra acquired in order to underline the importance and potentiality of the milk metabolomics studies. Segato et al. also used NMR to discriminate the metabolic profiles of different pasture-based alpine Asiago PDO cheeses [72]. To conclude the NMR overview, Yanibada [73] reported the application of OPLS-DA, preceded by an explorative PCA, to classify two groups of cows by NMR metabolomics. In Table 3 a synthesis of the more relevant papers identified has been reported.

To summarize, excluding PCA (mainly used to preliminarily study the problem), PLS-DA and OPLS-DA were the most used methods for classification in the recent papers on milk classification. The second most used have been SIMCA and DD-SIMCA, followed by many other various methods. The use of some classification techniques more than others could be attributed to different reasons: PLS-DA and OPLS-DA, the more used in the reviewed articles, are more known compared to some other more specific methods. The main reason for their popularity is probably linked to the fact that they are implemented in a lot of user-friendly commercial software, mainly used by non-expert users. It is advisable to use PLS-DA in place of LDA when the number of variables is higher than the number of samples and when the predictors are correlated. When classes are not balanced (i.e., the number of samples for each class is very different), better results are often obtained by class-modeling techniques, such as SIMCA. The choice of the proper classification method should also be influenced by their parametric or non-parametric nature: the former, such as LDA, assumes that the data follow a particular statistical distribution, so the model calculation becomes the calculation of the parameters of these distributions. The disadvantage of parametric techniques is that they can lead to big mistakes when starting assumptions fail to be verified. The advantage is that they make it easier to obtain the probability of obtaining a correct classification. On the other hand, non-parametric methods do not explicitly assume no statistical distribution (e.g., SIMCA, kNN, etc.).

Table 3. Recent studies (since 2018) involving classification methods related to milk adulteration.

Type of Milk	Target	Analytical Method(s)	Classification Method(s)	Reference
Cow	Classification	NIRS	EELM	Chen [67]
Cow	Organic milk geographical indication	Isotope ratio	OPLS-DA	Chung [58]
Cow	Authenticity	NMR	CDA	Segato [72]
Goat	Adulteration detection	Image analysis	OC-Classifier, OC-PLS, DD-SIMCA	dos Santos Pereira [66]
Cow	Quality	Chemical analysis, NIRS	PCA, SIMCA, PLS-DA	Ejeahalaka [62]
Various	Authenticity	NIRS, EDXRF	DD-SIMCA, PLS-DA, C-SVC	Galvan [61]
Cow	Adulteration	IR	LS-SVM	Jin [59]
Cow	Adulteration	NIRS	PCA, PLS	Kamboj [57]
Milk powder	Adulteration	Raman	PCA, SIMCA	Karunathilaka [60]
Cow	Geographical origin	Isotope ratio	ANOVA, DA, OPLS-DA, DD-SIMCA	Potočnik [68]
Cow	Authentication	Chemical analysis	PCA, OPLS-DA	Vargas [56]
Cow	Authentication	FTIR	PCA, kNN, SVM, RF, LDA	Wang [65]
Cow	Traceability	Chemical analysis, isotope ratio,	PCA, OPLS-DA	Xie [69]
Cow	Quality, breed classification	NMR	PLS, PLS-DA	Tomassini [70]

Table 3. *Cont.*

Type of Milk	Target	Analytical Method(s)	Classification Method(s)	Reference
Cow	Quality	NMR	PCA, PLS-DA, OPLS-DA, HCA, RF	Sundekilde [71]
Cow	Quality	NMR	PCA, OPLS-DA	Yanibada [73]
Donkey	Authentication	NIRS	PLS-DA, VSN, ASCA	Di Donato [63]

Abbreviations: ANOVA = analysis of variance, ASCA = ANOVA simultaneous component analysis, CDA = canonical discriminant analysis, C-SVC = C-classification support vector classifier, DA = discriminant analysis, DD-SIMCA = data-driven soft independent modeling of class analogy, EELM = ensemble of extreme learning machine, HCA = hierarchical cluster analysis, k-NN = k-nearest neighbors, LS-SVM = least squares support vector machine, LDA = linear discriminant analysis, OC = one-class classifier, OC-PLS = one-class partial least Squares, OPLS-DA = orthogonal partial least squares-discriminant analysis, PCA = principal component analysis, PLS = partial least squares, PLS-DA = partial least squares-discriminant analysis, RF = random forest, SVM = support vector machine, VSN = variable sorting for normalization.

2.3. Regression

Multivariate regression is widely used to quantify the concentration of adulterants in food matrices. In Table 4, the papers presented for this review in the last five years, with reference to regression methods, are listed.

The most popular multivariate regression method is certainly partial least squares (PLS) [74], as it is relatively simple to use and is implemented in a lot of statistical software, including instruments software (e.g., Opus). For this reason, in the last five years, PLS regression was used in more than three-quarters of the works on milk adulteration. The main advantage of PLS is its ability to handle data with many more variables than samples, specifically when these variables co-vary. The algorithm performs a simultaneous decomposition of both X (descriptors matrix) and Y (response matrix) matrices with the aim to maximize the covariance between the two matrices, computing at the same time latent variables (LVs) that explain the maximum variability of X. Due to its features, PLS is often used to treat spectral data, especially in the infrared region. In fact, with respect to other methods, such as chromatography, near- and mid-infrared spectroscopies (NIR and MIR, respectively) offer numerous practical advantages: they are fast, non-destructive, non-invasive, and relatively cheap techniques. Moreover, sample preparation is usually absent or extremely simple. The only drawback is the complex interpretation of the spectra, especially for NIR spectra, where differences in overtones and combination bands are difficult to detect and interpret. For this reason, the use of a simple multivariate tool for the extraction of relevant information is essential.

NIR spectroscopy is used to detect and quantify different kinds of adulterants: the most common and simple ones, such as water [57], urea [75–77], melamine [76–78], and sugar [79], and less common ones, such as sodium dodecyl sulfate (a milk surfactant) [80] or different vegetable oils added to yogurt [81]. Moreover, NIR spectroscopy is also used to detect specific adulterants for particular matrices as showed by Pandiselvam et al., where coconut milk residue was used to adulterate desiccated coconut powder [82], or by Di Donato et al., which used cow's milk as an adulterant in goat's milk samples [63].

MIR spectroscopy is also widely used coupled with PLS regression to detect and quantify adulterants in different milk samples. In several works, MIR was used to quantify the amount of cow's milk in more expensive milk types: buffalo [83,84], goat [85], and horse [86]. It was used to analyze coconut milk samples adulterated with water [87]. MIR spectrometers equipped with an ATR cell were employed to detect soya bean oil and common sugar [88], sucrose [89], and formalin [13] in cow's milk. The use of an ATR cell allows for minimizing sample preparation, as the penetration depth in the sample of IR radiation does not depend on sample thickness. Obviously, NIR and MIR spectra have to be properly pre-processed to minimize noise, scattering, and other undesirable contributions. Hence, it is good practice to build PLS models applying different combinations of pre-processing methods and compare the results to see which one provides the best prediction performance. For instance, Temizkan et al. [81] tried different preprocessing options: normalization, smoothing, first derivative, second derivative, multiplicative scatter correction (MSC), and standard normal variate (SNV). These, together with the baseline correction, are the most common row pre-processing method used to treat NIR and MIR spectra.

Another spectroscopic technique coupled with PLS in the milk adulteration field is Raman spectroscopy, whose spectra rely on the light scattering of vibrating molecules. Raman spectroscopy was employed to find maltodextrin, sodium carbonate, and whey in bovine milk [90,91], as well as margarine, palm oil, and corn oil in cheeses made using adulterated milk samples [92,93].

Although in the majority of papers PLS regression is applied to vibrational spectroscopic data, in recent literature, there are also many applications with different techniques. Cyclic voltammetry, using a graphite/SiO_2 hybrid-working electrode, was employed to quantify reconstituted skim milk in cow's milk [11], electrochemical impedance spectroscopy was used to measure urea [36] whereas face fluorescence spectroscopy and laser-induced breakdown spectroscopy assessed the amount of bovine milk in buffalo milk [90] and ovine and caprine milk [94], respectively. Moreover, time-domain NMR [12] and opto-electronic nose [40] quantified formaldehyde in bovine milk. The versatility of this technique is one of the reasons why its presence is predominant among papers that deal with multivariate regression. Actually, in many papers, PLS is frequently compared with other two multivariate regression methods, i.e., multiple linear regression (MLR) [95] and principal component regression (PCR) [96]. Jaiswal et al. [85] and Gonçalves et al. [84] showed comparable results between PLS and MLR in quantifying adulterants with MIR spectroscopy. Conceição et al. [97] used MLR coupled with MIR spectroscopy to assess the amount of sodium bicarbonate, sodium hydroxide, hydrogen peroxide, starch, sucrose, and urea in cow's milk. However, the use of MLR is not recommended if the data matrix is ill-conditioned, namely has more variables (e.g., wavenumbers) than samples, and if those variables co-vary, as the regression model would be unstable. On the other hand, PCR is a more reliable method, since the variables are orthogonal (the ill-conditioned matrix problem has been overcome) and only relevant information in the original data matrix is considered, being based on PCA. Unlike PLS, in PCR the information in the response matrix (Y) is not taken into account when choosing the number of PCs. Moreover, for this reason, PLS has been habitually preferred to PCR. In some of the papers inspected for this review, these two methods were compared: on three occasions PLS provided the best prediction performances [13,86,89], whereas in one case the results obtained by the two methods were similar [87].

Throughout the years, the PLS algorithm has been modified by many authors to add features and make it more suitable for specific tasks (e.g., multiblock analysis, locally weighted models, etc.). One of the most famous extensions of PLS is orthogonal PLS (OPLS) [98], which removes the systematic variation from X that is not correlated (orthogonal) to Y. It was used by Delatour et al. [99] on data collected from eight different NIR and MIR miniature sensors to measure the amount of semicarbazide hydrochloride, ammonium sulfate, and cornstarch in skimmed milk powder [96]. Another different use of PLS regression, synergy interval PLS (siPLS) [100], has been used by Vinciguerra et al. to quantify cheese whey in cow's milk samples through MIR spectroscopy [38]. In this method, the MIR spectra were divided into different intervals (8, 16, 32, 64, and 128) with the same number of variables, applying a PLS on each interval. Furthermore, combinations of these intervals (two by two, three by three, and four by four) were also explored and PLS was performed for each combination. Hosseini et al. used the genetic algorithm PLS (GA-PLS) in order to perform an efficient variable selection before calculating the regression models [80]. Lastly, unfolded PLS with residual bilinearization (U-PLS/RBL) [101] coupled with fluorescence spectroscopy was used by Barreto et al. to quantify melamine in bovine milk [102]. Actually, U-PLS/RBL belongs to the family of multiway methods, similar to other techniques such as parallel factor analysis (PARAFAC) and multivariate curve resolution-alternating least squares (MCR-ALS), all based on obtaining pure profiles of the components present in a mixture system. They are also called second-order calibration algorithms, as they can operate by decomposing the 3-way data matrix and then performing a regression between the resolved relative concentration of the constituents of interest and the corresponding reference concentration. Fluorescence spectroscopy provides excitation-emission matrices (EEMs) that can be resolved by those algorithms. According to de Araújo Gomes et al., U-PLS/RBL is particularly suitable to deal with fluorescence data, as it is able to model the inner filter effect that occurs in chemical fluorescence spectroscopy analysis systems [103]. Barreto et al. also used PARAFAC to quantify melamine, obtaining slightly better results than the ones achieved with U-PLS/RBL. PARAFAC [104] is a generalization of PCA to higher-order matrices, and its models furnish parameters (loadings) that describe the variability in the samples. Hence, MCR-ALS [105] was used by Zhao et al. on NIR data to compute calibration models for the simultaneous quantification of multiple adulterants (urea, melamine, and starch) [77]. In this case, MCR-ALS was used on classical 2-way data (i.e., NIR spectra), but the assumptions made earlier are valid. In general, MCR decomposes the data matrix into a bilinear model constraining the components' profiles to assure that the solution makes sense not only from a statistical point of view, but also chemically. ALS optimization explores the possible solutions through an iterative least square calculation until convergence is achieved.

Moving forward, some other less popular (but no worse) applications of multivariate regression techniques employed in the area of milk adulteration than PLS and its extensions can be found in the literature. Artificial neural network (ANN) regression methods, namely generalized regression-NN [106] and back propagation-ANN [107], were used to assess the amount of melamine, wheat flour, and corn flour in milk powder samples [108] and acidity in cow's milk samples [109], respectively, both through Raman spectroscopy. Least squares support vector machine (LS-SVM) [110] was applied on both NIR and dielectric spectroscopic data to quantify mature bovine milk in colostrum samples [111] and on MIR data to assess cheese whey in bovine milk [38], providing better results than PLS. A generalized linear model with lasso regularization (GLM-Lasso) [112] coupled with MALDI-TOF mass spectroscopy provides better results than PLS too, in this case, to detect bovine milk in caprine and ovine milk [113]. Ehsani et al. applied boosted regression tree (BRT) [114] on NIR spectra collected by a portable spectrometer for a fast water quantification in cow's milk [115]. The presence of water in cow's milk was also inspected by Asefa et al. [116], who proposed a procedure based on digital image analysis coupled with extreme gradient boosting (XGBoost) [117].

To sum up, the most-used technique for multivariate regression in the field of milk adulteration is by far PLS, as it is relatively simple to use and is present in much commercial software. In most cases, proper use of PLS regression is enough to obtain good prediction performances, but in the case of a more complex data structure, it is worth trying more advanced techniques. The use of the many extensions of PLS can be useful to increase the signal-to-noise ratio, to compute prediction models only with the most relevant variables, or to deal with 3-way data. More expert users sometimes use other kinds of multivariate regression methods, such as ANN or SVM. In some cases, they provide slightly better results than PLS, but in many other cases, the results are comparable.

Table 4. Recent studies (2018–2022) involving regression methods related to milk adulteration.

Type of Milk	Target	Analytical Method(s)	Regression Method(s)	Reference
Cow milk	Water	NIR	PLS	[57]
Cow milk	Urea	NIR	PLS	[75]
Fat-filled milk powder	Melamine, urea	NIR	PLS	[76]
Goat milk powder	Melamine, urea, starch	NIR	PLS, MCR-ALS	[77]
Milk powder—infant formula	Melamine, vanillin	NIR HSI	PLS	[78]
Cow milk	Sugar	NIR	PLS	[79]
Cow milk	Anionic surfactant (SDS)	NIR, MIR (ATR)	PLS, GA-PLS	[80]
Yogurt	Margarine, sunflower oil, corn oil, hydrogenated vegetable oil	NIR, MIR	PLS	[81]
Desiccated coconut powder	Coconut milk	Vis-NIR	PLS	[82]
Donkey milk	Cow milk	NIR	PLS	[73]
Buffalo milk	Cow milk	MIR	PLS	[83]
Buffalo milk	Cow milk	MIR	PLS, MLR	[84]
Goat milk	Cow milk	MIR, Raman	PLS	[85]
Horse milk	Cow milk, goat milk	MIR	PLS, PCR	[86]
Coconut milk	Water	MIR	PLS, PCR	[87]
Cow milk	Soya bean oil, sugar	MIR (ATR)	PLS, MLR	[88]
Cow milk	Sucrose	MIR (ATR)	PLS, PCR	[89]
Cow milk	Formalin	MIR (ATR)	PLS, PCR	[13]
Cow milk	Maltodextrin, sodium carbonate, whey	Raman	PLS	[90]
Cow milk	Whey	Raman	PLS	[91]
White ultra-filtered cheese	Margarine, palm oil, and corn oil	Raman	PLS	[92]
Cow milk	Reconstituted skim milk powder	Cyclic voltammetry	PLS	[11]
Cow milk	Urea	Electrochemical impedance spectroscopy	PLS	[36]
Buffalo milk	Cow milk	Face fluorescence spectroscopy	PLS	[93]
Ovine and caprine milk	Cow milk	Laser-induced breakdown spectroscopy	PLS	[94]
Cow milk	Formaldehyde	TD-NMR	PLS	[12]
Cow milk	Formaldehyde	Opto-electronic nose	PLS	[40]
Cow milk	Sodium bicarbonate, sodium hydroxide, hydrogen peroxide, starch, sucrose, urea	MIR (ATR)	MLR	[97]
Skimmed milk powder	Semicarbazide hydrochloride, ammonium sulfate, cornstarch	NIR (miniature spectral devices)	OPLS	[99]

Table 4. *Cont.*

Type of Milk	Target	Analytical Method(s)	Regression Method(s)	Reference
Cow milk	Whey	MIR	PLS, siPLS, LS-SVM	[38]
Cow milk	Melamine	Fluorescence spectroscopy	PARAFAC, U-PLS/RBL	[102]
Milk powder	Melamine, wheat flour, corn flour	Raman	GRNN	[108]
Cow milk	Acidity	Raman	PLS, BP-ANN	[109]
Colostrum	Mature cow milk	NIR, dielectric spectroscopy	PLS, LS-SVM	[111]
Ovine milk and caprine milk	Cow milk	MALDI-TOF-MS	PLS, GLM-Lasso	[113]
Cow milk	Water	NIR (portable)	BRT	[115]
Cow milk	Water	Digital image analysis	XGBoost	[116]

Abbreviations: ATR = attenuated total reflection, BP-ANN = back propagation artificial neural networks, BRT = boosted regression trees, GA-PLS = genetic-algorithm partial least squares, GLM-Lasso = generalized linear model with lasso regularization, GR-NN = generalized regression neural networks, HSI = hyperspectral imaging, LS-SVM = least squares support vector machine, MALDI-TOF-MS = matrix-assisted laser desorption ionization time-of-flight mass spectrometry, MCR-ALS = multivariate curve resolution alternating least squares, MIR = mid-infrared, MLR = multiple linear regression, NIR = near-infrared, OPLS = orthogonal partial least squares, PARAFAC = parallel factor analysis, PCR = principal component regression, PLS = partial least squares, siPLS = synergy interval partial least squares, TD-NMR = time-domain nuclear magnetic resonance, U-PLS/RBL = unfolded partial least squares with residual bilinearization, Vis = visible, XGBoost = extreme gradient boosting.

3. Steps for Development and Validation of a Chemometric Approach

It is difficult to define a precise pipeline for the correct development and validation of a chemometric approach for authentication purposes. This chapter tries to face the fundamental steps, covering the sampling procedure, considering the analytical source of data, the model calibration and validation, and the main figure of merits useful for model evaluation (Figure 2).

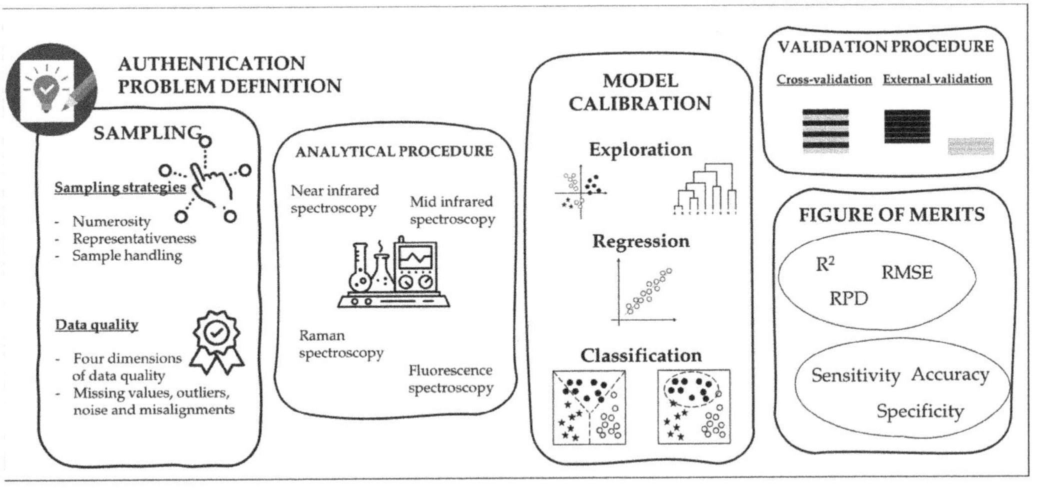

Figure 2. Schematic representation of the main steps useful to develop and validate a chemometric approach.

3.1. Correct Sampling Procedure

3.1.1. Sampling Strategies

No matter the chemometric model to be performed, according to the developed strategy goal, it is mandatory to perform a proper sampling strategy. Behind the word "proper" there are a set of extremely challenging standpoints that should consider the nature of the sample, the statistical representativeness, the analytical chemistry principles, and the quality and the management of the obtained datasets. Sampling procedures are very important to assure the robustness and reliability of the developed chemometric models. However, no well-defined sampling protocols exist so far for fingerprint techniques.

When addressing the nature of the sample, a relevant emphasis should be placed on the heterogeneity of a food matrix, together with the wide possibility of frauds, from the adulteration, i.e., the presence of an undeclared substance to the geographical origin discrimination, passing through the substitution of ingredients or commodities. In any case, the source of the samples, i.e., the provider, must be extremely reliable when addressing an authentication issue. They must be of provable provenance to assure they are authentic or not; thus, it would be advisable to obtain them from the producer rather than buying at retail markets [21].

For instance, the collection of commercial samples from local grocery stores to study goat's milk adulteration by cow's milk [85] could be inappropriate. Indeed, the commercial milk already passed to technological operation (heat treatments, fat separation, homogenization); thus, it would be more representative of real fraud to mix the different types of milk before any unit operation. This is what was done by Spina et al. [83], who described in detail the farmers, the breeds, and the sampling period and batches. Furthermore, they strengthened their experimental plan by planning a randomized pairing of cow and buffalo milk to obtain 17 adulteration levels.

Pandiselvam et al. [82] also adopted the strategy of ad hoc sample preparation. They prepared different adulterated samples by adding coconut milk residue to desiccated coconut powder. Even though the sample numerosity was quite high, i.e., 20 samples prepared for ten adulteration levels (from 0 to 100% *w/w*), it seems that the raw materials used to prepare the standard samples were always the same, thus not covering all the possible sources of variability. The variability of simulated adulterated samples was better covered by de Oliveira Mendes et al. [88], who considered six samples of milk from different producers to be adulterated with sweet whey prepared at a laboratory scale at eight adulteration levels.

From a statistical standpoint, the size and the representativeness of the sample collection must be considered [32] to obtain samples spanning all the sources of variability associated with the application of the model [118]. Different strategies described by the theory of sampling (ToS) could be followed to guarantee representative sampling and appropriate analytical quality [119]. A power analysis could be performed to establish the adequate number of samples required and to reduce the technical and biological variability. When a wide variability should be covered in a limited set of measures, design of experiments (DoE) techniques could be applied to obtain statistically valid data; the advantages of these approaches are well described by Peris-Díaz & Krężel [120].

In the literature there are examples of poor sampling strategies; for example, there are works considering a number of samples that is too low to be representative from both a technological/chemical and statistical point of view [12,38,87].

From an analytical point of view, the sample handling in terms of conservation prior to analysis, preparation, and analytical replicates should be faced to circumscribe the intrinsic variability. This is quite a challenging issue which has been clearly pointed out by Kemsley, et al. [121], and too often poorly described in the revised literature.

Finally, to sum up the useful sampling strategy to be adopted, the approach proposed by the "Five Ws" iterative interrogative technique could be winning. The first W to be clearly set is the goal of the developed approach, i.e., why, and the definition of the authentication issue to be addressed. Then, it is appropriate to cover the personnel and instrument variability (who), together with the definition of sample unit, the number of samples, handling procedure, representativeness, balanced/not balanced datasets, and possible development/availability of trusted samples (what). Moreover, the range of time (when)—which could refer to seasons, harvesting years, vintages, product aging, and so on—should be adequately covered. Finally, the investigation of the effects of the area of origin and/or the processing steps (where) should be faced.

3.1.2. Data Quality

The quality of the collected raw data strongly influences the data processing and the model quality. This is highly dependent on the instrumentation characteristics and related analytical methodology. The review by Szymanska [122] deeply described the four main dimensions of data quality (accuracy, completeness, timeliness, and consistency) and their characteristics. The most common artifacts generated by quality collection failures are missing values, outliers, noise, and misalignments. According to the type, there are strategies for their detection and deletion, substitution, or correction [122]. However, in most of the literature, little attention is given to the description of these strategies, which are hopefully applied to assess and monitor the quality of the collected data before the chemometric model construction.

3.2. Pre-Processing

An exception is the description of data pre-processing, which is generally reported as a winning strategy to remove irrelevant sources of variation, such as instrumental and experimental artifacts due to the employed analytical method. However, there are still authors who miss the preprocessing description, such as Kamboj et al. (2020) [57], or just mention an automatic strategy applied by the software. Different preprocessing strategies are available; in-depth information is given by Engel et al. [123]. Every specific dataset has specific features; thus, the definition of a rule of thumb to define which preprocessing strategy is more appropriate is impossible.

In any case, the spectroscopic data requires a pre-processing step before the statistical data analysis to remove or minimize variability in the spectra not related to the sample's characteristics. It will be clear that pre-processing cannot generate information, but only help to extract proper information already existing in the data. Moreover, incorrect use of pre-processing may cause a loss of information. Pre-treatment should be well calibrated to minimize the effects of "noise" such as optical phenomena, effects of temperature changes, light scattering, baseline shift or trends, and so on.

Most of the revised works, especially the ones dealing with infrared data, apply different preprocessing strategies, such as smoothing, standard normal variate (SNV) or multiplicative scatter correction (MSC), and derivatives alone or in combination [87,99,124]. Later on, they select the most appropriate one to solve the specific adulteration issue based on the performance criteria obtained in the developed models. However, it is important not to apply all of them by default without looking back at their effect on the data. Indeed, it should be considered that an inappropriate transformation can cause alterations to data quality, driving relevant consequences on model outcomes. A must-read tutorial concerning pre-processing has been written by Oliveri et al. [125].

Between the papers explored, some different approaches have been found in NIR pre-processing. Ejeahalaka [62] performed a comparison between two different approaches: first, no pre-processing at all, and second, extended multiplicative signal correction (EMSC) on a selected part of the spectrum. In Galvan [61] some different pre-processing methods were tested before a mean centering for all: (1) raw data, (2) Savitzky–Golay smoothing (third-order polynomial and 21 window points), (3) standard normal variate (SNV), (4) multiplicative scatter correction (MSC), (5) first and second derivative with Savitzky–Golay smoothing, (6) SNV plus first and second derivative, and (7) MSC plus first and second derivative. At the end, the best performance (evaluated by RMSE of the calculated models) was obtained by the application of the first derivative with smoothing (pre-processing 5).

Wang [65] used three pre-processing steps: (1) mean centering, (2) first, and (3) second-order Savitzky–Golay derivative, selecting at the end the first-order derivative as the better pre-processing method.

Kamboj [57] did not indicate which pre-processing was used. Not mentioning the pre-processing step should be avoided because this step implies some assumptions on the nature of the data set variability, and it is crucial that these assumptions are well understood and appropriate. An innovative approach was reported by Di Donato [63] in a study on donkey milk. NIR data were used to identify and quantify cow adulteration in expensive donkey milk. In this case, the pre-processing was done by variable sorting for normalization (VSN), a recent scatter correction technique [126] that estimates the weight of wavelengths that are or are not related to scattering effects instead of that related to the response of interest. Not-related wavelengths were not considered in the successive step. In this way, it is possible to obtain an improvement in signal and model interpretation.

Karunathilaka [60] in an application of Raman spectroscopy cites different spectral pre-processing to remove fluorescence and laser fluctuations, including Savitzky–Golay first and second derivatives and standard normal variate (SNV), choosing at the end the second derivative.

3.3. Data Reduction

The analysis of spectroscopic results is a typical example in which the dimension of the analytical part of the dataset (n columns) is much higher than the number of samples (m rows), normally thousands of columns vs. tens or hundreds of rows. So, to avoid elaboration problems and to select just the variables relevant for the statistical analysis, a variable selection step is often evaluated. Reviewing in detail all the possible algorithms is out of scope, considering their relevant number; thus, here we only report the ones used in the evaluated papers. Between them, just a few used a data reduction algorithm. For example, Chen [67] on NIRS data used an extension of the ReliefF filter algorithm [74]. ReliefF filter works on multiple classes, building a weight vector that indicates for each feature (wavelengths in the NIRS case) how important it is to explain the differences between samples of different classes. Wang [65] instead used just an observation of the first two PCA loadings as the criterion to understand relevant wavelengths, but it was unclear if just the relevant wavelengths in the subsequent classification step were used.

3.4. Use of Robust Validation Procedures

Before detailing the possible validation procedures, it is essential to consider the quality of the calibration. Taking for granted that the data representativeness and numerosity must be guaranteed according to the defined purpose, it is relevant not to overfit or underfit the model calibration.

Model validation is frequently addressed by iterative validation procedures, such as cross-validation. In the considered papers, the most used cross-validation strategy is leave-one-out, to whatever degree it should be avoided for its over-optimistic results, especially in the case of exhaustive sampling procedures [13,75,83]. Indeed, it means that during the iterative recalculation of the model just one sample at a time is removed; this way the robustness of the model is poorly investigated. None of the work internally validates models with other iterative procedures such as Monte Carlo, Jackknife, holdout, or bootstrapping.

The use of internal validation is often justified when a low number of samples is at disposal. In these situations, it can be unaffordable to exclude 30–40% of the collected data to be used as a test set. Westad and Marini [127] suggest this strategy when the number of samples is smaller than 40.

Moreover, the internal validation procedures are fundamental insights to study the model stability, identify the main sources of variation, and improve model performance, i.e., by setting model dimensionality [128]. This was the approach followed by Ejeahalaka et al. [76] for both SIMCA and PLS model development. It is important to notice that the correct model dimensionality is fundamental for predicting the test set; if the model dimensionality is incorrect, the performance criteria/figure of merit may not be a good estimate of future samples, as reported by Westad and Marini [127]. For instance, the results obtained from internal validation give insights about model overfitting due to the selection of a huge number of components/variables, which means fitting too much of the data so that also the measurement noise is interpreted as a relevant effect.

Then, it is the time to use robust, mandatory validation procedures in order to guarantee reliable and reproducible results. Usually, the available samples are divided into two subsets: a training (or calibration) set to be used for building the model, and a test set used to evaluate its validity [20] in terms of quality and generalization ability [129]. The division should guarantee that the calibration set covers the whole variability domain to obtain reliable results. The dataset split could be performed arbitrarily—according to the acquired knowledge of the data, randomly, or designed by sampling strategies—such as the Kennard and Stone algorithm, Duplex, D-optimality criterion, and K-means or Kohonen mapping; for more details about the differences among the strategies and their effects refer to Westad and Marini [127].

Infrequently, the experimental structure is considered for data splitting. This was the case for Genis et al. [92] who considered 15 concentrations of fat in the calibration set, and 11 concentrations of fat as validation data set when developing methods for the identification of foreign lipid types and adulteration ratio in milk. Most of the revised papers apply random sample selection to build the test set considering from 40 to 20% of the whole data. Among the designed sampling strategies, the Kennard and Stone algorithm is the one mostly used. However, in many cases no information is provided for dataset splitting, thus making the model robustness evaluation difficult.

In any case, it would be advisable to use a fully independent set of data to test the model; for example, considering a different production batch, a different time of the year, or a different harvesting year.

This option will represent the ideal procedure for model validation, anyway it should be set to guarantee the samples' diversity if possible, or at least their mutual independence [130].

If someone argues it is still not enough, we can reply as suggested by Westad and Marini [127]: "Another way to overcome the problem of using the same criterion to select a subset of variables and the error (i.e., cross-validation) is to divide the objects into a calibration, a validation and a verification set, where the verification set is the 'proof of the pudding'".

Each step of model development (i.e., calibration, cross-validation, and external validation) should be properly evaluated by diagnostic metrics (i.e., Figures of Merits), which are discussed in the next session.

3.5. *Performance Criteria/Figure of Merits*

Before mentioning the performance criteria useful for regression evaluation, it is important to have enough information to evaluate the quality of the collected data. In particular specific information must be reported about the numerosity of the data, their variability (i.e., mean, median, and standard deviation), the nature of the measure (instrumentation used), the removed outliers (and adopted strategy), the regression algorithm employed (mainly PLS, OPLS, PCR, MLR, LSSVM, SWM, ANN, GLM-Lasso, and so on), or the classification approach (mainly PLS-DA and OPLS-DA for the pure classification, and SIMCA and DD-SIMCA for the class-modeling techniques), the characteristics of the model development steps (calibration, internal- and external validation), the potential data pre-treatments, and the selected components/latent variables [131]. Last but not least, the information about the reference method employed to determine the specific compound and the associated error, i.e., the standard error of the laboratory (SEL), or the standard error of the test (SET), must be reported [131]. Having a clear idea of the variance covered by the data and the error of the reference analysis would be crucial to judge the results obtained by the regression model obtained. Indeed, the accuracy of chemometric model predictors depends on the repeatability of the reference methods and it combines both the error of the reference measure and the error of the fingerprint analysis [132].

3.5.1. R^2 (Coefficient of Determination) and RMSE (Root Mean Squared Error)

The main effective tests used to evaluate multivariate regression models are R^2, SEP, and the RPD. R^2, the coefficient of determination, is commonly used to evaluate regression models in every development step. It is quite relevant to compare the different coefficients of determination obtained in calibration, cross-validation, and prediction to understand the model stability. It would be better to evaluate the R^2 adjusted, which corrects for the number of explanatory terms in relation to the number of data points.

The coefficient of determination (R^2) is, in its most general definition, computed by:

$$R^2 = 1 - \frac{SS_{res}}{SS_{tot}} \tag{1}$$

where SS_{res} is the sum of squares of residuals for measurements y_i and mean of observed data ($\bar{Y}$) and SS_{tot} is the total sum of squares.

The R^2 adjusted is:

$$R^2_{adj} = 1 - \frac{n-1}{1-k-1}\frac{SS_{res}}{SS_{tot}} \tag{2}$$

where n is the number of observations and k is the number of independent variables.

However, the evaluation of R^2 alone is not exhaustive: there may be models with high coefficient values, thus describing high data variability, but with high error, expressed as root mean square error. To determine the reasonability of RMSE value it should be compared to measurement errors such as reference method, reproducibility error, historical data, and so on.

The RMSE is computed as:

$$RMSE = \sqrt{\frac{1}{n}\sum_{i=1}^{n}(y_i - \hat{y})^2} \tag{3}$$

where n is the number of observations, y_i is the predicted value and $\hat{y}$ is the actual value.

If divided by the standard deviation of the experimental values it is obtained the normalized RMSE (nRMSE), which is an unbiased measurement for model predictions.

Good error estimation was performed for the models developed by Genis [92]. They calculated the relative error of standard deviation (RSD) and relative error of prediction (REP) together with the limit of detection (LOD) and the limit of quantification (LOQ) in the regression model intended for fat authenticity in milk for ultra-filtered white cheese.

The use of both criteria, R^2 and RMSE, is relevant especially in cases of high range of variability of the considered compound; in this case, it could be plausible to obtain a model with higher R^2, but accompanied by higher RMSE, if compared with a dataset with limited range of variability. Generally speaking, "wide" calibration could be less precise, but more dangerous is a too-narrow calibration which will be valid just for the case understudy [132].

The ratio between the SD and the RMSE is referred to as ratio percentage deviation (RPD). It can be seen as a performance criterion like R^2, even if RPD is a ratio of SD, whereas R^2 is a ratio of variance. Its calculation is present in few papers dealing with milk adulteration [13,82,84,89,91], but its use can give an immediate insight to evaluate the predictions as well as to compare models predicting different compounds [132]. There are different papers that give an interpretation of model performance according to RPD values, among them the one of Williams [133] which defines six levels of performance. In the considered works the RPD was always quite high. Indeed, very good prediction capabilities were reached by the MLR model for buffalo's milk authenticity verification developed by Gonçalves et al. [84]; the RPD was 7.9. When developing a PLS regression on the same data it improved to 9.0, thus demonstrating the excellent performance of mid-infrared spectroscopy to assess cow's milk levels in buffalo's milk. The model developed by

Pandiselvam et al. [82] for the detection of adulteration with coconut powder also achieved excellent performance, resulting in an RPD of 11. Excellent performances were found by Balan et al. [13] when developing a PLS model to predict formalin in cow's milk, reaching an RPD above 8. Also, the RPD of the PLS models developed by Balan et al. [89] was high (13.4), demonstrating an excellent prediction capability of sucrose in milk, thus being able to detect sucrose addition intended to increase total solid content as well as the sweet taste. Similarly, de Oliveira Mendes et al. [91] developed a PLS model for whey quantification in raw milk by Raman spectroscopy obtaining an RPD of 13.9.

In any case, where RPD is not reported as a model parameter, it can be calculated directly from the R^2 such as $1/-(1-R^2)$.

Bellon-Maurel et al. [134] proposed to substitute RPD with a new index, RPIQ (ratio of performance to IQ). The index is based on quartiles, thus better representing the population distribution. They found out that, in sample sets with skewed distribution, the RPD is not a good approach for SEP standardization according to population spread, whereas the RPIQ index, in which standard deviation is replaced by IQ (=Q3 − Q1), better considers the spread of the population. However, none of the works considered here applied this figure of merit.

3.5.2. Specificity and Sensitivity, and Graphical Representations

The performance of classification models is assessed by verifying if samples belonging to the class of interest are designated as true positives (TP) or false negatives (FN), as well as if samples not belonging to the class of interest are labeled as false positives (FP) or true negatives (TN) [20]. Just to recall the theory, TP defines the samples recognized to belong to the class a priori assigned, FN are samples erroneously rejected, FP are samples erroneously assigned to the class, and TN are samples correctly refused.

From their assignments, it is possible to calculate the sensitivity and sensibility of the method. Sensitivity is the true positive rate (TPR), computed as TP/(TP + FN). Specificity is the true negative rate (TNR), computed as TN/(TN + FP).

The graphical tool used to represent the performance criteria of a discriminant model is the receiver operating characteristic (ROC) curves (Figure 3a). The plot represents a two-axis Cartesian space, with the horizontal axis reporting FPR, and the vertical axis the TPR. The dashed diagonal represents the performance of a random classifier. Two examples of classifiers (green and red) are shown, representing good and scarce results, respectively. The curves are built by connecting with a line the experimental outcomes. This tool is useful to compare the performances of models obtained with different parameter settings, such as the threshold value. A detailed analysis of ROC curves is discussed by Oliveri [20].

If discriminant methods can be applied only to solve multi-class situations, class modeling can be used to address both multi-class and one-class problems.

When performing a class-modeling analysis it could be useful to evaluate the results with a graphical representation, so Coomans' plots (Figure 3b). In a two-class problem, the two axes represent the distances of samples from the models of Class 1 (○) and Class 2 (star), respectively. The two dashed lines correspond to the critical acceptance levels for each model at the defined confidence level (normally 95%). Samples of the two classes are projected as scatter points, with coordinates indicating the relative similarity with the two models in the four sectors defined in the plot. In sector 1 it is possible to find samples accepted only by Class 1 (○); in sector 2 it is possible to find samples accepted only by Class 2 (star). Both sectors include samples defined as TP for the a priori defined class.

In sector 3 are positioned samples accepted by both models; indeed, since models for each class are independently built, class spaces may overlap. Lastly, in sector 4 it is possible to observe samples rejected by both models, which highlights that the used variables do not completely resolve the class space. They prevent the forced (but possibly wrong) classification of samples that may occur in discriminant approaches [20].

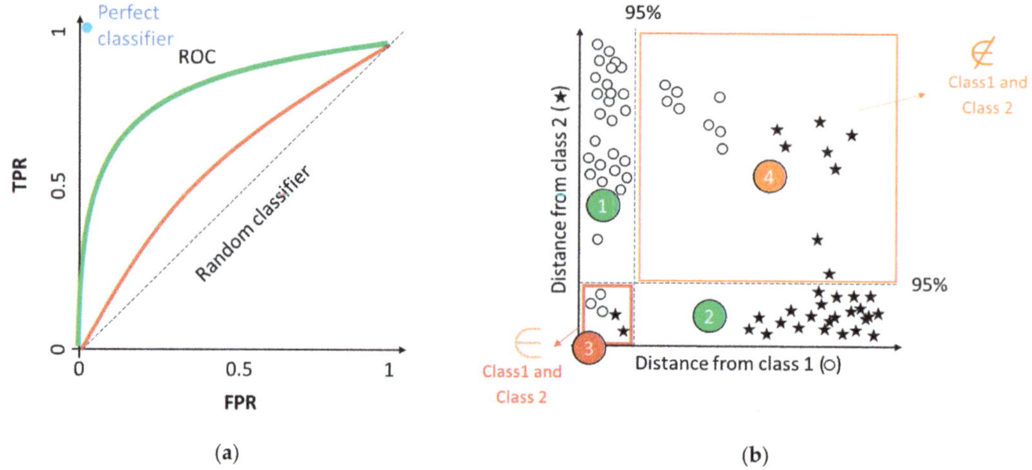

Figure 3. Graphical tools to represent classification model performance. (**a**) Receiver operating characteristic (ROC) curves; (**b**) Coomans' plot.

4. Methods for Rapid and On-Site Detection to Combat Milk Adulteration

The dairy industry as well as regulatory bodies are looking for simple and rapid methods for the detection of milk adulteration [135]. Lateral flow immunoassays (LFIAs) have been used as in situ screening tools to monitor food raw material quality as they provide rapid results [136]. LFIAs have been developed, among other applications [137], for the detection and quantification of mycotoxins [138], such as aflatoxin M1 [139]. LFIAs have been also used for the detection of adulteration of milk with melamine [140]. In a very recent study adulteration of cow's milk with buffalo's milk was detected by an on-site carbon nanoparticle-based lateral flow immunoassay in 10 min, with the sensitivity of the test being 5%, i.e., 5% adulteration of cow's milk with buffalo's milk, proving that this tool is suitable for rapid detection of adulteration [135].

Another novel technology for the rapid detection of milk adulteration is DNAFoil. It is a portable, fully self-administered, on-site DNA test that does not require the use of expensive PCR equipment or laboratory setups to confirm the detection of milk adulteration within a short period of time. The efficiency of the DNAFoil kit used to detect the vegetable material in milk products (DNAFoil UniPlant) was confirmed using real-time PCR assays. The results showed that using 24 μL of DNAFoil UniPlant master mix, a 17.5 min reaction time allowed the detection of 10% adulteration of liquid cow's milk by wheat flour [141].

Moreover, an electronic nose (e-nose) system is being evolved for the falsification detection of milk and dairy products in a reliable and rapid way [142]. This technology avoids the disadvantages of chromatography, spectrometry, and chemical methods with high costs and long cycle times [143]. Adulteration of bovine milk with formaldehyde, based on aldehydes and ketones, was examined by electronic nose by Mostafapour et al. [40]. In another investigation, the identification of trace amounts of detergent powder in raw milk using a customized low-cost electronic nose was achieved [144].

5. Conclusions

An overview of the different chemometric techniques (from clustering to classification and regression applied to several analytical data) has been presented along with spectroscopy, chromatography, and electrochemical sensors as well as rapid and on-site detection devices in the fight against milk adulteration and fraud. HCA is the main representative of the clustering methods. The classification of objects in groups depending on their characteristics expressed as results of a set of measurements is one of the applications of chemometrics. Classification methods were distinguished into "discriminant" and "class-modeling" techniques. The classification statistical techniques mostly employed in the last few years for milk applications were PLS-DA as a pure classification technique and SIMCA as a class-modeling approach. Multivariate regression is widely used to quantify the concentration of adulterants in food matrices and was deeply described.

Finally, the steps which should be followed to develop a chemometric model to face adulteration issues were carefully presented with the required critical discussion describing sampling strategies, pre-processing, data reduction, and use of robust validation procedures along with performance criteria/figure of merits.

All chemometric methods, supervised and unsupervised, had fundamental results in order to serve the goals of each research study. It cannot be concluded which chemometric method is the best, as each dataset is unique and different. Robustness is usually more related to supervised methods, but unsupervised methods are also important in the field. Usually, the availability and access to each chemometric method are the variables that influence their specific selection. With regard to the field of milk adulteration, it is clear that, in most cases, the simplest methods are enough to obtain good results. However, even the simplest methods are in some cases used improperly, making the results obtained inconsistent.

Author Contributions: Conceptualization, S.G., M.T., A.D., S.A., T.V. and L.S.; methodology, S.G., M.T., A.D., S.A., T.V. and L.S.; writing—original draft preparation, S.G., M.T., A.D., S.A., T.V. and L.S.; writing—review and editing, S.G., M.T., A.D., S.A., T.V. and L.S. All authors have read and agreed to the published version of the manuscript.

Funding: This research received no external funding.

Acknowledgments: The authors would like to thank Foods for the financial support provided.

Conflicts of Interest: The authors declare no conflict of interest.

References

1. Uncu, A.O.; Uncu, A.T. A barcode-DNA analysis method for the identification of plant oil adulteration in milk and dairy products. *Food Chem.* **2020**, *326*, 126986. [CrossRef] [PubMed]
2. Zhang, T.; Wu, X.; Wu, B.; Dai, C.; Fu, H. Rapid authentication of the geographical origin of milk using portable near-infrared spectrometer and fuzzy uncorrelated discriminant transformation. *J. Food Process. Eng.* **2022**, *45*, e14040. [CrossRef]
3. Ye, H.; Yang, J.; Xiao, G.; Zhao, Y.; Li, Z.; Bai, W.; Zeng, X.; Dong, H. A comprehensive overview of emerging techniques and chemometrics for authenticity and traceability of animal-derived food. *Food Chem.* **2023**, *402*, 134216. [CrossRef] [PubMed]
4. Nascimento, C.F.; Santos, P.M.; Pereira-Filho, E.R.; Rocha, F.R. Recent advances on determination of milk adulterants. *Food Chem.* **2017**, *221*, 1232–1244. [CrossRef] [PubMed]
5. Moore, J.C.; Spink, J.; Lipp, M. Development and application of a database of food ingredient fraud and economically motivated adulteration from 1980 to 2010. *J. Food Sci.* **2012**, *77*, R118–R126. [CrossRef]
6. Giglioti, R.; Polli, H.; Azevedo, B.T.; Katiki, L.M.; Vercesi Filho, A.E. Detection and quantification of adulteration in milk and dairy products: A novel and sensitive qPCR-based method. *Food Chem. Mol. Sci.* **2022**, *4*, 100074. [CrossRef]
7. Teixeira, J.L.D.P.; Carames, E.T.D.S.; Baptista, D.P.; Gigante, M.L.; Pallone, J.A.L. Vibrational spectroscopy and chemometrics tools for authenticity and improvement the safety control in goat milk. *Food Control* **2020**, *112*, 107105. [CrossRef]
8. Du, L.; Lu, W.; Cai, Z.J.; Bao, L.; Hartmann, C.; Gao, B.; Yu, L.L. Rapid detection of milk adulteration using intact protein flow injection mass spectrometric fingerprints combined with chemometrics. *Food Chem.* **2018**, *240*, 573–578. [CrossRef]
9. Motta, T.C.; Hoff, R.B.; Barreto, F.; Andrade, R.B.S.; Lorenzini, D.M.; Meneghini, L.Z.; Pizzolato, T.M. Detection and confirmation of milk adulteration with cheese whey using proteomic-like sample preparation and liquid chromatography–electrospray–tandem mass spectrometry analysis. *Talanta* **2014**, *120*, 498–505. [CrossRef]
10. Qin, C.; Liu, L.; Wang, Y.; Leng, T.; Zhu, M.; Gan, B.; Xie, J.; Yu, Q.; Chen, Y. Advancement of omics techniques for chemical profile analysis and authentication of milk. *Trends Food Sci. Technol.* **2022**, *127*, 114–128. [CrossRef]
11. Nikolaou, P.; Deskoulidis, E.; Topoglidis, E.; Kakoulidou, A.T.; Tsopelas, F. Application of chemometrics for detection and modeling of adulteration of fresh cow milk with reconstituted skim milk powder using voltammetric fingerpriting on a graphite/SiO_2 hybrid electrode. *Talanta* **2020**, *206*, 120223. [CrossRef] [PubMed]
12. Coimbra, P.T.; Bathazar, C.F.; Guimarães, J.T.; Coutinho, N.M.; Pimentel, T.C.; Neto, R.P.C.; Esmerino, E.A.; Freitas, M.Q.; Silva, M.C.; Tavares, M.I.B.; et al. Detection of formaldehyde in raw milk by time domain nuclear magnetic resonance and chemometrics. *Food Control* **2020**, *110*, 107006. [CrossRef]
13. Balan, B.; Dhaulaniya, A.S.; Jamwal, R.; Sodhi, K.K.; Kelly, S.; Cannavan, A.; Singh, D.K. Application of Attenuated Total Reflectance-Fourier Transform Infrared (ATR-FTIR) spectroscopy coupled with chemometrics for detection and quantification of formalin in cow milk. *Vib. Spectrosc.* **2020**, *107*, 103033. [CrossRef]

14. Wasnik, P.G.; Menon, R.R.; Sivaram, M.; Nath, B.S.; Balasubramanyam, B.V.; Manjunatha, M. Development of mathematical model for prediction of adulteration levels of cow ghee with vegetable fat using image analysis. *J. Food Sci. Technol.* **2019**, *56*, 2320–2325. [CrossRef]
15. Roy, M.; Doddappa, M.; Yadav, B.K.; Jaganmohan, R.; Sinija, V.R.; Manickam, L.; Sarvanan, S. Detection of soybean oil adulteration in cow ghee (clarified milk fat): An ultrafast study using flash gas chromatography electronic nose coupled with multivariate chemometrics. *J. Sci. Food Agric.* **2022**, *102*, 4097–4108. [CrossRef]
16. Vatavali, K.; Kosma, I.; Louppis, A.; Gatzias, I.; Badeka, A.V.; Kontominas, M.G. Characterisation and differentiation of geographical origin of Graviera cheeses produced in Greece based on physico-chemical, chromatographic and spectroscopic analyses, in combination with chemometrics. *Int. Dairy J.* **2020**, *110*, 104799. [CrossRef]
17. Aleixandre-Tudo, J.L.; Castello-Cogollos, L.; Aleixandre, J.L.; Aleixandre-Benavent, R. Chemometrics in food science and technology: A bibliometric study. *Chemom. Intell. Lab. Syst.* **2022**, *222*, 104514. [CrossRef]
18. Kamal, M.; Karoui, R. Analytical methods coupled with chemometric tools for determining the authenticity and detecting the adulteration of dairy products: A review. *Trends Food Sci. Technol.* **2015**, *46*, 27–48. [CrossRef]
19. Gómez-Caravaca, A.M.; Maggio, R.M.; Cerretani, L. Chemometric applications to assess quality and critical parameters of virgin and extra-virgin olive oil. A review. *Anal. Chim. Acta* **2016**, *913*, 1–21. [CrossRef]
20. Oliveri, P. Class-modelling in food analytical chemistry: Development, sampling, optimisation and validation issues—A tutorial. *Anal. Chim. Acta* **2017**, *982*, 9–19. [CrossRef]
21. McGrath, T.F.; Haughey, S.A.; Patterson, J.; Fauhl-Hassek, C.; Donarski, J.; Alewijn, M.; van Ruth, S.; Elliott, C.T. What are the scientific challenges in moving from targeted to non-targeted methods for food fraud testing and how can they be addressed? Spectroscopy case study. *Trends Food Sci. Technol.* **2018**, *76*, 38–55. [CrossRef]
22. Cubero-Leon, E.; Penalver, R.; Maquet, A. Review on metabolomics for food authentication. *Food Res. Int.* **2014**, *60*, 95–10711.
23. Hanganu, A.; Chira, N. When detection of dairy food fraud fails: An alternate approach through proton nuclear magnetic resonance spectroscopy. *J. Dairy Sci.* **2021**, *104*, 8454–8466. [CrossRef] [PubMed]
24. Souhassou, S.; Bassbasi, M.; Hirri, A.; Kzaiber, F.; Oussama, A. Detection of camel milk adulteration using Fourier transformed infrared spectroscopy FT-IR coupled with chemometrics methods. *Int. Food Res. J.* **2018**, *25*, 1213–1218.
25. Wang, X.; Esquerre, C.; Downey, G.; Henihan, L.; O'Callaghan, D.; O'Donnell, C. Feasibility of discriminating dried dairy ingredients and preheat treatments using mid-infrared and Raman Spectroscopy. *Food Anal. Methods* **2018**, *11*, 1380–1389. [CrossRef]
26. Karunathilaka, S.R.; Yakes, B.J.; He, K.; Chung, J.K.; Mossoba, M. Non-targeted NIR spectroscopy and SIMCA classification for commercial milk powder authentication: A study using eleven potential adulterants. *Heliyon* **2018**, *4*, e00806. [CrossRef]
27. Da Silva Dias, L.; da Silva Junior, J.C.; Felício, A.L.D.S.M.; de França, J.A. A NIR photometer prototype with integrating sphere for the detection of added water in raw milk. *IEEE Trans. Instrum. Meas.* **2018**, *67*, 2812–2819. [CrossRef]
28. Windarsih, A.; Rohman, A.; Irnawati; Riyanto, S. The Combination of Vibrational Spectroscopy and Chemometrics for Analysis of Milk Products Adulteration. *Int. J. Food Sci.* **2021**, *2021*, 8853358. [CrossRef]
29. de Lima, A.B.S.; Batista, A.S.; de Jesus, J.C.; de Jesus Silva, J.; de Araújo, A.C.M.; Santos, L.S. Fast quantitative detection of black pepper and cumin adulterations by near-infrared spectroscopy and multivariate modeling. *Food Control* **2020**, *107*, 106802. [CrossRef]
30. Jiménez-Carvelo, A.M.; González-Casado, A.; Bagur-González, M.G.; Cuadros-Rodríguez, L. Alternative data mining/machine learning methods for the analytical evaluation of food quality and authenticity—A review. *Food Res. Int.* **2019**, *122*, 25–39. [CrossRef]
31. Teixeira, J.L.d.P.; Caramês, E.T.d.S.; Baptista, D.P.; Gigante, M.L.; Pallone, J.A.L. Rapid adulteration detection of yogurt and cheese made from goat milk by vibrational spectroscopy and chemometric tools. *J. Food Compost. Anal.* **2021**, *96*, 103712. [CrossRef]
32. Ramirez-Lopez, L.; Schmidt, K.; Behrens, T.; Van Wesemael, B.; Demattê, J.A.; Scholten, T. Sampling optimal calibration sets in soil infrared spectroscopy. *Geoderma* **2014**, *226*, 140–150. [CrossRef]
33. Jain, A.K.; Murty, M.N.; Flynn, P.J. Data clustering: A review. *ACM Comput. Surv. CSUR* **1999**, *31*, 264–323. [CrossRef]
34. Bratchell, N. Chapter 6 Cluster Analysis. *Data Handl. Sci. Technol.* **1992**, *9*, 179–208.
35. Saxena, A.; Prasad, M.; Gupta, A.; Bharill, N.; Patel, O.P.; Tiwari, A.; Er, M.J.; Ding, W.; Lin, C. A review of clustering techniques and developments. *Neurocomputing* **2017**, *267*, 664–681. [CrossRef]
36. Minetto, T.A.; França, B.D.; da Silva Dariz, G.; Veiga, E.A.; Galvão, A.C.; da Silva Robazza, W. Identifying adulteration of raw bovine milk with urea through electrochemical impedance spectroscopy coupled with chemometric techniques. *Food Chem.* **2022**, *385*, 132678. [CrossRef]
37. Cirak, O.; Icyer, N.C.; Durak, M.Z. Rapid detection of adulteration of milks from different species using Fourier Transform Infrared Spectroscopy (FTIR). *J. Dairy Res.* **2018**, *85*, 222–225. [CrossRef]
38. Vinciguerra, L.L.; Marcelo, M.C.; Motta, T.; Meneghini, L.Z.; Bergold, A.M.; Ferrão, M.F. Chemometric tools and FTIR-ATR spectroscopy applied in milk adulterated with cheese whey. *Química Nova* **2019**, *42*, 249–254. [CrossRef]
39. Ezhilan, M.; Gumpu, M.B.; Ramachandra, B.L.; Nesakumar, N.; Babu, K.J.; Krishnan, U.M.; Rayappan, J.B.B. Design and development of electrochemical biosensor for the simultaneous detection of melamine and urea in adulterated milk samples. *Sens. Actuators B Chem.* **2017**, *238*, 1283–1292. [CrossRef]

40. Mostafapour, S.; Gharaghani, F.M.; Hemmateenejad, B. Converting electronic nose into opto-electronic nose by mixing MoS_2 quantum dots with organic reagents: Application to recognition of aldehydes and ketones and determination of formaldehyde in milk. *Anal. Chim. Acta* **2021**, *1170*, 338654. [CrossRef]
41. Li, Q.; Yu, Z.; Zhu, D.; Meng, X.; Pang, X.; Liu, Y.; Frew, R.; Chen, H.; Chen, G. The application of NMR-based milk metabolite analysis in milk authenticity identification. *J. Sci. Food Agric.* **2017**, *97*, 2875–2882. [CrossRef] [PubMed]
42. Sowmya, N.; Ponnusamy, V. Development of spectroscopic sensor system for an IoT application of adulteration identification on milk using machine learning. *IEEE Access* **2021**, *9*, 53979–53995. [CrossRef]
43. Souza, S.O.; Santos, V.S.; Santos, E.S.; Ávila, D.V.L.; Nascimento, C.C.; Costa, S.S.L.; Garcia, C.A.B.; Araujo, R.G.O. Evaluation of the mineral content in milk and yogurt types using chemometric tools. *Microchem. J.* **2018**, *143*, 1–8. [CrossRef]
44. Rodríguez-Bermúdez, R.; López-Alonso, M.; Miranda, M.; Fouz, R.; Orjales, I.; Herrero-Latorre, C. Chemometric authentication of the organic status of milk on the basis of trace element content. *Food Chem.* **2018**, *240*, 686–693. [CrossRef]
45. Zain, S.M.; Behkami, S.; Bakirdere, S.; Koki, I.B. Milk authentication and discrimination via metal content clustering–A case of comparing milk from Malaysia and selected countries of the world. *Food Control* **2016**, *66*, 306–314. [CrossRef]
46. Xu, S.; Zhao, C.; Deng, X.; Zhang, R.; Qu, L.; Wang, M.; Ren, S.; Wu, H.; Yue, Z.; Niu, B. Determining the geographical origin of milk by multivariate analysis based on stable isotope ratios, elements and fatty acids. *Anal. Methods* **2021**, *13*, 2537–2548. [CrossRef]
47. Amenzou, N.; Hamid, M.; Fouad, T.; Elyahyaoui, A.; Elghali, T.; Elmoqrani, L.; Mahmoud, E. *Stable Isotope Ratios in Dairy Products (Milk) as New Tool to Determine Their Different Origins in Morocco*; Joint FAO/IAEA Centre of Nuclear Techniques in Food and Agriculture, Food Safety and Control Section: Vienna, Austria, 2022; pp. 60–69, 128, ISBN 978-92-0-124822-0. ISSN 1011-4289. CONTRACT MOR 18051.
48. Podkolzin, I.; Solovev, A. *Application of Stable Isotope Techniques and Elemental Analysis to Confirm Geographical Origin of Milk Produced in the Russian Federation*; IAEA: Vienna, Austria, 2022.
49. Karrar, E.; Mohamed Ahmed, I.A.; Huppertz, T.; Wei, W.; Jin, J.; Wang, X. Fatty acid composition and stereospecificity and sterol composition of milk fat from different species. *Int. Dairy J.* **2022**, *128*, 105313. [CrossRef]
50. Bhumireddy, S.R.; Rocchetti, G.; Pallerla, P.; Lucini, L.; Sripadi, P. A combined targeted/untargeted screening based on GC/MS to detect low-molecular-weight compounds in different milk samples of different species and as affected by processing. *Int. Dairy J.* **2021**, *118*, 105045. [CrossRef]
51. Tan, D.; Zhang, X.; Su, M.; Jia, M.; Zhu, D.; Kebede, B.; Wu, H.; Chen, G. Establishing an untargeted-to-MRM liquid chromatography–mass spectrometry method for discriminating reconstituted milk from ultra-high temperature milk. *Food Chem.* **2021**, *337*, 127946. [CrossRef]
52. Couvreur, S.; Hurtaud, C. Relationships between milks differentiated on native milk fat globule characteristics and fat, protein and calcium compositions. *Animal* **2017**, *11*, 507–518. [CrossRef]
53. Dhankhar, J.; Sharma, R.; Indumathi, K. A comparative study of sterols in milk fat of different Indian dairy animals based on chemometric analysis. *J. Food Meas. Charact.* **2020**, *14*, 2538–2548. [CrossRef]
54. Marini, F. Classification methods in chemometrics. In Proceedings of the Mediterraneum Meeting, Ventotene, Italy, 1–4 June 2008.
55. Derde, M.P.; Massart, D.L. UNEQ: A disjoint modelling technique for pattern recognition based on normal distribution. *Anal. Chim. Acta* **1986**, *184*, 33–51. [CrossRef]
56. Vargas-Bello-Pérez, E.; Gomez-Cortes, P.; Geldsetzer-Mendoza, C.; Sol Morales, M.; Toro-Mujica, P.; Fellenberg, M.A.; Ibanez, R.A. Authentication of retail cheeses based on fatty acid composition and multivariate data analysis. *Int. Dairy J.* **2018**, *85*, 280–284. [CrossRef]
57. Kamboj, U.; Kaushal, N.; Mishra, S.; Munjal, N. Application of Selective Near Infrared Spectroscopy for Qualitative and Quantitative Prediction of Water Adulteration in Milk. *Mater. Today Proc.* **2020**, *24*, 2449–2456. [CrossRef]
58. Chung, I.M.; Kim, J.K.; Yang, Y.J.; An, Y.J.; Kim, S.Y.; Kwon, C.; Kim, S.H. A case study for geographical indication of organic milk in Korea using stable isotope ratios-based chemometric analysis. *Food Control* **2020**, *107*, 106755. [CrossRef]
59. Jin, H.; Dong, G.M.; Wu, H.Y.; Yang, Y.R.; Huang, M.Y.; Wang, M.Y.; Yang, R.J. Identification of adulterated milk based on auto-correlation spectra. *Spectrochim. Acta Part A Mol. Biomol. Spectrosc.* **2023**, *286*, 121987. [CrossRef]
60. Karunathilaka, S.R.; Yakes, B.J.; He, K.; Brückner, L.; Mossoba, M.M. First use of handheld Raman spectroscopic devices and on-board chemometric analysis for the detection of milk powder adulteration. *Food Control* **2018**, *92*, 137–146. [CrossRef]
61. Galvan, D.; Lelis, C.A.; Effting, L.; Melquiades, F.L.; Bona, E.; Conte-Junior, C.A. Low-cost spectroscopic devices with multivariate analysis applied to milk authenticity. *Microchem. J.* **2022**, *181*, 107746. [CrossRef]
62. Ejeahalaka, K.K.; On, S.L.W. Chemometric studies of the effects of milk fat replacement with different proportions of vegetable oils in the formulation of fat-filled milk powders: Implications for quality assurance. *Food Chem.* **2019**, *295*, 198–205. [CrossRef]
63. Di Donato, F.; Biancolillo, A.; Ferretti, A.; D'Archivio, A.A.; Marini, F. Near Infrared Spectroscopy coupled to Chemometrics for the authentication of donkey milk. *J. Food Compos. Anal.* 2022, *in press*. [CrossRef]
64. Zontov, Y.V.; Rodionova, O.Y.; Kucheryavskiy, S.V.; Pomerantsev, A.L. DD-SIMCA—A MATLAB GUI tool for data driven SIMCA approach. *Chemom. Intell. Lab. Syst.* **2017**, *167*, 23–28. [CrossRef]
65. Wang, Y.T.; Ren, H.B.; Liang, W.Y.; Jin, X.; Yuan, Q.; Liu, Z.R.; Chen, D.M.; Zhang, Y.H. A novel approach to temperature-dependent thermal processing authentication for milk by infrared spectroscopy coupled with machine learning. *J. Food Eng.* **2021**, *311*, 110740. [CrossRef]

66. Dos Santos Pereira, E.V.; de Sousa Fernandes, D.D.; de Almeida, L.F.; Sucupira Maciel, M.I.; Gonçalves Dias Diniz, P.H. Goat milk authentication by one-class classification of digital image-based fingerprint signatures: Detection of adulteration with cow milk. *Microchem. J.* **2022**, *180*, 107640. [CrossRef]
67. Chen, H.; Tan, C.; Lin, Z.; Wua, T. Classification of different liquid milk by near-infrared spectroscopy and ensemble modelling. *Spectrochim. Acta Part A Mol. Biomol. Spectrosc.* **2021**, *251*, 119460. [CrossRef]
68. Potočnik, D.; Nečemer, M.; Perišić, I.; Jagodic, M.; Mazej, D.; Camin, F.; Eftimov, T.; Strojnik, L.; Ogrinc, N. Geographical verification of Slovenian milk using stable isotope ratio, multielement and multivariate modelling approaches. *Food Chem.* **2020**, *326*, 126958. [CrossRef]
69. Xie, L.; Zhao, S.; Rogers, K.M.; Xia, Y.; Zhang, B.; Suo, R.; Zhao, Y. A case of milk traceability in small-scale districts-Inner Mongolia of China by nutritional and geographical parameters. *Food Chem.* **2020**, *316*, 126332. [CrossRef]
70. Tommasini, A.; Curone, G.; Solè, M.; Capuani, G.; Sciubba, F.; Conta, G.; Miccheli, G.; Vigo, D. NMR-based metabolomics to evaluate the milk composition from Friesian and autochtonous cows of Northern Italy at different lactation times. *Nat. Prod. Res.* **2019**, *33*, 1085–1091. [CrossRef]
71. Sundekilde, U.K.; Larsen, L.B.; Bertram, C. NMR-Based milk metabolomics. *Metabolites* **2013**, *3*, 204–222. [CrossRef]
72. Segato, S.; Caligiani, A.; Contiero, B.; Galaverna, G.; Bisutti, V.; Cozzi, G. 1H NMR metabolic profile to discriminate pasture based alpine Asiago PDO cheeses. *Animals* **2019**, *9*, 722. [CrossRef]
73. Yanibada, B.; Boudra, H.; Debrauwer, L.; Martin, C.; Morgavi, D.P.; Canlet, C. Evaluation of sample preparation methods for NMR-based metabolomics of cow milk. *Heliyon* **2018**, *4*, e00856. [CrossRef]
74. Wold, S.; Martens, H.; Wold, H. The multivariate calibration problem in chemistry solved by PLS method. In *Matrix Pencils*; Springer: Berlin/Heidelberg, Germany, 1983; pp. 286–293.
75. Mabood, F.; Ali, L.; Boque, R.; Abbas, G.; Jabeen, F.; Haq, Q.M.I.; Hussain, J.; Hamaed, A.M.; Naureen, Z.; Al-Nabhani, M.; et al. Robust Fourier transformed infrared spectroscopy coupled with multivariate methods for detection and quantification of urea adulteration in fresh milk samples. *Food Sci. Nutr.* **2020**, *8*, 5249–5258. [CrossRef] [PubMed]
76. Ejeahalaka, K.K.; On, S.L. Effective detection and quantification of chemical adulterants in model fat-filled milk powders using NIRS and hierarchical modelling strategies. *Food Chem.* **2020**, *309*, 125785. [CrossRef] [PubMed]
77. Zhao, X.; Wang, Y.; Liu, X.; Jiang, H.; Zhao, Z.; Niu, X.; Li, C.; Pang, B.; Li, Y. Single-and Multiple-Adulterants Determinations of Goat Milk Powder by NIR Spectroscopy Combined with Chemometric Algorithms. *Agriculture* **2022**, *12*, 434. [CrossRef]
78. Zhao, X.; Li, C.; Zhao, Z.; Wu, G.; Xia, L.; Jiang, H.; Wang, T.; Chu, X.; Liu, J. Generic models for rapid detection of vanillin and melamine adulterated in infant formulas from diverse brands based on near-infrared hyperspectral imaging. *Infrared Phys. Technol.* **2021**, *116*, 103745. [CrossRef]
79. Kamboj, U.; Kaushal, N.; Jabeen, S. Near Infrared Spectroscopy as an efficient tool for the Qualitative and Quantitative Determination of Sugar Adulteration in Milk. *J. Phys. Conf. Ser.* **2020**, *1531*, 12024. [CrossRef]
80. Hosseini, E.; Ghasemi, J.B.; Daraei, B.; Asadi, G.; Adib, N. Application of genetic algorithm and multivariate methods for the detection and measurement of milk-surfactant adulteration by attenuated total reflection and near-infrared spectroscopy. *J. Sci. Food Agric.* **2021**, *101*, 2696–2703. [CrossRef]
81. Temizkan, R.; Can, A.; Dogan, M.A.; Mortas, M.; Ayvaz, H. Rapid detection of milk fat adulteration in yoghurts using near and mid-infrared spectroscopy. *Int. Dairy J.* **2020**, *110*, 104795. [CrossRef]
82. Pandiselvam, R.; Mahanti, N.K.; Manikantan, M.R.; Kothakota, A.; Chakraborty, S.K.; Ramesh, S.V.; Beegum, P.S. Rapid detection of adulteration in desiccated coconut powder: Vis-NIR spectroscopy and chemometric approach. *Food Control* **2022**, *133*, 108588. [CrossRef]
83. Spina, A.A.; Ceniti, C.; Piras, C.; Tilocca, B.; Britti, D.; Morittu, V.M. Mid-Infrared (MIR) Spectroscopy for the quantitative detection of cow's milk in buffalo milk. *J. Anim. Sci. Technol.* **2022**, *64*, 531–538. [CrossRef]
84. Gonçalves, B.H.R.; Silva, G.J.; Jesus, J.C.D.; Conceição, D.G.; Santos, L.S.; Ferrão, S.P. Fast verification of buffalo's milk authenticity by mid-infrared spectroscopy, analytical measurements and multivariate calibration. *J. Braz. Chem. Soc.* **2020**, *31*, 1453–1460. [CrossRef]
85. Yaman, H. A rapid method for detection adulteration in goat milk by using vibrational spectroscopy in combination with chemometric methods. *J. Food Sci. Technol.* **2020**, *57*, 3091–3098. [CrossRef] [PubMed]
86. Arifah, M.F.; Nisa, K.; Windarsih, A.; Rohman, A. The Application of FTIR Spectroscopy and Chemometrics for the Authentication Analysis of Horse Milk. *Int. J. Food Sci.* **2022**, *2022*, 7643959. [CrossRef] [PubMed]
87. Sitorus, A.; Muslih, M.; Cebro, I.S.; Bulan, R. Dataset of adulteration with water in coconut milk using FTIR spectroscopy. *Data Br.* **2021**, *36*, 107058. [CrossRef]
88. Jaiswal, P.; Jha, S.N.; Kaur, J.; Hg, R. Rapid detection and quantification of soya bean oil and common sugar in bovine milk using attenuated total reflectance–fourier transform infrared spectroscopy. *Int. J. Dairy Technol.* **2018**, *71*, 292–300. [CrossRef]
89. Balan, B.; Dhaulaniya, A.S.; Jamwal, R.; Yadav, A.; Kelly, S.; Cannavan, A.; Singh, D.K. Rapid detection and quantification of sucrose adulteration in cow milk using Attenuated total reflectance-Fourier transform infrared spectroscopy coupled with multivariate analysis. *Spectrochim. Acta Part A Mol. Biomol. Spectrosc.* **2020**, *240*, 118628. [CrossRef] [PubMed]
90. Tian, H.; Chen, S.; Li, D.; Lou, X.; Chen, C.; Yu, H. Simultaneous detection for adulterations of maltodextrin, sodium carbonate, and whey in raw milk using Raman spectroscopy and chemometrics. *J. Dairy Sci.* **2022**, *105*, 7242–7252. [CrossRef] [PubMed]

91. De Oliveira Mendes, T.; Rodrigues, B.V.M.; Porto, B.L.S.; da Rocha, R.A.; de Oliveira, M.A.L.; de Castro, F.K.; de Carvalho dos Anjos, V.; Bell, M.J.V. Raman Spectroscopy as a fast tool for whey quantification in raw milk. *Vib. Spectrosc.* **2020**, *111*, 103150. [CrossRef]
92. Genis, D.O.; Sezer, B.; Durna, S.; Boyaci, I.H. Determination of milk fat authenticity in ultra-filtered white cheese by using Raman spectroscopy with multivariate data analysis. *Food Chem.* **2021**, *336*, 127699. [CrossRef]
93. Ullah, R.; Khan, S.; Ali, H.; Bilal, M. Potentiality of using front face fluorescence spectroscopy for quantitative analysis of cow milk adulteration in buffalo milk. *Spectrochim. Acta Part A Mol. Biomol. Spectrosc.* **2020**, *225*, 117518. [CrossRef]
94. Sezer, B.; Durna, S.; Bilge, G.; Berkkan, A.; Yetisemiyen, A.; Boyaci, I.H. Identification of milk fraud using laser-induced breakdown spectroscopy (LIBS). *Int. Dairy J.* **2018**, *81*, 1–7. [CrossRef]
95. Lai, T.L.; Robbins, H.; Wei, C.Z. Strong consistency of least squares estimates in multiple regression II. *J. Multivar. Anal.* **1979**, *9*, 343–361. [CrossRef]
96. Mandel, J. Use of the singular value decomposition in regression analysis. *Am. Stat.* **1982**, *36*, 15–24.
97. Conceição, D.G.; Gonçalves, B.H.R.; Hora, F.F.D.; Faleiro, A.S.; Santos, L.S.; Ferrão, S.P. Use of FTIR-ATR spectroscopy combined with multivariate analysis as a screening tool to identify adulterants in raw milk. *J. Braz. Chem. Soc.* **2019**, *30*, 780–785. [CrossRef]
98. Trygg, J.; Wold, S. Orthogonal projections to latent structures (O-PLS). *J. Chemom. J. Chemom. Soc.* **2002**, *16*, 119–128. [CrossRef]
99. Delatour, T.; Becker, F.; Krause, J.; Romero, R.; Gruna, R.; Längle, T.; Panchaud, A. Handheld spectral sensing devices should not mislead consumers as far as non-authentic food is concerned: A case study with adulteration of milk powder. *Foods* **2021**, *11*, 75. [CrossRef]
100. Nørgaard, L.; Hahn, M.T.; Knudsen, L.B.; Farhat, I.A.; Engelsen, S.B. Multivariate near-infrared and Raman spectroscopic quantifications of the crystallinity of lactose in whey permeate powder. *Int. Dairy J.* **2005**, *15*, 1261–1270. [CrossRef]
101. Öhman, J.; Geladi, P.; Wold, S. Residual bilinearization. Part 1: Theory and algorithms. *J. Chemom.* **1990**, *4*, 79–90. [CrossRef]
102. Barreto, M.C.; Braga, R.G.; Lemos, S.G.; Fragoso, W.D. Determination of melamine in milk by fluorescence spectroscopy and second-order calibration. *Food Chem.* **2021**, *364*, 130407. [CrossRef] [PubMed]
103. De Araújo Gomes, A.; Schenone, A.V.; Goicoechea, H.C.; de Araújo, M.C.U. Unfolded partial least squares/residual bilinearization combined with the Successive Projections Algorithm for interval selection: Enhanced excitation-emission fluorescence data modeling in the presence of the inner filter effect. *Anal. Bioanal. Chem.* **2015**, *407*, 5649–5659. [CrossRef]
104. Bro, R. PARAFAC. Tutorial and applications. *Chemom. Intel. Lab. Syst.* **1997**, *38*, 149–171. [CrossRef]
105. De Juan, A.; Tauler, R. Multivariate curve resolution (MCR) from 2000: Progress in concepts and applications. *Crit. Rev. Anal. Chem.* **2006**, *36*, 163–176. [CrossRef]
106. Kelwade, J.P.; Salankar, S.S. Radial basis function neural network for prediction of cardiac arrhythmias based on heart rate time series. In Proceedings of the 2016 IEEE First International Conference on Control, Measurement and Instrumentation (CMI), Kolkata, India, 8–10 January 2016; IEEE: New York, NY, USA, 2016; pp. 454–458.
107. Yabunaka, K.I.; Hosomi, M.; Murakami, A. Novel application of a back-propagation artificial neural network model formulated to predict algal bloom. *Water Sci. Technol.* **1997**, *36*, 89–97. [CrossRef]
108. Wang, Y.; Liu, Q. Fast identification of powdered milk adulteration by generalized regression neural network algorithm. In *International Conference on Computer Graphics, Artificial Intelligence, and Data Processing (ICCAID 2021)*; SPIE: Bellingham, WA, USA, 2022; Volume 12168, pp. 717–724.
109. Huang, W.; Fan, D.; Li, W.; Meng, Y.; Liu, T.C.Y. Rapid evaluation of milk acidity and identification of milk adulteration by Raman spectroscopy combined with chemometrics analysis. *Vib. Spectrosc.* **2022**, *123*, 103440. [CrossRef]
110. Suykens, J.A.K.; Vandewalle, J. Least square support vector machine classifiers. *Neural Process. Lett.* **1999**, *9*, 293–300. [CrossRef]
111. Yang, K.; An, C.; Zhu, J.; Guo, W.; Lu, C.; Zhu, X. Comparison of near-infrared and dielectric spectra for quantitative identification of bovine colostrum adulterated with mature milk. *J. Dairy Sci.* **2022**, *105*, 8638–8649. [CrossRef] [PubMed]
112. Tibshirani, R. Regression shrinkage and selection via the lasso. *J. R. Stat. Soc. B Methodol.* **1996**, *58*, 267–288. [CrossRef]
113. Rysova, L.; Cejnar, P.; Hanus, O.; Legarova, V.; Havlik, J.; Nejeschlebova, H.; Nemeckova, I.; Jedelska, R.; Bozik, M. Use of MALDI-TOF MS technology to evaluate adulteration of small ruminant milk with raw bovine milk. *J. Dairy Sci.* **2022**, *105*, 4882–4894. [CrossRef]
114. Friedman, J.; Hastie, T.; Tibshirani, R. Additive logistic regression: A statistical view of boosting (with discussion and a rejoinder by the authors). *Ann. Stat.* **2000**, *28*, 337–407. [CrossRef]
115. Ehsani, S.; Dastgerdy, E.M.; Yazdanpanah, H.; Parastar, H. Ensemble classification and regression techniques combined with portable near infrared spectroscopy for facile and rapid detection of water adulteration in bovine raw milk. *J. Chemom.* 2022, *Early View*.
116. Asefa, B.G.; Hagos, L.; Kore, T.; Emire, S.A. Feasibility of Image Analysis Coupled with Machine Learning for Detection and Quantification of Extraneous Water in Milk. *Food Anal. Methods* **2022**, *15*, 3092–3103. [CrossRef]
117. Chen, T.; Guestrin, C. Xgboost: A scalable tree boosting system. In Proceedings of the 22nd ACM Sigkdd International Conference on Knowledge Discovery and Data Mining, San Francisco, CA, USA, 13–17 August 2016; pp. 785–794.
118. Schlesier, K.; Fauhl-Hassek, C.; Forina, M.; Cotea, V.; Kocsi, E.; Schoula, R.; van Jaarsveld, F.; Wittkowaki, R. Characterisation and determination of the geographical origin of wines. Part I: Overview. *Eur. Food Res. Technol.* **2009**, *230*, 13. [CrossRef]
119. Esbensen, K.H.; Paoletti, C.; Thiex, N. Representative sampling for food and feed materials: A critical need for food/feed safety. *J. AOAC Int.* **2015**, *98*, 249–251. [CrossRef] [PubMed]

120. Peris-Díaz, M.D.; Krężel, A. A guide to good practice in chemometric methods for vibrational spectroscopy, electrochemistry, and hyphenated mass spectrometry. *Trends Anal. Chem.* **2021**, *135*, 116157. [CrossRef]
121. Kemsley, E.K.; Defernez, M.; Marini, F. Multivariate statistics: Considerations and confidences in food authenticity problems. *Food Control* **2019**, *105*, 102–112. [CrossRef]
122. Szymańska, E. Modern data science for analytical chemical data—A comprehensive review. *Anal. Chim. Acta* **2018**, *1028*, 1–10. [CrossRef] [PubMed]
123. Engel, J.; Gerretzen, J.; Szymańska, E.; Jansen, J.J.; Downey, G.; Blanchet, L.; Buydens, L.M. Breaking with trends in pre-processing? *Trends Anal. Chem.* **2013**, *50*, 96–106. [CrossRef]
124. Yang, Y.; Hettinga, K.A.; Erasmus, S.W.; Pustjens, A.M.; van Ruth, S.M. Opportunities for fraudsters: When would profitable milk adulterations go unnoticed by common, standardized FTIR measurements? *Food Res. Int.* **2020**, *136*, 109543. [CrossRef]
125. Oliveri, P.; Malegori, C.; Simonetti, R.; Casale, M. The impact of signal pre-processing on the final interpretation of analytical outcomes—A tutorial. *Anal. Chim. Acta* **2019**, *1058*, 9–17. [CrossRef] [PubMed]
126. Rabatel, G.; Marini, F.; Walczak, B.; Roger, J.M. VSN: Variable sorting for normalization. *J. Chemom.* **2020**, *34*, e3164. [CrossRef]
127. Westad, F.; Marini, F. Validation of chemometric models–a tutorial. *Anal. Chim. Acta* **2015**, *893*, 14–24. [CrossRef]
128. Rajamanickam, V.; Babel, H.; Montano-Herrera, L.; Ehsani, A.; Stiefel, F.; Haider, S.; Presser, B.; Knapp, B. About model validation in bioprocessing. *Processes* **2021**, *9*, 961. [CrossRef]
129. Smilde, A.; Bro, R.; Geladi, P. *Multi-Way Analysis in Chemistry and Related Fields*; John Wiley & Sons, Ltd.: New York, NY, USA, 2004; Volume 240, pp. 260–280.
130. Daszykowski, M.; Walczak, B.; Massart, D.L. Representative subset selection. *Anal. Chim. Acta* **2022**, *468*, 91–103. [CrossRef]
131. Williams, P.; Dardenne, P.; Flinn, P. Tutorial: Items to be included in a report on a near infrared spectroscopy project. *J. Near Infrared Spectrosc.* **2017**, *25*, 85–90. [CrossRef]
132. Bittante, G.; Patel, N.; Cecchinato, A.; Berzaghi, P. Invited review: A comprehensive review of visible and near-infrared spectroscopy for predicting the chemical composition of cheese. *J. Dairy Sci.* **2022**, *105*, 1817–1836. [CrossRef]
133. Williams, P.C. Implementation of near-infrared technology. In *Near-Infrared Technology in the Agricultural and Food Industries*; American Association of Cereal Chemist Press: St. Paul, Minnesota, USA, 2001; pp. 145–169.
134. Bellon-Maurel, V.; Fernandez-Ahumada, E.; Palagos, B.; Roger, J.M.; McBratney, A. Critical review of chemometric indicators commonly used for assessing the quality of the prediction of soil attributes by NIR spectroscopy. *Trends Anal. Chem.* **2010**, *29*, 1073–1081. [CrossRef]
135. Sharma, R.; Verma, A.; Shinde, N.; Mann, B.; Gandhi, K.; Wichers, J.H.; van Amerongen, A. Adulteration of cow's milk with buffalo's milk detected by an on-site carbon nanoparticles-based lateral flow immunoassay. *Food Chem.* **2021**, *351*, 129311. [CrossRef] [PubMed]
136. Agriopoulou, S.; Stamatelopoulou, E.; Varzakas, T. Advances in analysis and detection of major mycotoxins in foods. *Foods* **2020**, *9*, 518. [CrossRef]
137. Di Nardo, F.; Chiarello, M.; Cavalera, S.; Baggiani, C.; Anfossi, L. Ten Years of Lateral Flow Immunoassay Technique Applications: Trends, Challenges and Future Perspectives. *Sensors* **2021**, *21*, 5185. [CrossRef]
138. Agriopoulou, S.; Stamatelopoulou, E.; Varzakas, T. Advances in Occurrence, Importance, and Mycotoxin Control Strategies: Prevention and Detoxification in Foods. *Foods* **2020**, *9*, 137. [CrossRef]
139. Wang, C.; Peng, J.; Liu, D.-F.; Xing, K.-Y.; Zhang, G.-G.; Huang, Z.; Cheng, S.; Zhu, F.F.; Duan, M.L.; Zhang, K.Y.; et al. Lateral flow immunoassay integrated with competitive and sandwich models for the detection of aflatoxin M1 and *Escherichia coli* O157:H7 in milk. *J. Dairy Sci.* **2018**, *101*, 8767–8777. [CrossRef]
140. Yue, X.; Pan, Q.; Zhou, J.; Ren, H.; Peng, C.; Wang, Z.; Zhang, Y. A simplified fluorescent lateral flow assay for melamine based on aggregation induced emission of gold nanoclusters. *Food Chem.* **2022**, *385*, 132670. [CrossRef]
141. El Sheikha, A.F. DNAFoil: Novel technology for the rapid detection of food adulteration. *Trends Food Sci. Technol.* **2019**, *86*, 544–552. [CrossRef]
142. Roy, M.; Yadav, B.K. Electronic nose for detection of food adulteration: A review. *J. Food Sci. Technol.* **2022**, *59*, 846–858. [CrossRef] [PubMed]
143. Tian, H.; Chen, B.; Lou, X.; Yu, H.; Yuan, H.; Huang, J.; Chen, C. Rapid detection of acid neutralizers adulteration in raw milk using FGC E-nose and chemometrics. *J. Food Meas. Charact.* **2022**, *16*, 2978–2988. [CrossRef]
144. Tohidi, M.; Ghasemi-Varnamkhasti, M.; Ghafarinia, V.; Mohtasebi, S.S.; Bonyadian, M. Identification of trace amounts of detergent powder in raw milk using a customized low-cost artificial olfactory system: A novel method. *Measurement* **2018**, *124*, 120–129. [CrossRef]

foods

MDPI

Article

Investigating the Tocopherol Contents of Walnut Seed Oils Produced in Different European Countries Analyzed by HPLC-UV: A Comparative Study on the Basis of Geographical Origin

Petros D. Mitsikaris [1], Lambros Kokokiris [1], Agathi Pritsa [1], Athanasios N. Papadopoulos [1] and Natasa P. Kalogiouri [1,2,*]

[1] Laboratory of Chemical Biology, Department of Nutritional Sciences and Dietetics, International Hellenic University, Sindos, 57400 Thessaloniki, Greece
[2] Laboratory of Analytical Chemistry, Department of Chemistry, Aristotle University of Thessaloniki, 54124 Thessaloniki, Greece
* Correspondence: kalogiourin@chem.auth.gr

Abstract: A rapid HPLC-UV method was developed for the determination of tocopherols in walnut seed oils. The method was validated and the LODs ranged between 0.15 and 0.30 mg/kg, while the LOQs were calculated over the range of 0.50 to 1.00 mg/kg. The accuracy values ranged between 90.8 and 97.1% for the within-day assay ($n = 6$) and between 90.4 and 95.8% for the between-day assay ($n = 3 \times 3$), respectively. The precision of the method was evaluated and the RSD% values were lower than 6.1 and 8.2, respectively. Overall, 40 samples of walnuts available on the Greek market, originating from four different European countries (Greece, Ukraine, France, and Bulgaria), were processed into oils and analyzed. One-way ANOVA was implemented in order to investigate potential statistically significant disparities between the concentrations of tocopherols in the walnut oils on the basis of the geographical origin, and Tukey's post hoc test was also performed to examine exactly which varieties differed. The statistical analysis of the results demonstrated that the Ukrainian walnut seed oils exhibited significantly higher total concentrations compared to the rest of the samples.

Keywords: walnut; seed oil; tocopherols; HPLC-UV; geographical origin

Citation: Mitsikaris, P.D.; Kokokiris, L.; Pritsa, A.; Papadopoulos, A.N.; Kalogiouri, N.P. Investigating the Tocopherol Contents of Walnut Seed Oils Produced in Different European Countries Analyzed by HPLC-UV: A Comparative Study on the Basis of Geographical Origin. *Foods* **2022**, *11*, 3719. https://doi.org/10.3390/foods11223719

Academic Editor: Cristina Alamprese

Received: 15 September 2022
Accepted: 15 November 2022
Published: 19 November 2022

1. Introduction

Walnuts (*Juglans regia* L.) are a prominent member of the culinary nuts family. Their chemical composition, both macronutrient- and micronutrient-wise [1,2], suggests that they are a useful choice in a healthy and balanced diet. According to the FAO [3], 3,323,964 tons of walnuts were produced worldwide in 2020, while China, USA, Iran, Turkey, and Mexico were the top producers. The bioactive compounds in walnuts and walnuts oils have been linked with several beneficial effects to human health, such as anti-inflammatory [4], cardioprotective [5], and anti-proliferative [6] effects, among others. Furthermore, walnut consumption also aids in slowing down degenerative brain diseases [7,8] and has also been associated with hunger suppression [9], rendering it an effective tool in a weight-loss plan.

Tocopherols (α-, β-, γ-, and δ-) are part of the commonly known vitamin E complex, along with the corresponding tocotrienols. Their molecular structure consists of a chromanol ring and a saturated phytyl side chain located at the C2 position. Within the human organism they act as powerful antioxidants, as has been proven by both in vivo [10] and in vitro [11] studies. Specifically, their main function is to protect against the non-enzymatic peroxidation of polyunsaturated fatty acids that construct the bilayer cellular membrane [12]. Tocopherols are plant-derived [13], meaning they have to be provided to humans through diet or supplementation. One of their main characteristics is that they are lipid-soluble, and this is the reason why they can only be found in high-fat plant foods [14,15], such as walnuts.

The exploration of the tocopherol content in walnut oil samples is a broadly analyzed topic in the literature. There are an array of studies available discussing walnut oil's tocopherol content on the basis of the variety, maturity level, crop year, harvest time, and storage conditions [16–27]. However, there is no information available regarding the walnut oil tocopherol profile on the basis of the geographical origin. This is a gap that has to be filled, taking into consideration that domestic and imported products are both widely available in the Western world and consumers are concerned about which to select.

Several analytical protocols have been proposed in the literature for tocopherol analysis, employing high-pressure liquid chromatography coupled to diode array (HPLC-DAD) [27–29], UV–Vis [30,31], fluorescence (FLD) [32], or mass spectrometric (MS) detectors [33–36]. Undoubtedly, UV detectors allow a fast, reliable, and cost-effective analysis, especially when it comes to such a small number of compounds, where separation and identification can be achieved without much adversity. The flagship of the analytical workflow techniques, however, is sample preparation. Several prolonged and laborious protocols have been suggested, including some that involve the use of substantial volumes of organic solvents in a Soxhlet apparatus [37]. Solid liquid extraction (SLE) is also frequently used, but in most cases requires the use of large amounts of solvents [37,38]. The use of solid-phase extraction (SPE) allows for a reduction in the amount of toxic solvents but demands the fulfillment of various steps throughout the sample preparation process, as well as the acquisition of cartridges [27]. The objective is to choose the right approach in order to determine the bioactive analytes that are present in the samples, eliminate the use of time-consuming extraction steps, and minimize the use of organic solvents.

The next step after the chromatographic analysis involves data mining. The statistical analysis of the experimental data enables the interpretation of the results. Statistical and chemometric tools are widely used in authenticity studies to support the outcomes that are acquired from the experimental results [30,31,39,40].

The goal of this study is to propose a rapid HPLC-UV analytical protocol that could be applied in the investigation of the authenticity of walnut seed oils prepared from walnut seeds that are available in the Greek market, originating from four European countries, namely Greece, Ukraine, France, and Bulgaria. The concentrations of tocopherols among the analyzed samples of different geographical origins were examined using a one-way ANOVA. Tukey's post-hoc test was also performed to examine exactly which samples from different geographical origins differed from each other.

2. Materials and Methods

2.1. Chemicals and Reagents

The HPLC-grade acetonitrile (ACN) and HPLC-grade methanol (MeOH) were acquired from Carl-Roth (Carlsruhe, Germany), while the 2-propanol (IPA) and heptane were purchased from Panreac-AppliChem (Darmstadt, Germany). The ultrapure water was provided by a Milli-Q water purification system (Millipore, Bedford, MA, USA). The standard compounds that were utilized were all purchased by Sigma-Aldrich (Steinheim, Germany), namely α-tocopherol (96%), β-tocopherol (96%), γ-tocopherol (96%), and δ-tocopherol (96%). For each one of the standard compounds, stock solutions at a concentration of 1000 μg/mL were prepared in MeOH and they were then stored in dark brown bottles at −20 °C.

2.2. Walnut Samples

Forty samples of dried walnut kernels (500 g each) were used in the study, and they were all available on the Greek market. Specifically, 10 samples that originated from various regions of Macedonia and Thessaly in Greece, 10 samples of walnuts produced in Bulgaria, 10 samples produced in Ukraine, and 10 samples produced in France were provided by traders. All samples were acquired in 2021. The walnut kernels were separated from their shells and subsequently chopped and homogenized in a mixer. All samples were stored at −20 °C until further processing and analysis.

2.3. Instrumentation

The chromatographic analysis was performed using an Agilent 1220 Infinity HPLC-UV system (Agilent Technologies, Santa Clara, CA, USA). The system comprised of: (i) the degasser, (ii) the column oven, (iii) the manual injector, and (iv) the UV detector. The OpenLAB software (Agilent Technologies, Santa Clara, CA, USA) and the Method and Run control package were employed to monitor the analysis. The sonication of the samples was performed in a MRC:DC-150-H ultrasonic bath provided by MRC (Essex, UK). A vortex mixer purchased from VELP Scientifica (Usmate Velate, Italy) was used for agitation. Centrifugation was carried out using a 3-16PK centrifuge system supplied by Sigma (Osterode am Harz, Germany). Prior to the injection in the chromatographic system, all samples were filtered through QMax RR 25 mm 0.22 μm PTFE syringe filters purchased from Frisenette ApS (Knebel, Germany).

2.4. Chromatographic Analysis

For the separation of the analytes, a reversed-phase (RP) Kromasil C18 (4.6 × 250 mm, 5 μm) analytical column was used, provided by Macherey-Nagel (Dueren, Germany). A binary gradient elution program consisting of MeOH (A) and ACN (B) was used for the separation of tocopherols. The temperature in the column oven was maintained at 28 °C. The elution program lasted for 15 min in total. The gradient program started with 50% A and remained stable for seven min, then gradually increased to 100% A until the 12 min mark, and remained stable for the last 3 min [31]. The flow rate was set to 1 mL/min and the absorbance was measured at 295 nm.

2.5. Sample Preparation

For sample preparation, a slightly altered version of the extraction procedure previously introduced by Martakos et al. [41] was employed. First of all, the walnut seeds were thoroughly homogenized in a porcelain mortar. After, 1 g of solid sample was weighed in a falcon tube and 10 mL of heptane was added. The solution was vigorously agitated in a vortex mixer for 1 min and was then placed in an ultrasonic bath at 40 °C for 20 min. The samples were subsequently centrifuged at 8000 rpm for 10 min. In a following step, the organic layer was transferred into the rotary evaporator and evaporated under vacuum to obtain the pure oil. Then, 100 mg of oil was weighed and extracted with 400 μL of isopropanol. The mixture was vortexed during 1 min, centrifuged at 8000 rpm for 10 min, and finally, the aliquot was collected and filtered through 0.22 μm PTFE syringe filters prior to the injection in the chromatographic system.

2.6. Method Validation

Linearity, accuracy, precision, limits of detection (LODs), and limits of quantification (LOQs) were assessed to validate the method. Linearity studies were conducted at the concentration range LOQ—50 mg/kg, by plotting the peak area versus the concentration of the standard compounds for 7 calibration points. The r^2 values of the standard calibration curves were calculated in order to assess the linearity of the method. The LOQs were considered to be the lower point of the calibration curve that corresponded to a signal-to-noise ratio (S/N) higher than 10. In order to calculate the LOD values, the LOQ of each analyte was divided by 3.3 [42]. Accuracy and precision were evaluated using a pool sample spiked at three different concentration levels (0.5, 25 and 50 mg/kg). The relative recoveries (%R) were calculated by means of recovery percentage, by comparing the found and added concentrations of the examined analytes, expressing accuracy. The relative standard deviations (%RSDs) were calculated to assess the precision of the method. Repeatability was assessed by measuring within-day precision using six replicates (n = 6), and reproducibility was evaluated by performing triplicate analysis of a pool sample spiked at three different concentrations (0.5, 25 and 50 mg/kg) within three consecutive days.

2.7. Statistical Analysis

The statistical analysis was performed using IBM SPSS Statistics 21. As the assumption of normality was not met by tocopherol concentration values in some groups, the concentrations of each tocopherol were compared between groups using the Kruskal–Wallis non parametric test at $\alpha = 0.05$. One-way ANOVA was performed to explore potential significant disparities between the samples, both in the concentration of every single tocopherol and in their total sum, as well. The confidence level was set at $\alpha = 0.05$. To discover this, a post hoc analysis was carried out. There isa large variety of post hoc tests available, all of which have their own pros and cons [43]. In this study, we chose to perform Tukey's test, which is the most commonly used one [44]. The MetaboAnalyst 5.0 package was used to create box plots and present the concentrations of the determined tocopherols in the analyzed samples [45].

3. Results

3.1. Method Validation Results

Table 1 presents the analytical parameters of the HPLC-UV method. The LODs were found to range between 0.15 to 0.30 mg/kg, and the LOQ ranged between 0.50 to 1.00 mg/kg. The method precision was good since the %RSD values of the within-day ($n = 6$) and between-day assays ($n = 3 \times 3$) were lower than 6.1 and 8.2, respectively. The accuracy was assessed by means of relative percentage of recovery (%R) were calculated at three concentration levels (0.5, 25, 50 mg/kg), and ranged between 90.8 and 97.1% for the within-day assay ($n = 6$) (Table 2), and between 90.4 and 95.8% for between-day assay ($n = 3 \times 3$) (Table 3).

Table 1. HPLC-UV analytical parameters.

Compound	Calibrationequation	Linear Range (mg/kg)	r^2	LOD (mg/kg)	LOQ (mg/kg)
α-tocopherol	y = 5.68x − 1.23	1–50	0.999	0.30	1.00
β+γ-tocopherol	y = 6.94x + 0.19	0.5–50	0.997	0.15	0.50
δ-tocopherol	y = 5.89x + 0.33	0.8–50	0.996	0.27	0.80

LOD: limit of detection; LOQ: limit of quantitation.

Table 2. Recovery (%R) rates for the evaluation of the repeatability.

Compound	Low Concentration (%R, $n = 6$)	%RSD	Medium Concentration (%R, $n = 6$)	%RSD	High Concentration (%R, $n = 6$)	%RSD
α-tocopherol	92.2	5.4	90.8	5.8	96.5	6.1
β+γ-tocopherol	94.1	3.6	92.5	2.9	97.1	5.8
δ-tocopherol	93.5	6.5	96.9	4.3	93.4	5.3

Table 3. Recovery (%R) rates for the evaluation of the reproducibility.

Compound	Low Concentration (%R, $n = 3 \times 3$)	%RSD	Medium Concentration (%R, $n = 3 \times 3$)	%RSD	High Concentration (%R, $n = 3 \times 3$)	%RSD
α-tocopherol	93.7	6.5	93.3	6.5	95.5	8.2
(β+γ)-tocopherol	94.3	7.3	94.1	8.3	91.2	6.9
δ-tocopherol	95.8	5.8	90.4	7.5	92.4	7.5

3.2. Walnut Seed Oil Analysis

Overall, 40 samples originating from 4 different countries were analyzed in triplicate, and the contentrations of α-tocopherol, (β+γ)-tocopherol, and δ-tocopherol were determined. The separation of the tocopherols was accomplished within twelve minutes, as it is shown in the characteristic chromatogram of a standard mixture of tocopherols (α-tocopherol: 5 mg/kg; (β+γ)-tocopherol: 5 mg/kg; δ-tocopherol: 5 mg/kg) monitored at 295 nm, presented in Figure 1. The β- and γ-tocopherol are isomers and were quantified as a sum [46]. Table 4 presents the determined analytes along with the molecular formulas, molecular structures, and retention times (RTs). The β- and γ-tocopherol are isomers, and subsequently they eluted at the same RT [27,30,31].

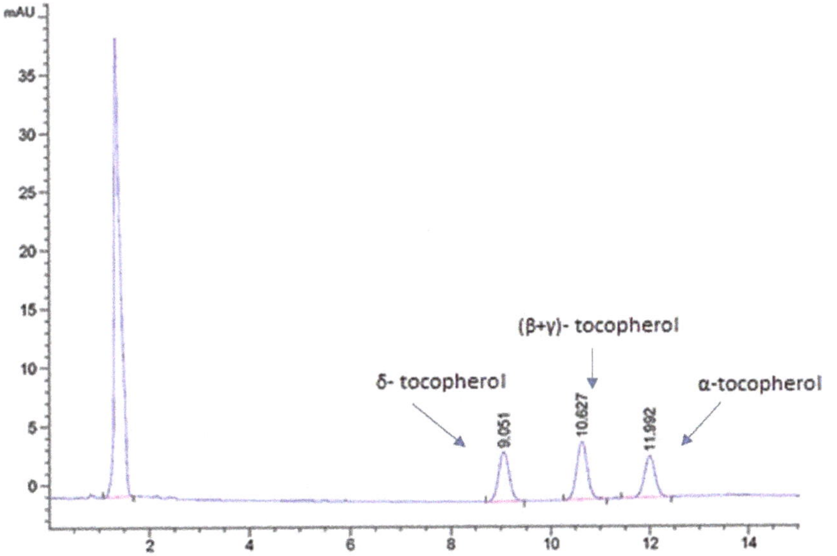

Figure 1. Characteristic chromatogram of a standard mixture of tocopherols (α-tocopherol: 5 mg/kg; (β+γ)-tocopherol: 5 mg/kg; δ-tocopherol: 5 mg/kg) monitored at 295 nm.

Table 4. Molecular formulas, molecular structures, and retention times.

Compound	Molecular Formula	Molecular Structure	RT (min)
α-tocopherol	$C_{29}H_{50}O_2$	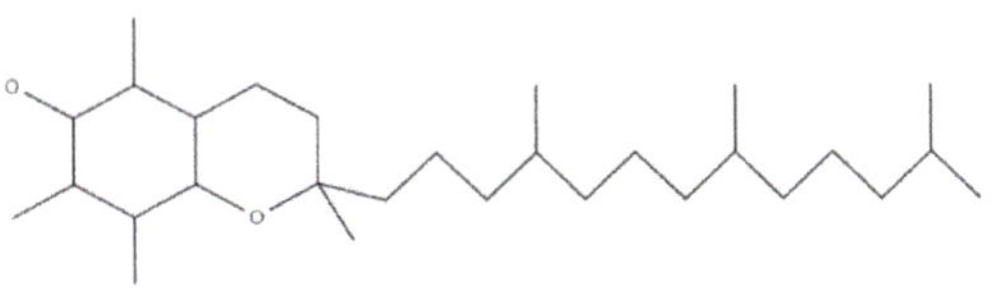	12.0
β-tocopherol	$C_{28}H_{48}O_2$	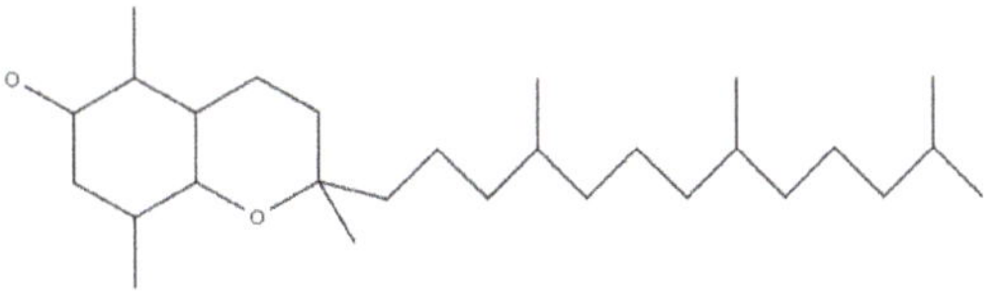	10.6

Table 4. *Cont.*

Compound	Molecular Formula	Molecular Structure	RT (min)
γ-tocopherol	$C_{28}H_{48}O_2$	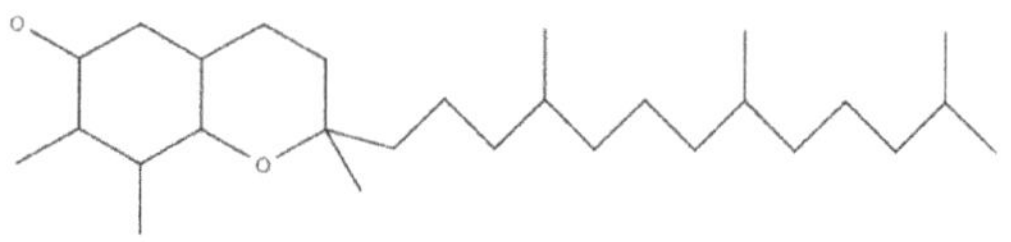	10.6
δ-tocopherol	$C_{27}H_{46}O_2$	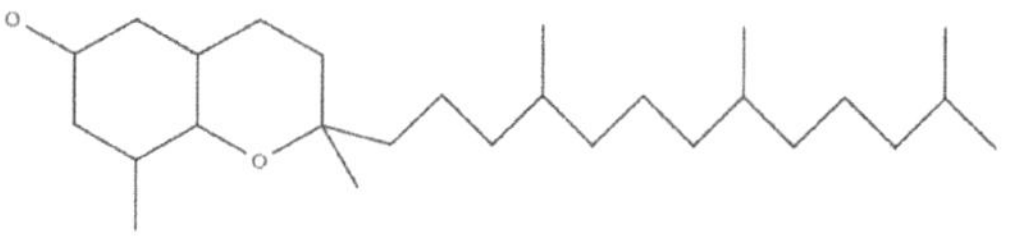	9.1

RT: retention time.

3.3. *Quantitative Analysis of Tocopherols*

The concentration ranges (minimum and maximum values) of the tocopherols expressed in mg per kg of walnut seed oil, are presented in Table 5. The average concentration ranges and their mean values (±SD), as well as their average total concentrations are presented in Table 6. The sum of β- and γ-tocopherol was proven to be the most abundant in all samples, owing to the high concentrations of γ-tocopherol in the walnuts [16]. The second most abundant tocopherol was δ-tocopherol, whileα-tocopherol was the least abundant.

Table 5. Concentration ranges (mg per kg of walnut seed oil) of the samples.

Tocopherol	Greek Walnuts	French Walnuts	Ukrainian Walnuts	Bulgarian Walnuts
α-	5.3–18.4	2.9–10.9	7.2–22.1	4.1–7.2
β- and γ-	100.3–189	92.0–142.5	166.9–229.0	105.9–160.8
δ-	11.2–19.6	14.5–20.9	18.3–24.8	12.5–20.9
Total	117.5–220.8	110.6–174.3	192.4–263.2	124.4–187.0

Table 6. Mean values ± SD (mg per kg of walnut seed oil) of samples.

Tocopherol	Greek Walnuts (n = 10)	French Walnuts (n = 10)	Ukrainian Walnuts (n = 10)	Bulgarian Walnuts (n = 10)
α-	10± 4	6 ± 3	12 ± 6	5.5 ± 1.0
β- and γ-	150 ± 30	122 ± 17	204 ± 21	138.0 ± 17.8
δ-	15± 3	17 ± 2	23 ± 2	16 ± 2
Total	176 ± 34	145 ± 20	239± 23	159 ± 20

Box and whisker plots were created for each compound to graphically depict their concentrations among the different European regions. The concentration levels of α-tocopherol did not differ significantly between Greek and Ukrainian walnuts. However, these concentration levels were significantly higher compared to those determined in Bulgarian and French walnuts (x2 (3) = 19.9, $p < 0.001$, Figure 2A). The sum of (β and γ)-tocopherols and the concentration levels of δ- tocopherol were determined in significantly higher levels in Ukrainian walnuts (x2 (3) = 23.6, $p < 0.001$ for sum of β- and γ-tocopherols, Figure 2B; x2 (3) = 20.6, $p < 0.001$, for δ tocopherol, Figure 2C). Similarly, the average total concentrations of tocopherols were found to be significantly higher in Ukrainian walnuts (x2 (3) = 22.9, $p < 0.001$, Figure 3). δ-Tocopherol was found to be the most abundant tocopherol in Ukrainian walnuts with a mean concentration equal to 23 ± 2 mg per kg of oil. The second most abundant concentration was determined in French samples (17 ± 2 mg/kg). The Bulgarian and Greek walnuts mean values were 16 ± 2 mg/kg and 15 ± 3 mg/kg, respectively. The total tocopherol content was highest in Ukrainian samples, presenting a mean value equal to 239 ± 23 mg per kg of oil. The Greek walnuts had the second highest mean value of 176 ± 34 mg/kg. The Bulgarian walnuts mean concentration was 159 ± 20 mg/kg, while the French walnuts demonstrated a mean value of 145 ± 20 mg/kg.

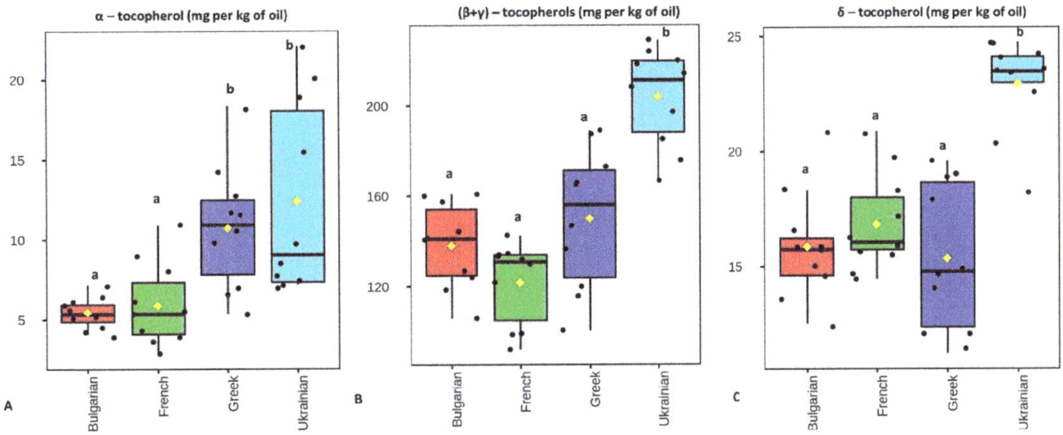

Figure 2. Box plot of α-tocopherol (**A**) and the sum of β- and γ-tocopherols (**B**) and δ-tocopherol in oils (**C**) from Bulgarian, French, Greek, and Ukrainian walnuts. Box plot values highlighted with different Latin letters differ significantly at α = 0.001.

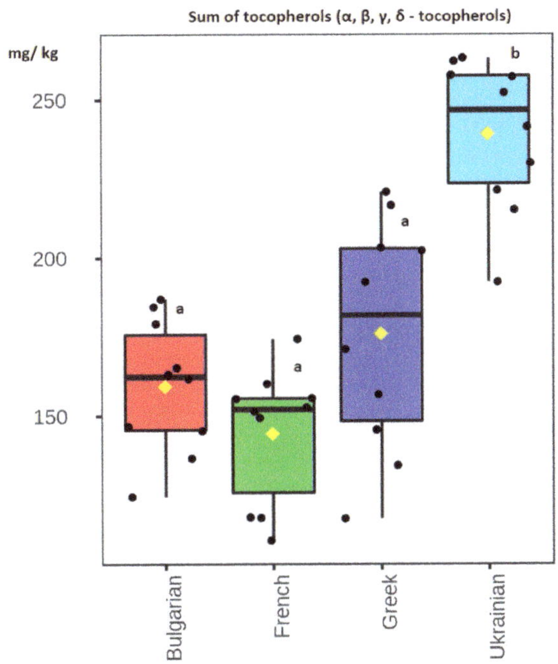

Figure 3. Box plot of total tocopherol contents (sum of α-, β-, γ- and δ-tocopherols) in oils from Bulgarian, French, Greek, and Ukrainian walnuts. Box plot values highlighted with different Latin letters differ significantly at α = 0.001.

The concentration ranges reported were in accordance with studies published in the literature. There is a wide variety of studies available that confirm that γ-tocopherol is of the highest concentration in walnut seed oils, no matter the cultivar or the geographical origin of the sample [19,20,22–25,47–50].

Concerning α-tocopherol, in particular, some of the aforementioned studies have reported higher values compared to those reported in this study, at up to 45 mg per kg of walnut oil [20,21,24]. The range of concentrations is even wider for the sum contents of β- and γ-tocopherols, since some studies have reported the sum of β- and γ-tocopherols at the lowest limit of 35 mg per kg of oil, which is significantly lower compared to the concentrations found in this study [20]. On the other hand, higher concentrations have been reported in crops originating from Argentina and Turkey, supporting the idea that there is a correlation between the tocopherol content and the geographical origin [18,24]. As for δ-tocopherol, the concentration ranges reported in the literature are similar to those determined in the present study.

4. Conclusions

A rapid protocol was developed to determine the concentrations of tocopherol homologues in walnut samples originating from four different European regions (Greece, Ukraine, France, and Bulgaria) using HPLC-UV. According to the results, the highest concentrations were determined in Ukrainian walnuts, the second largest were found in Greek walnuts, and then the Bulgarian and the French walnuts followed. The results were analyzed with one-factor ANOVA, and it was revealed that the samples differed significantly from each other (p-value < 0.001), indicating that big discrepancies do exist on the basis of the samples' geographical origin. Furthermore, the Ukrainian walnut seed oils presented a significantly higher concentration compared to the rest of the samples. Finally, the Greek walnuts possessed a significantly higher mean concentration of α-tocopherol compared to the French and Bulgarian walnuts and a significantly higher mean concentration of β- plus γ-tocopherols compared to the French walnuts.

Author Contributions: Conceptualization, N.P.K.; methodology, N.P.K. and P.D.M.; validation, P.D.M.; formal analysis, P.D.M.; investigation, N.P.K., P.D.M. and L.K.; resources, A.N.P.; data curation, P.D.M.; writing—original draft preparation, P.D.M., L.K., A.P. and N.P.K.; writing—review and editing, N.P.K. and L.K.; visualization, N.P.K. and A.N.P.; supervision, N.P.K. and A.P.; project administration, N.P.K. and A.P. All authors have read and agreed to the published version of the manuscript.

Funding: This research received no external funding.

Conflicts of Interest: The authors declare no conflict of interest.

References

1. Rébufa, C.; Artaud, J.; Le Dréau, Y. Walnut (*Juglans Regia* L.) Oil Chemical Composition Depending on Variety, Locality, Extraction Process and Storage Conditions: A Comprehensive Review. *J. Food Compos. Anal.* **2022**, *110*, 104534. [CrossRef]
2. Kalogiouri, N.P.; Manousi, N.; Rosenberg, E.; Zachariadis, G.A.; Samanidou, V.F. Advances in the Chromatographic Separation and Determination of Bioactive Compounds for Assessing the Nutrient Profile of Nuts. *Curr. Anal. Chem.* **2021**, *17*, 495–511. [CrossRef]
3. Food and Agricultural Association (FAO). Available online: https://www.fao.org/faostat/en/#data/QC (accessed on 18 August 2022).
4. Zhang, Y.G.; Kan, H.; Chen, S.X.; Thakur, K.; Wang, S.; Zhang, J.G.; Shang, Y.F.; Wei, Z.J. Comparison of Phenolic Compounds Extracted from Diaphragma Juglandis Fructus, Walnut Pellicle, and Flowers of *Juglans Regia* Using Methanol, Ultrasonic Wave, and Enzyme Assisted-Extraction. *Food Chem.* **2020**, *321*, 126672. [CrossRef] [PubMed]
5. Ros, E.; Izquierdo-Pulido, M.; Sala-Vila, A. Beneficial Effects of Walnut Consumption on Human Health: Role of Micronutrients. *Curr. Opin. Clin. Nutr. Metab. Care* **2018**, *21*, 498–504. [CrossRef] [PubMed]
6. Park, J.M.; An, J.M.; Han, Y.M.; Surh, Y.J.; Hwang, S.J.; Kim, S.J.; Hahm, K.B. Walnut Polyphenol Extracts Inhibit Helicobacter Pylori-Induced STAT3Tyr705 Phosphorylation through Activation of PPAR-γ and SOCS1 Induction. *J. Clin. Biochem. Nutr.* **2020**, *67*, 248–256. [CrossRef]
7. Muthaiyah, B.; Essa, M.M.; Lee, M.; Chauhan, V.; Kaur, K.; Chauhan, A. Dietary Supplementation of Walnuts Improves Memory Deficits and Learning Skills in Transgenic Mouse Model of Alzheimer's Disease. *J. Alzheimer's Dis.* **2014**, *42*, 1397–1405. [CrossRef] [PubMed]

8. Pandareesh, M.D.; Chauhan, V.; Chauhan, A. Walnut Supplementation in the Diet Reduces Oxidative Damage and Improves Antioxidant Status in Transgenic Mouse Model of Alzheimer's Disease. *J. Alzheimer's Dis.* **2018**, *64*, 1295–1305. [CrossRef]
9. Wilson, T.; Devaan, L.S.; Lacasse, M.E.; Gile, E.M.; Weis, M.J.; Ahmann, M.D.; Schnellman, G.I.; Lenz, M.T.; Hooks, T.L. Effect of Walnut Predinner Snack on Mealtime Hunger and Nutrient Intake Among University Students. *J. Med. Food* **2022**, *25*, 89–96. [CrossRef] [PubMed]
10. Niki, E.; Noguchi, N. Antioxidant Action of Vitamin E in Vivo as Assessed from Its Reaction Products with Multiple Biological Oxidants. *Free Radic. Res.* **2021**, *55*, 352–363. [CrossRef]
11. Basta-kaim, A.; Trela-makowej, A.; Le, M.; Kruk, J.; Andrzej, Z. Antioxidant and Neuroprotective Activity of Vitamin E Homologues: In Vitro Study. *Metabolites* **2022**, *8*, 608.
12. Niki, E. Lipid Oxidation That Is, and Is Not, Inhibited by Vitamin E: Consideration about Physiological Functions of Vitamin E. *Free Radic. Biol. Med.* **2021**, *176*, 1–15. [CrossRef] [PubMed]
13. Fritsche, S.; Wang, X.; Jung, C. Recent Advances in Our Understanding of Tocopherol Biosynthesis in Plants: An Overview of Key Genes, Functions, and Breeding of Vitamin E Improved Crops. *Antioxidants* **2017**, *6*, 99. [CrossRef]
14. Aksoz, E.; Korkut, O.; Aksit, D.; Gokbulut, C. Vitamin E (α-, $\beta + \gamma$- and δ-Tocopherol) Levels in Plant Oils. *Flavour Fragr. J.* **2020**, *35*, 504–510. [CrossRef]
15. Matthäus, B.; Özcan, M.M.; Al Juhaimi, F.; Adiamo, O.Q.; Alsawmahi, O.N.; Ghafoor, K.; Babike, E.E. Effect of the Harvest Time on Oil Yield, Fatty Acid, Tocopherol and Sterol Contents of Developing Almond and Walnut Kernels. *J. Oleo Sci.* **2018**, *67*, 39–45. [CrossRef] [PubMed]
16. Nguyen, T.H.D.; Vu, D.C. A Review on Phytochemical Composition and Potential Health-Promoting Properties of Walnuts. *Food Rev. Int.* **2021**. [CrossRef]
17. Grilo, F.S.; Wang, S.C. Walnut (Juglans Regia L.) Volatile Compounds Indicate Kernel and Oil Oxidation. *Foods* **2021**, *10*, 329. [CrossRef] [PubMed]
18. Cittadini, M.C.; Martín, D.; Gallo, S.; Fuente, G.; Bodoira, R.; Martínez, M.; Maestri, D. Evaluation of Hazelnut and Walnut Oil Chemical Traits from Conventional Cultivars and Native Genetic Resources in a Non-Traditional Crop Environment from Argentina. *Eur. Food Res. Technol.* **2020**, *246*, 833–843. [CrossRef]
19. Pycia, K.; Kapusta, I.; Jaworska, G.; Jankowska, A. Antioxidant Properties, Profile of Polyphenolic Compounds and Tocopherol Content in Various Walnut (*Juglans Regia* L.) Varieties. *Eur. Food Res. Technol.* **2019**, *245*, 607–616. [CrossRef]
20. Pycia, K.; Kapusta, I.; Jaworska, G. Impact of the Degree of Maturity of Walnuts (*Juglans Regia* L.) and Their Variety on the Antioxidant Potential and the Content of Tocopherols and Polyphenols. *Molecules* **2019**, *24*, 2936. [CrossRef] [PubMed]
21. Mohamed Ahmed, I.A.; Al-Juhaimi, F.Y.; Özcan, M.M.; Osman, M.A.; Gassem, M.A.; Salih, H.A.A. Effects of Cold-Press and Soxhlet Extraction Systems on Antioxidant Activity, Total Phenol Contents, Fatty Acids, and Tocopherol Contents of Walnut Kernel Oils. *J. Oleo Sci.* **2019**, *68*, 167–173. [CrossRef] [PubMed]
22. Ojeda-Amador, R.M.; Salvador, M.D.; Gómez-Alonso, S.; Fregapane, G. Characterization of Virgin Walnut Oils and Their Residual Cakes Produced from Different Varieties. *Food Res. Int.* **2018**, *108*, 396–404. [CrossRef] [PubMed]
23. Gao, P.; Jin, J.; Liu, R.; Jin, Q.; Wang, X. Chemical Compositions of Walnut (*Juglans Regia* L.) Oils from Different Cultivated Regions in China. *JAOCS J. Am. Oil Chem. Soc.* **2018**, *95*, 825–834. [CrossRef]
24. Kafkas, E.; Burgut, A.; Ozcan, H.; Ozcan, A.; Sutyemez, M.; Kafkas, S.; Türemis, N. Fatty Acid, Total Phenol and Tocopherol Profiles of Some Walnut Cultivars: A Comparative Study. *Food Nutr. Sci.* **2017**, *8*, 1074–1084. [CrossRef]
25. Gao, P.; Liu, R.; Jin, Q.; Wang, X. Comparative Study of Chemical Compositions and Antioxidant Capacities of Oils Obtained from Two Species of Walnut: *Juglans regia* and *Juglans sigillata*. *Food Chem.* **2019**, *279*, 279–287. [CrossRef] [PubMed]
26. González-Gómez, D.; Ayuso-Yuste, M.C.; Blanco-Roque, C.; Bernalte-García, M.J. Optimization of Enzyme-Assisted Aqueous Method for the Extraction of Oil from Walnuts Using Response Surface Methodology. *J. Food Process. Preserv.* **2019**, *43*, e14218. [CrossRef]
27. Kalogiouri, N.P.; Kabir, A.; Olayanju, B.; Furton, K.G.; Samanidou, V.F. Development of Highly Hydrophobic Fabric Phase Sorptive Extraction Membranes and Exploring Their Applications for the Rapid Determination of Tocopherols in Edible Oils Analyzed by High Pressure Liquid Chromatography-Diode Array Detection. *J. Chromatogr. A* **2022**, *1664*, 462785. [CrossRef]
28. Xie, Q.; Xia, M.; Sun, D.; Cao, J.; Xiao, Y.; Lin, M.; Hou, B.; Jia, L.; Li, D. Deep Eutectic Solvent-Based Liquid-Phase Microextraction Coupled with Reversed-Phase High-Performance Liquid Chromatography for Determination of α-, β-, γ-, and δ-Tocopherol in Edible Oils. *Anal. Bioanal. Chem.* **2021**, *413*, 577–584. [CrossRef] [PubMed]
29. Bakir, D.S.; Yalcin, G.; Cucu, A.K. Isolation and Determination of Tocopherols and Tocotrienols from the Seed of *Capparis Ovata* Grown in Turkey by Reversed-Phase High-Performance Liquid Chromatography. *Chromatographia* **2020**, *83*, 77–86. [CrossRef]
30. Kalogiouri, N.P.; Mitsikaris, P.D.; Papadopoulos, A.N.; Samanidou, V.F. Microwave-Assisted Extraction Coupled to HPLC-UV Combined with Chemometrics for the Determination of Bioactive Compounds in Pistachio Nuts and the Guarantee of Quality and Authenticity. *Molecules* **2022**, *27*, 1435. [CrossRef]
31. Kalogiouri, N.P.; Mitsikaris, P.D.; Klaoudatos, D.; Papadopoulos, A.N.; Samanidou, V.F. A Rapid Hplc-Uv Protocol Coupled to Chemometric Analysis for the Determination of the Major Phenolic Constituents and Tocopherol Content in Almonds and the Discrimination of the Geographical Origin. *Molecules* **2021**, *26*, 5433. [CrossRef] [PubMed]
32. Rotondo, A.; La Torre, G.L.; Gervasi, T.; Di Matteo, G.; Spano, M.; Ingallina, C.; Salvo, A. A Fast and Efficient Ultrasound-Assisted Extraction of Tocopherols in Cow Milk Followed by HPLC Determination. *Molecules* **2021**, *26*, 4645. [CrossRef]

33. Liu, S.; Hu, H.; Yu, Y.; Zhao, J.; Liu, L.; Zhao, S.; Xie, J.; Li, C.; Shen, M. Simultaneous Determination of Tocopherols, Phytosterols, and Squalene in Vegetable Oils by High Performance Liquid Chromatography-Tandem Mass Spectrometry. *Food Anal. Methods* **2021**, *14*, 1567–1576. [CrossRef]
34. Gachumi, G.; Demelenne, A.; Poudel, A.; Dallal Bashi, Z.; El-Aneed, A. Novel Fast Chromatography-Tandem Mass Spectrometric Quantitative Approach for the Determination of Plant-Extracted Phytosterols and Tocopherols. *Molecules* **2021**, *26*, 1402. [CrossRef]
35. Dugo, L.; Russo, M.; Cacciola, F.; Mandolfino, F.; Salafia, F.; Vilmercati, A.; Fanali, C.; Casale, M.; De Gara, L.; Dugo, P.; et al. Determination of the Phenol and Tocopherol Content in Italian High-Quality Extra-Virgin Olive Oils by Using LC-MS and Multivariate Data Analysis. *Food Anal. Methods* **2020**, *13*, 1027–1041. [CrossRef]
36. Zhang, L.; Wang, S.; Yang, R.; Mao, J.; Jiang, J.; Wang, X.; Zhang, W.; Zhang, Q.; Li, P. Simultaneous Determination of Tocopherols, Carotenoids and Phytosterols in Edible Vegetable Oil by Ultrasound-Assisted Saponification, LLE and LC-MS/MS. *Food Chem.* **2019**, *289*, 313–319. [CrossRef] [PubMed]
37. Bodoira, R.; Maestri, D. Phenolic Compounds from Nuts: Extraction, Chemical Profiles, and Bioactivity. *J. Agric. Food Chem.* **2020**, *68*, 927–942. [CrossRef] [PubMed]
38. Indyk, H. Total Vitamin E in Dairy Products, Foods and Tissues By. *Analyst* **1988**, *113*, 1217–1221. [CrossRef] [PubMed]
39. Kalogiouri, N.P.; Manousi, N.; Rosenberg, E.; Zachariadis, G.A.; Paraskevopoulou, A.; Samanidou, V. Exploring the Volatile Metabolome of Conventional and Organic Walnut Oils by Solid-Phase Microextraction and Analysis by GC-MS Combined with Chemometrics. *Food Chem.* **2021**, *363*, 130331. [CrossRef] [PubMed]
40. Kalogiouri, N.P.; Samanidou, V.F. HPLC Fingerprints for the Characterization of Walnuts and the Detection of Fraudulent Incidents. *Foods* **2021**, *10*, 2145. [CrossRef] [PubMed]
41. Martakos, I.; Kostakis, M.; Dasenaki, M.; Pentogennis, M.; Thomaidis, N. Simultaneous determination of pigments, tocopherols, and squalene in Greek olive oils: A study of the influence of cultivation and oil-production parameters. *Foods* **2019**, *9*, 31. [CrossRef] [PubMed]
42. Manousi, N.; Alampanos, V.; Priovolos, I.; Kabir, A.; Furton, K.G.; Rosenberg, E.; Zachariadis, G.A.; Samanidou, V.F. Talanta Designing a Moderately Hydrophobic Sol-Gel Monolithic Carbowax 20 M Sorbent for the Capsule Phase Microextraction of Triazine Herbicides from Water Samples Prior to HPLC Analysis. *Talanta* **2021**, *234*, 122710. [CrossRef] [PubMed]
43. McHugh, M.L. Multiple Comparison Analysis Testing in ANOVA. *Biochem. Med.* **2011**, *21*, 203–209. [CrossRef]
44. Ruxton, G.D.; Beauchamp, G. Time for Some a Priori Thinking about Post Hoc Testing. *Behav. Ecol.* **2008**, *19*, 690–693. [CrossRef]
45. Pang, Z.; Chong, J.; Zhou, G.; de Lima Morais, D.A.; Chang, L.; Barrette, M.; Gauthier, C.; Jacques, P.-É.; Li, S.; Xia, J. MetaboAnalyst 5.0: Narrowing the Gap between Raw Spectra and Functional Insights. *Nucleic Acids Res.* **2021**, *49*, 388–396. [CrossRef] [PubMed]
46. Gliszczynska-Swiglo, A.; Sikorska, E.; Khmelinskii, I.; Sikorski, M. Tocopherol Content in Edible Plant Oils. *Pol. J. Food Nutr. Sci.* **2007**, *57*, 157–161.
47. Rabrenovic, B.; Dimic, E.; Maksimovic, M. Determination of Fatty Acid and Tocopherol Compositions. *Czech J. Food Sci.* **2011**, *29*, 74–78. [CrossRef]
48. Amaral, J.S.; Alves, M.R.; Seabra, R.M.; Oliveira, B.P.P. Vitamin E Composition of Walnuts (*Juglans Regia* L.): A 3-Year Comparative Study of Different Cultivars. *J. Agric. Food Chem.* **2005**, *53*, 5467–5472. [CrossRef]
49. Abdallah, I.B.; Tlili, N.; Martinez-Force, E.; Rubio, A.G.P.; Perez-Camino, M.C.; Albouchi, A.; Boukhchina, S. Content of Carotenoids, Tocopherols, Sterols, Triterpenic and Aliphatic Alcohols, and Volatile Compounds in Six Walnuts (*Juglans Regia* L.) Varieties. *Food Chem.* **2015**, *173*, 972–978. [CrossRef]
50. Wu, S.; Ni, Z.; Wang, R.; Zhao, B.; Han, Y.; Zheng, Y.; Liu, F.; Gong, Y.; Tang, F.; Liu, Y. The Effects of Cultivar and Climate Zone on Phytochemical Components of Walnut (*Juglans Regia* L.). *Food Energy Secur.* **2020**, *9*, e196. [CrossRef]

foods

Review

Detection of Saffron's Main Bioactive Compounds and Their Relationship with Commercial Quality

Raul Avila-Sosa [1], Guadalupe Virginia Nevárez-Moorillón [2], Carlos Enrique Ochoa-Velasco [1], Addí Rhode Navarro-Cruz [1], Paola Hernández-Carranza [1] and Teresa Soledad Cid-Pérez [1,*]

1 Facultad de Ciencias Químicas, Benemérita Universidad Autónoma de Puebla, Edificio 105E, 14 Sur y Av. San Claudio, Ciudad Universitaria, Col. San Manuel, Puebla 72420, Mexico
2 Facultad de Ciencias Químicas, Universidad Autónoma de Chihuahua, Circuito Universitario s/n Campus, Universitario II, Chihuahua 31125, Mexico
* Correspondence: teresa.cid@correo.buap.mx

Abstract: This review aims to evaluate the state of saffron's main bioactive compounds and their relationship with its commercial quality. Saffron is the commercial name for the dried red stigmas of the *Crocus sativus* L. flower. It owes its sensory and functional properties mainly to the presence of its carotenoid derivatives, synthesized throughout flowering and also during the whole production process. These compounds include crocin, crocetin, picrocrocin, and safranal, which are bioactive metabolites. Saffron's commercial value is determined according to the ISO/TS3632 standard that determines their main apocatotenoids. Other techniques such as chromatography (gas and liquid) are used to detect the apocarotenoids. This, together with the determination of spectral fingerprinting or chemo typing are essential for saffron identification. The determination of the specific chemical markers coupled with chemometric methods favors the discrimination of adulterated samples, possible plants, or adulterating compounds and even the concentrations at which these are obtained. Chemical characterization and concentration of various compounds could be affected by saffron's geographical origin and harvest/postharvest characteristics. The large number of chemical compounds found in the by-products (flower parts) of saffron (catechin, quercetin, delphinidin, etc.) make it an interesting aromatic spice as a colorant, antioxidant, and source of phytochemicals, which can also bring additional economic value to the most expensive aromatic species in the world.

Keywords: saffron; bioactive compounds; chemometric methods; adulteration; by-products

Citation: Avila-Sosa, R.; Nevárez-Moorillón, G.V.; Ochoa-Velasco, C.E.; Navarro-Cruz, A.R.; Hernández-Carranza, P.; Cid-Pérez, T.S. Detection of Saffron's Main Bioactive Compounds and Their Relationship with Commercial Quality. *Foods* **2022**, *11*, 3245. https://doi.org/10.3390/foods11203245

Academic Editor: Theodoros Varzakas

Received: 9 August 2022
Accepted: 8 October 2022
Published: 18 October 2022

1. Introduction

Plants and vegetables are major sources of food bioactives. Spices and herbs are plant materials that provide a wide range of biologically active compounds. In addition to being used as sources of aroma, flavor, and color and as preservatives, spices and herbs have been used for medicinal purposes and health and wellness for centuries. Aromatic spices can be added to food in their natural state as a powder or extract [1]. In the food industry, it is not only the active parts of vegetables or plants that are important since there are several uses for their waste or by-products as ingredients in different food formulations [2].

Saffron is the commercial name for the dried red stigmas of the *Crocus sativus* L. flower. It is appreciated for adding color, flavor, and a particular aroma to different food dishes or drinks (paella in Spain, Milanese risotto in Italy, lussekatter buns in Sweden, and alcoholic beverages). It is considered a high-priced condiment (1500–2200 euro/kg) due to the considerable labor involved in its production since it requires manual harvesting as well as a laborious handling process (sorting, drying, and storage) [3–6]. Saffron's principal producers are Iran and Spain, whereas the leading importers are Spain, Hong Kong, and the United States [3]. Saffron's quality is essential for consumers in the food industry [7] and is based on the concentration of its apocarotenoids and their respective sensory attributes: crocin's coloring strength, picrocrocin's bitter taste, and safranal's aromatic intensity. Saffron contains over 150 volatile and non-volatile compounds including proteins, carbohydrates, vitamins, amino acids, minerals, gums, and other compounds [8,9]. However, the apocarotenoids (crocin, picrocrocin, and safranal) are responsible for saffron's sensorial attributes and are the major bioactive compounds used as markers for its quality. Furthermore, the quality and, consequently, the commercial value of saffron are based on the estimation of its coloring power, bitter taste, and aroma [10].

Reductions in saffron's commercial quality can be attributed to inappropriate harvesting methods, insufficient dehydration processing, exposure to direct sunlight, improper storage, and adulteration [4,5]. Saffron fraud is related to unfair competition, including (a) by adding substances (parts of other, cheaper plants or synthetic dyes) to produce low-cost spices [5,7,11] or (b) spices that carry the Protected Designations of Origin (PDO) logo without being produced or processed in the specified geographic area [11,12]. To prevent adulteration, it is necessary to establish a precise chemical identification protocol to protect producers' and consumers' interests [13]. Metabolic and chemical profiling is a valuable tool for product standardization and for detecting mislabeled or fraudulent samples [4]. This review aims to evaluate the state of saffron's main bioactive compounds and their relationship with its commercial quality. To fully achieve this purpose, the following topics are addressed: (i) we describe the *C. sativus* characteristics for obtaining saffron and its uses in the food industry; (ii) we present and discuss saffron's chemical composition, its main bioactive compounds, and their determinations; (iii) we explain saffron's quality compounds related to color, odor, and flavor; (iv) we differentiate the saffron authentication techniques and their relationships with chemical compounds and chemometric methods as a critical parameter of its commercial quality; and (v) we consider the saffron by-products and their applications in the food industry.

2. *C. sativus*

C. sativus belongs to the Iridaceae family and is considered a sterile herb from the *Crocus* genus [14–18]. It is a perennial plant; therefore, soil fertility must be carefully controlled to achieve high production. Its cultivation is adapted to arid and semi-arid lands. It grows abundantly in regions with cold winters and abundant rain in spring and autumn and low rainfall in summer; it can also grow in temperate and subtropical climates with sandy or clay soils with good drainage. Saffron is grown in Iran, Spain, India, and Greece. The plant is small, with a height of up to 30 cm. Predominantly, it consists of leaves (deep green), flowers, and a globular underground corm or bulb measuring 3 to 5 cm in diameter [14,19–21]. The flowers are composed of six tepals; inside the flower, three stamens are present, and a filiform white style terminates in a stigma divided into three threads. During its development and growth, the stigma changes color from white to scarlet [16,18,22–26]. The stigma constitutes between 7 and 7.4% of the flower and the remaining 93% is composed of the petals, stamens, and style. Stigmas represent the unique, marketable part; the rest of the plant is called the floral biomass [14]. Saffron flowers are sterile; therefore, they do not produce viable seeds and must be propagated manually by planting corms that grow underground. Flowering occurs approximately 40 days after sowing and lasts from 20 to 30 days [16–18,25,26]. Corms remain dormant during summer and grow at the end of the season [16,17]. The geographical origin and their respective environmental conditions (altitude, temperature, rainfall, irrigation cycles, harvest season, humidity, and properties or type of soil) influence plant growth and development, exerting strong effects on the production of secondary metabolites [6,27].

3. From *C. sativus* to Saffron

There are a variety of methodologies and techniques for obtaining saffron from the *C. sativus* flower. The main phases are described in the following subsections.

3.1. Harvesting

Harvesting begins in the morning (deep red stigma); flowers are cut before the tepals open to prevent them from wilting in the sun (causing loss of color and concentration of apocarotenoids). This break is made in the lower part of the corolla. Fresh-cut flowers should be kept in good storage conditions, with high humidity, a low temperature, and moderate airflow. This is due to their short shelf life, rapid senescence, high water loss, and high likelihood of contamination by bacteria and fungi [3,28,29].

3.2. Post-Harvest

Cut flowers are transferred in baskets or sacks to the processing area, avoiding pressure or deformation of the stigmas (in Greece, stigmas are cut on the plant). Next, they are placed on a table for "monda" (separation of tepals and removal of the styles). The flower is opened and the stigmas are separated from the tepals and stamens (the stigma is cut at the base of the filaments and the style is removed). The whole, manually performed operation takes around 4 s per flower (a step responsible for the high cost). The stigmas are collected manually to preserve the bioactive compounds. However, easy degradation in the presence of light or oxidizing agents means that few stigmas are classified as high-quality saffron. Poor hygiene, transportation, bulk storage, manual harvesting, monda, sudden rain during flowering, and prolonged and inadequate storage temperatures are critical factors of quality and contamination. Accelerated stigma separation after flower harvesting is recommended to reduce these factors [3,17,28–31].

3.3. Drying

As described above, fresh stigmas do not transmit the typical color, flavor, and aroma so a drying treatment is necessary. This step is crucial and essential to convert *C. sativus* stigmas into the aromatic spice saffron. In most cases, it can be stated that the drying method affects the color, morphological characteristics, bioactive composition, flavor, and aroma of saffron; this is explained by the fact that, during the process, a series of biochemical and enzymatic changes occur, generating volatile and non-volatile compounds. There are several techniques for carrying out the drying process (conventional: room or moderate temperatures of 35–45 °C over long periods of time are recommended; non-conventional: very short periods of time at high temperatures of 60–70 °C) and each method has its own variables (place, temperature, relative humidity, raw material load, etc.). The variables differ between countries and according to the experience, available resources, and climate of each region, which results in variations in saffron quality [32–36].

3.4. Storage

Dried stigmas are packed in sealed containers away from moisture and light at temperatures between 5 and 25 °C. Saffron is marketed as strands or ground saffron. One kilogram of dry saffron requires between 110,000 and 165,000 flowers, which implies around 50 h of labor to pick the flowers plus 200 h to peel the stigmas from them. Storage favors the oxidative and hydrolytic decomposition of the secondary metabolites (crocin and picrocrocin). However, inadequate storage can affect the properties of the finished product [30,36,37].

4. Saffron in the Food Industry

Saffron's aroma develops during the drying and storage stages. However, the loss of apocarotenoid quality occurs due to poor harvesting, inadequate drying and storage conditions, the mixing of stigmas with other parts of the plant, etc. [10,36,38]. It is traditionally used in industry as a medicine, textile dye [8], cosmetic raw material, ornamental flower [35], and aphrodisiac [39]. Specifically, it is used in the food industry and cuisines worldwide as a spice or seasoning [8,40], acting as a flavoring and coloring agent [39,41–43]. However, it is also recognized as a medicinal plant [41], carrying various beneficial health properties such as analgesic, sedative, antioxidant, anticancer, and other therapeutic properties [39].

Saffron has been used in the food industry for culinary purposes as an aromatic, flavoring, and coloring agent, in many products. The chemical composition of saffron makes it a valuable functional ingredient for various products in the food industry [14]. Saffron has been added to several formulations for the development of functional foods as a preservative, colorant, flavoring, antioxidant, base for bioactive compounds, etc.

Regarding the bakery and confectionery industries, Gani et al. (2021) produced fortified cookies with encapsulated bioactive compounds from saffron. The additions enhanced its antioxidant activity, providing a better color and suitable stability. In addition, in vitro digestibility showed a low glycemic index [44]. Moreover, Bhat et al. (2018) designed whole-wheat flour cookies with saffron extracts. They reported acceptable sensory characteristics (except texture), antioxidant properties, and a suitable shelf life [45]. Bhat et al. (2022) produced functional cakes from whole-wheat flour combined with saffron or tomato extracts. The addition of saffron extract to the cakes produced desserts with improved antioxidant properties, without affecting the product's sensory quality [46]. Armellini et al. (2018) evaluated the qualities (texture, physicochemical, and sensory) of dough enriched with saffron powder. The results showed that saffron provided better textural properties, higher sensory acceptability (visual appearance, color, aroma, flavor, chewiness, hardness, gumminess, and overall acceptability), and improved antioxidant activity (higher values of crocin) in the saffron-enriched dough [47]. The same research group [43] studied the effect of saffron extract addition on starch digestibility and crocin fate and release at different cooking times in fresh pasta. The results showed that the saffron extract affected the digestibility and glycemic index. The higher the saffron concentration and the shorter the cooking time, the higher the amount of crocin released in the digestive fluids.

Sena-Moreno et al. (2018) used a saffron extract (rich in safranal) as a flavoring agent in olive oil. They reported that small concentrations of safranal led to organoleptic improvements in the oils. In addition, positive values were obtained for oxidative stability, indicating this product's potential in the charcuterie market [48]. Almodóvar et al. (2018) compared the advantages and culinary applications of a natural commercial saffron extract (affron®eye) vs. saffron stigmas in refrigerated foods. They demonstrated that affron®eye has advantages in terms of microbiological safety, ease of dissolution, quick application, and simple mixing of ingredients [49]. Finally, Moghaddam et al. (2018) developed a probiotic beverage (*Lactobacillus*, *Lactococcus*) fermented using saffron petals. They reported its physicochemical, antioxidant, rheological, and sensory properties, showing overall benefits in terms of antioxidant and phenolic activity after fermentation [50].

5. Saffron's Chemical Composition

Saffron contains more than 150 compounds (volatile and non-volatile) including carotenoids (crocetin, crocin, β-carotene, lycopene, and zeaxanthin), monoterpene aldehydes (picrocrocin and safranal), monoterpenoids, and isopherones [8,28]. However, it also contains other compounds such as flavonoids, vitamins, proteins, and amino acids [51]. Saffron owes its sensory and functional properties mainly to the presence of its carotenoid derivatives, synthesized throughout flowering but also during the whole production process [43]. These compounds include crocin, crocetin, picrocrocin, and safranal, which are the secondary or bioactive metabolites [8,43,44,52]. Saffron's quality depends on its chemical profile and is directly related to the geographic area, climate variability, environmental practices, genetic traits, soil composition, cultivation conditions, and processing and storage methods [53,54]. Nevertheless, according to the ISO standards (3632-1:2011 and ISO 3632-2:2010), the value and quality of the stigma are measured based on the content of the color components (crocin and crocetin), the bitter taste component (picrocrocin), and the volatile compounds responsible for the odor and aroma (safranal). These specific parameters are influenced by the environmental conditions, extraction method, purification, etc. [14,28,55–57]. Some studies have been conducted on the extraction of bioactive compounds from saffron using the concept of green chemistry [58]. Some research on saffron stability demonstrates that temperature and humidity exert a strong influence on the degradation of the principal active ingredients [8].

5.1. Saffron's Important Apocarotenoids

Crocin: The main bioactive compound of saffron was isolated by Aschoff in 1818, reporting a family of yellowish-red water-soluble carotenoids (mono-glycosyl or di-glycosyl-polyene esters) of 20 carbons [8,34,58–60]. In other words, this was a group of compounds formed by crocetin esterification (dicarboxylic carotenoid), which were classified according to their sugar fractions [59]. The abbreviations used in this review are as follows. The cis/trans-X-R1R2 crocin abbreviation system is used based on three main characteristics: (a) cis/trans isomers, (b) X: number of glucose components (1–5), and (c) type of structure in R1 and R2 (acid form: H; glucose: g; gentiobiose: G; Neapolitan: n; or triglucose: t.) (Suchareau et al. (2021)). The most represented crocins are trans-4-GG, trans-3-Gg, trans-2-G, trans-2-gg, trans-5-tG, and trans-1-g, among others [19,59,61–70].

Crocins are unusual apocarotenoids since their terminal glycoside rings confer high solubility. These pigments are detected in the red lobes of the stigmas of the *Crocus sativus* flower [14,19] and their content is proportional to the color and quality index. However, it should be noted that zeaxanthin (fat-soluble carotenoid) can also influence the color [35]. Crocins as such have low stability and lose their functionality during exposure to heat, oxygen, light absorption, acidic environments, and/or due to the presence of additives [43]. Therefore, the drying and storage temperatures are important for proper color development [68]; poor storage conditions lead to color pigment degradation [71]. Several factors are related to the concentration of these pigments in saffron stigmas, which are mainly the geographical growing region, crop conditions, type of soil, plant genetic traits, climate, planting time (rate), seed/crown rate, planting depth, corm size/weight, crop density, nutrient management, weed management, growth regulators, harvest and postharvest management, and drying conditions [49,72]. Finally, crocin (digentiobiose ester of crocetin) is recognized as a natural food-grade dye that displays biological activity such as antigenotoxic, cytotoxic, antioxidant, anti-inflammatory, anti-atherosclerotic, anti-diabetic, hypotensive, hypolipidemic, hypoglycemic, and antidepressant properties [14,28,55].

Crocetins are lipophilic carotenoids derived from the hydrolysis of crocin glycosides, which is a crocin aglycone [67]. It contains a carboxyl group at each end of the polyene chain [19]; these groups of compounds (α-crocetin or crocetin I, crocetin II, β-crocetin, γ-crocetin) are produced from the degradation of zeaxanthin [73].

Picrocrocin's structure was established by Khun and Winterstein in 1934 [60]. It is a colorless and odorless glycoside monoterpene (4-hydroxy-2,6,6-trimethyl-1-cyclohexene-1-carboxaldehyde or hydroxy-β-cyclocitral: HTCC and glucose), a product of the degradation of zeaxanthin, and is responsible for saffron's bitter taste [8,28,34,52,58,74]. Picrocrocin is the second most abundant component in dry matter content [66,73,75]. During the drying process (35–50 °C for 4–7 h), picrocrocin's temperature and/or hydrolysis form an aglycone [73,76]. Therefore, picrocrocin decreases during dehydration, whereas safranal is absent before drying [17].

Safranal is an aldehyde monoterpene and the volatile component responsible for saffron essential oil. HTCC (hydroxy-β-cyclocitral or 4-hydroxy-2,6,6-trimethyl-1-cyclohexene-1-carboxaldehyde) is regarded by many authors as a safranal precursor. This compound is obtained by chemical or enzymatic hydrolysis (dissociation) or when the vegetal material is dehydrated and transformed into safranal, but this also happens due to the handling and storage processes [8,53,58,63,75,77,78]. The safranal content changes according to the duration and intensity of drying, causing quality fluctuations [34], whereas its concentration increases with the storage and timely harvesting of flowers. However, heat and sunlight decrease the final quality and price [28].

5.2. Hypotheses on the Method of Obtaining Apocarotenoids

There are various hypotheses on the method of obtaining these important apocarotenoids from saffron. The first theory focuses on synthesizing these compounds in the plant from protocrocin (glycosyl derivative of zeaxanthin), the substrate of an oxidative enzyme that produces a molecule of crocin and two molecules of picrocrocin. Regarding safranal, it has been described that only a minimal concentration is detected in the fresh spice [79]. Fallahi et al. [80] described another pathway wherein apocarotenoids, which are commercially important, are obtained by the cleavage of carotenoids (zeaxanthin and β-carotene) by the carotenoid dioxygenase enzyme, giving rise to crocetin and hydroxy-β-cyclocitral as products. Later, they propose a glycosylation (glycosyltransferases) step, which produces crocins and picrocrocin, respectively. Finally, they describe that picrocrocin is hydrolyzed to form safranal. This hypothesis is consistent with that described by Sereshti et al. [81], who also describe other, more specific enzymes and substrates, as seen in Figure 1.

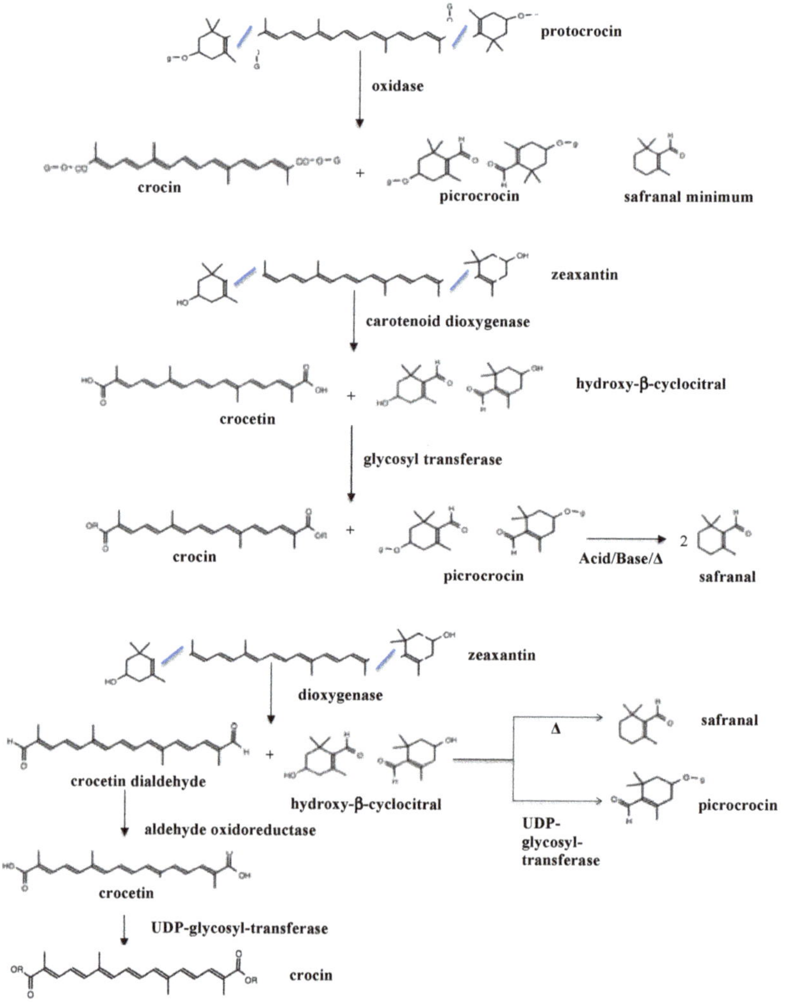

Figure 1. Possible pathways of commercial apocarotenoids in saffron.

The enzyme dioxygenase performs a 7–8C and 7′–8′C symmetric cleavage on the carotenoid zexanthin, converting it to 3-hydroxy-β-cyclocitral and dialdehyde crocetin. Crocetin dialdehyde undergoes oxidation by aldehyde dehydrogenase to crocetin. Crocetin further undergoes glycosylation at the carboxyl group by the enzyme UDP-glucuronosyl transferase, forming crocin. Picrocrocetin is obtained from 3-hydroxy-β-cyclocitral by glycosylation at the hydroxyl group by the enzyme UDP-glucuronosyl transferases. Picrocrocin is converted to safranal by the action of the enzyme β-glucosidase along with heat during drying [14].

6. Saffron Quality: Compounds Related to Color, Odor, and Flavor

Saffron's quality depends on its chemical profile, which provides the bitter taste, desirable aroma, and attractive yellowish-red color of this spice [29,82]. Several studies on saffron stability are related to temperature, humidity, pH, light, oxygen [76], geographical growth location, and drying and storage conditions [83]. Since 1980, a standard quality procedure has been employed for saffron classification according to the International Standard Organization (ISO/TS 3632), which was updated in subsequent years (2003, 2010, 2011). This regulation allows saffron to be classified into distinct categories based on physical and chemical criteria: Category I—high quality; Category II—±medium quality; and Category III—low quality [61,84,85]. The grouping parameters used are moisture content, flower residues, foreign material, ash, and coloring power. However, external parameters, such as the absence of other plants, biological micro-flora, and pesticide residues, are also used. The methodology to determine saffron's quality using these regulations is the spectrophotometric quantification of the stigmas' aqueous extracts (1%) at three maximum wavelengths, namely 257 nm to indicate flavor strength (picrocrocin), 330 nm related to aroma (safranal), and 440 nm for coloring force (crocins), using a 1 cm pathway quartz cell [85–89]. The results are reported according to Equation (1):

$$E_{1cm}^{1\%}(\lambda max) = \frac{(A \times 10,000)}{m \times (100 - H)} \quad (1)$$

where λmax is the wavelength (257, 330, or 420 nm), A is the absorbance, m is the saffron sample weight (g), and H is the moisture content (%) [20,79,88,90–92]. The color intensity is the most important characteristic related to quality and is used to establish the market price of saffron [93]. The crocin content (degraded carotene) [32] determines the market color specifications. Category I includes a minimum value of 200 units of coloring strength (ucs) and for Category III, the minimum value is 120 ucs [61]. Saffron merchants usually consider a 3-4-year shelf life for saffron when stored under suitable conditions (at room temperature without light exposure). The color intensity decreases by nearly 30 to 40 units per year and is a significant determinant of the final quality of saffron [94]. Diverse drying methods affect crocins, which may be related to the time, temperature, and resistance used [35]. Other factors that affect color are geographic location, harvest, storage, and mixing with additional non-colored parts of the plant (stems and other adulterating materials) [91]. Saffron's bitter taste is attributed to picrocrocin, a compound present in the plant's stigmas. The ISO standard determines the flavor strength with values of 70 (Category I), 55 (Category II), and 40 (Category III) [61]. The final picrocrocin content varies according to the dehydration process used [94]. The spice's flavor can suffer significant losses during processing [1]. Safranal is the active odor in this spice [18,94,95]. The ISO 3632 method determines three categories of aroma strength in safranal, with values within a range of 20–50 [61,96]. It is important to emphasize that during dehydration and storage, there are modifications in saffron's sensory characteristics [94,97].

Therefore, the chemical components of saffron quality are crocin, picrocrocin, and safranal. Lage and Cantrell [21] established that crocins are found in a more significant range (18–37%), followed by picrocrocin (4.2–28%) and, in a lower proportion, safranal (0.04–0.48%). This is consistent with the results described by various authors [21,64,72,90,96], who determined crocins as the major components, specifically trans-4-GG and trans-3-Gg crocins [61,64,98].

Concerning crocins, Chaouqi et al. [87] demonstrated that these coloring components are extracted in a more considerable proportion at 40 °C than at room temperature; the authors suggested the use of short dehydration times since an increase in temperature allows for the maximum crocin content, which also depends on the production [94]. However, Rocchi et al. [68] found that the use of elevated temperatures (125–200 °C) in the drying treatment can influence the pigments' degradation (glucose hydrolysis), and fresh samples (<1 year) retain a significant amount of glycosylated crocin, which is hydrolyzed after storage. Sereshti et al. [81] described that freshly dried samples have an intense color due to crocins since during storage, these pigments decrease (enzymes, temperature, light, hydrolysis), with a negative correlation with odor (the color is reduced, whereas the aroma increases). Saffron storage causes apocarotenoids' glycosidic bonds to break down (band at 1028 cm), which was confirmed using FT-IR spectroscopy, and is associated with the presence of glucose, together with intensities in the region of 1175–1157 cm linked with glucosidic bonds [99]. The second quality component in the percentage is picrocrocin, which increases with the dehydration temperature (40 °C) [21] but decreases with storage time [87]. Ordoudi et al. [78] determined that saffron produced under optimal processing and storage conditions retains its organoleptic characteristics for 1 to 4 years. Meanwhile, samples stored for more than four years produce low amounts of crocetin and picrocrocin esters. This is related to the findings described by Sereshti et al. [81], who determined that during storage, picrocrocin loses its sugar residues and becomes HTCC and safranal (fresh samples are more bitter). In other words, fresh samples contained a higher concentration of crocins and picrocrocins, whereas the level of safranal (the most abundant volatile component, but with a minimum total concentration in the aromatic spice) was higher in the stored samples; therefore, the relationship between time and safranal content was demonstrated by the higher concentration in the samples with extended storage. García-Rodríguez et al. [96] determined that the aged spice produces safranal from HTCC. The safranal concentration depends on the drying and storage conditions [97].

6.1. Quality Standards and Apocarotenoid Quantification

The ISO standard proposes a fast, economical, and easy-to-implement spectrophotometric UV-vis method for aqueous saffron extracts. However, this technique does not allow for the actual determination of the quality compounds [87]. ISO 3632 proposes the quantifications of picrocrocin, safranal, and crocins at a maximum of 257 nm, 330 nm, and 440nm, respectively. However, Cossignani et al. [88] and Aiello et al. [86] determined that crocins show an absorption spectrum between 250 and 470 nm that overlaps at various wavelengths between the compounds. Trans-crocin isomers showed two bands: the first at 260 nm (glycosidic ester bond) and the second band between 400 and 470 nm (typical of carotenoids). Meanwhile, the cis-crocin isomers showed three bands: two bands as previously described and a third band of medium intensity at 328 nm. This indicates that the amount of picrocrocin is affected by the concentration of cis and trans-crocins. Meanwhile, the safranal concentration obtained by UV-vis is not precise since cis-crocins interfere. In summary, overlapping causes quantification errors and limitations in this technique [57,90,96,100–102]. Another group of compounds that could interfere with saffron's quality is the kaempferol derivatives, which absorb UV-vis light at 264 and 344 nm [88,103]. Moreover, safranal is slightly soluble in water and therefore the use of hexane and chloroform has been determined as the best strategy for the extraction and detection of adulterants [101,104].

6.2. Apocarotenoids and Their Quantification by Chromatography

Color, flavor, and odor are the quality parameters for saffron aqueous extract according to ISO 3632. They are determined by a non-specific spectrophotometric technique, albeit with limitations in assessing the authenticity of saffron. In the search for a more effective technique, liquid chromatography (LC) or HPLC have been proposed to separate and identify the components contained in a sample [89]. Various studies have described the identification and detection of saffron metabolites by HPLC including safranal, crocins, picrocrocin, and kaempferol and its derivatives [86]. For its part, a mass spectrometry (MS) detector coupled to HPLC and/or DAD could improve quantification [105,106], and MS/MS could facilitate the identification of compounds through structural elucidation [107]. The key quality parameter of saffron is color and the compound to which it is attributed is crocin, which must be quantified in order to determine the market price. For the qualitative and quantitative determinations of crocins, it is necessary to implement standards (quantification by internal and external standards) such as trans-4-GG-crocin (high price and questionable purity ~80%) [53,67,102].

The MS detector has been of considerable help since the lack of suppliers and the high prices of the standards make the structural elucidation (fragmentation patterns) of each crocin important (the different crocins can be identified by the number of hexoses and the molecular weight provided by the mass spectra) to compare them with the patterns in the scientific literature [102,107]. Crocin determination was carried out by Aghhavani et al. [28]; they determined no correlation between the color indexes obtained with spectrophotometry and HPLC data. They concluded that one could use the most accurate, easiest, and low-cost method depending on the experimental conditions to evaluate the quality of saffron. Rocchi et al. [68], demonstrated a poor correlation between the total crocin content (quantification) obtained by the ISO method and by UHPLC-MS/MS.

García-Rodríguez et al. [96] and Kabiri et al. [90] found that the quantification of safranal obtained by UV-vis does not correlate with HPLC data due to the interferences (overestimation by interference) generated by cis-crocetin esters and other compounds with λmax 330 nm. They also demonstrated that crocins interfere with picrocrocin and safranal, resulting in overestimates of the latter compounds in samples with large amounts of crocin. They concluded that semipreparative HPLC could represent an efficient method for the quantification of apocarotenoids. Similar results were presented by Moras et al. [106]; they reported that safranal content is more accurately calculated using UHPLC-DAD-MS because it is not influenced by the overestimation of safranal (with cis-crocetin esters at λmax 310–330 nm), which is shown when using the ISO methodology. They recommend determining, separating, identifying, and quantifying the metabolite content using the UHPLC-DAD-MS method as a unique and rapid analysis technique. Maggi et al. [104] and Bononi et al. [100] reported a null correlation between safranal content obtained by ISO 3632 and the GC method, as many other saffron substances display absorbance at a maximum of 330 nm.

For this reason, several instruments and analytical methods have been developed for saffron quality control, including chromatography, spectroscopy, molecular biology, and biomimetic techniques, with varying degrees of success and benefits [89]. HPLC is used to isolate, identify, quantify, purify, and determine the quality or adulteration; reverse-phase chromatography is widely used as it is capable of detecting compounds of different polarities and molecular masses [108]. Some authors have pointed out that HPLC-DAD is a selective, precise, sensitive, and specific technique that could evaluate the commercial quality of saffron [27,109].

In Table 1, the major commercial-quality compounds in saffron quantified by HPLC, are shown. The extractant solvents used in the investigations (Table 1) are polar and are in agreement with the descriptions by Rahaiee et al. (2015), who suggested that solvents such as water, ethanol, and pure methanol can be used but that mixtures would be more appropriate for the extractions of bioactive compounds [110]. For many authors, ethanol is the most suitable solvent (compared to methanol, ethyl acetate, diethyl ether, hexane, and/or water) for extracting metabolites from saffron stamens [111]. Meanwhile, Rahaiee et al. (2015) showed that an ethanolic extract obtained higher yields compared to water and methanol [66]. Similarly, this solvent was better than methanol for obtaining qualitative and quantitative data from saffron extracts. Meanwhile, Kyriakoudi et al. (2012)

recommended the mixture of methanol: water (1:1, *v/v*) as a suitable solvent for industrial and analytical applications of saffron apocarotenoids [112]. Crocin isolation by solubility in a water–organic solvent mixture was tested by Zhang et al. (2004), who showed better results for methanol–water > ethanol–water > acetone–water extract [113]. Crocins are the most determined compound, followed by picrocrocin and safranal. In crocins, the ratios determined from highest to lowest were trans-4-GG, trans-3-Gg, cis-4-GG, trans-2-G, and trans-2-gg, respectively. An exception was Moratalla-López et al. [109], whose results did follow this relationship because the saffron samples used in their research were only of quality grade III. In general, ISO 3632 is used by researchers as a preliminary test. However, to perform the true quantification of saffron's commercial-quality compounds, more precise spectroscopic techniques are used (HPLC, GC-MS, etc.).

Table 1. Principal quality chemical components of saffrons obtained from different geographical origins and their concentrations.

Geographical Origin	Type of Extract	Compound	Concentration		Technique	Ref.
Azerbaijan	Methanol–water (50:50, *v/v*)	Trans-4-GG Trans-3-Gg Cis-4-GG Σ crocins Picrocrocin Safranal	39.08 27.25 7.49 77.16 3.34 0.98	mg/g	HPLC-PDA	[53]
China	Methanol–water (50:50, *v/v*)	Trans-4-GG Trans-3Gg Σ crocins Picrocrocin Safranal	6.29 2.44 8.73 0.53 0.22	mg/g	HPLC-PDA	[53]
Poitou, France	Methanol–water (50:50, *v/v*)	Trans-4-GG Trans-3-Gg Cis-4-GG Σ crocins Picrocrocin Safranal	38.43 27.74 5.89 75.07 5.97 0.81	mg/g	HPLC-PDA	[53]
Greece	Methanol–water (50:50, *v/v*)	Trans-4-GG Trans-3-Gg Cis-4-GG Σ crocins Picrocrocin Safranal	40.77 30.36 10.14 86.51 5.95 1.29	mg/g	HPLC-PDA	[53]
India	Methanol–water (50:50, *v/v*)	Trans-4-GG Trans-3-Gg Cis-4-GG Σ crocins Picrocrocin Safranal	37.54 22.13 9.12 75.68 7.87 0.47	mg/g	HPLC-PDA	[53]
Fars, Iran	Aqueous extracts	Trans-4-GG Trans-3-Gg Cis-4-GG Trans-2-gg Σ crocins Picrocrocin	56.16 48.72 12.53 12.49 153.81 77.29	mg/g	HPLC-DAD	[27]
Ghaen, Iran	Ethanol (70%)	Trans-4-GG Trans-3-Gg Cis-4-GG Trans-2-G Σ crocins Picrocrocin Safranal	197.84 71.56 26.88 24.86 338.87 43.82 1.35	mg/g	HPLC-DAD-MS	[64]
Gonabad, Iran	Ethanol (70%)	Trans-4-GG Trans-3-Gg Cis-4-GG Trans-2-G Σ crocins Picrocrocin Safranal	168.91 61.25 30.42 26 302.51 36.97 1.26	mg/g	HPLC-DAD-MS	[64]

Table 1. *Cont.*

Geographical Origin	Type of Extract	Compound	Concentration		Technique	Ref.
Isfahan, Iran	Aqueous extracts	Picrocrocin	150.64	mg/g	HPLC-DAD	[27]
		Trans-4-GG	46.86			
		Trans-3-Gg	43.51			
		Trans-2-G	14.53			
		Trans-2-gg	10.56			
		Σ crocins	137.05			
		Safranal	1.04			
Kerman, Iran	Aqueous extracts	Trans-4-GG	77.89	mg/g	HPLC-DAD	[27]
		Trans-3-Gg	46.69			
		Trans-2-G	12.79			
		Σ crocins	159.86			
		Picrocrocin	63.95			
		Safranal	1.31			
Razavi Khorasan, Iran	Aqueous extracts	Trans-4-GG	54.73	mg/g	HPLC-DAD	[27]
		Trans-3-Gg	34.51			
		Trans-2-G	9.35			
		Σ crocins	123.61			
		Picrocrocin	120.62			
		Safranal	2.13			
Tehran, Iran	Aqueous extracts	Trans-4-GG	59.7	mg/g	HPLC-DAD	[27]
		Trans-3-Gg	44.43			
		Cis-4-GG	12.39			
		Trans-2-gg	9.34			
		Σ crocins	146.66			
		Picrocrocin	131.61			
		Safranal	0.57			
Tehran, Iran	Aqueous extracts (1%) Freeze-Dried	Picrocrocin	33.88	mmol/100g	HPLC-DAD	[109]
		HTCC	20.2			
		Trans-3-Gg	3.81			
		Trans-4-GG	3.53			
		Trans-2-gg	1.17			
		Σ crocins	9.91			
		Safranal	0.84			
Tehran, Iran	Aqueous extracts (1%) Dark-Dried	HTCC	16.82	mmol/100g	HPLC-DAD	[109]
		Picrocrocin	15.14			
		Trans-4-GG	4.59			
		Trans-3-Gg	3.71			
		Σ crocins	11.95			
		Safranal	0.41			
Torbat, Iran	Ethanol (70%)	Trans-4-GG	238.02	mg/g	HPLC-DAD-MS	[64]
		Trans-3-Gg	85.36			
		Trans-2-G	24.3			
		Cis-4-GG	19.38			
		Σ crocins	388.23			
		Picrocrocin	67.95			
		Safranal	1.79			
Iran	Aqueous extracts	Trans-4-GG	42.24	%	HPLC	[70]
		Trans-3-Gg	24.76			
		Cis-4-GG	5.09			
		Trans-2-G	3.53			
		Trans-2-gg	3.18			
		Σ crocins	83.06			
		Picrocrocin	16.72			
		Safranal	0.22			
Iran	Methanol–water (50:50, *v*/*v*)	Trans-4-GG	38.41	mg/g	HPLC-PDA	[53]
		Trans-3-Gg	23.58			
		Cis-4-GG	4.73			
		Σ crocins	69.32			
		Picrocrocin	3.69			
		Safranal	0.65			
Iran	Ethanol 80%	Crocin	26.81	mg/0.1g	HPLC	[90]
		Picrocrocin	12.92			
		Safranal	0.042			
Cascia, Italy	Ethanol (70%)	Trans-4-GG	343.97	mg/g	HPLC-DAD-MS	[64]
		Trans-3-Gg	111.94			
		Trans-2-G	13.59			
		Σ crocins	494.42			
		Picrocrocin	127.83			
		Safranal	3.01			

Table 1. *Cont.*

Geographical Origin	Type of Extract	Compound	Concentration		Technique	Ref.
Città della Pieve, Italy	Ethanol (70%)	Trans-4-GG Trans-3-Gg Trans-2-G Σ crocins Picrocrocin Safranal	302.65 109.17 16.12 450.73 101.92 2.41	mg/g	HPLC-DAD-MS	[64]
Fiesole, Italy	Ethanol (70%)	Trans-4-GG Trans-3-Gg Trans-2-G Cis-4-GG Σ crocins Picrocrocin Safranal	372.49 123.15 21.24 12.55 548.84 130.35 2.01	mg/g	HPLC-DAD-MS	[64]
Fiesole, Italy	Ethanol (70%)—formic acid	Trans-4-GG Trans-3-Gg Trans-2-G Cis-4-GG Σ crocins Picrocrocin Safranal	238.91 65.64 16.96 4.95 342.02 111.14 2.27	mg/g	HPLC-DAD-MS	[98]
Navelli, Italy	Methanol–water (50:50, *v*/*v*)	Trans-4-GG Trans-3-Gg Σ crocins Picrocrocin Safranal	38.25 28.28 72.02 5.8 0.53	mg/g	HPLC-PDA	[53]
Perugia, Italy	Ethanol 70%—formic acid	Trans-4-GG Trans-3-Gg Trans-2-G Cis-4-GG Σ crocins Picrocrocin Safranal	148.5 46.2 14.8 14.1 231.1 68.9 2.6	mg/g	HPLC-DAD-MS	[98]
Italy	Aqueous extracts	Trans-4-GG Trans-3-Gg Cis-4-GG Trans-2-gg Σ crocins Picrocrocin Safranal	43.57 23.09 5.29 2.12 78.45 21.26 0.28	%	HPLC	[70]
Larache, Marruecos	Degassed methanol	Σ crocins Picrocrocin Safranal	17.9 11.92 0.21	%	HPLC-DAD	[21]
Safranier d'Ourika, Marruecos	Degassed methanol	Σ crocins Picrocrocin Safranal	37.23 28.78 0.24	%	HPLC-DAD	[21]
Rangiora, New Zealand	Methanol–water (50:50, *v*/*v*)	Trans-4-GG Trans-3-Gg Σ crocins Picrocrocin Safranal	41.21 31.26 74.61 7.94 0.47	mg/g	HPLC-PDA	[53]
La Mancha, Spain	Methanol–water (50:50, *v*/*v*)	Trans-4-GG Trans-3-Gg Cis-4-GG Σ crocins Picrocrocin Safranal	38.41 24.43 5.76 73.85 8.14 0.88	mg/g	HPLC-PDA	[53]
Turkey	Methanol–water (50:50, *v*/*v*)	Trans-4-GG Trans-3-Gg Cis-4-GG Σ crocins Picrocrocin Safranal	36.35 25.32 5.21 69.73 5.67 0.84	mg/g	HPLC-PDA	[53]

7. Saffron Authentication

Due to its high market price, saffron is the most adulterated spice in history, which is most frequently carried out by adding adulterants such as pulverized stigmas [114,115] since diverse plants with similar color and morphology to saffron function as adulterants when mixed [86]. Saffron adulteration can be classified into five common practices, as follows: (1) Adulteration using material from other plants such as calendula, arnica, gardenia, beet, pomegranate, turmeric, achiote, and safflower [93,106,115,116] or with other plant parts of C. *sativus* besides the stigmas; (2) Increasing saffron mass by moistening with honey, corn silk, sugar, fat, inorganic compounds, vegetable oils, or glycerin [18,116]; (3) Using natural or artificial food-grade colorants such as tartrazine, ponceau-4R, quinoline, methyl orange, sunset yellow, Sudan II, and Allura red [117,118]; and other less-used adulteration methods including (4) The addition of exogenous components mixed with food flavorings (erythrosine) and extracted spent saffron (recolored or old), and (5) Geographic origin tagging fraud [31,93,119,120].

The chemical composition of food is an indicator of quality, origin, authenticity, and/or adulteration. The chemical profile, also known as spectral fingerprinting or chemo typing, is considered a characteristic pattern [121]. In food, variations in a profile are related to alterations in production systems, the geographical origins of raw materials, storage conditions, or adulterant practices [122]. It should be emphasized that it is important to identify the adulterant and quantify the adulteration level [123]. Furthermore, the ISO/TS 3662 spectrophotometric technique does not differentiate between genuine and adulterated saffron [9,124]. Saffron authentication is based on a pharmacognostic analysis (microscopic examination of histomorphological features). It is time-consuming and requires the availability of trained and experienced personnel [115,125].

Regulatory systems evaluate saffron using sensory inspections (macroscopic and microscopic examinations) as well as conduct quantitative determinations of specific chemical compounds [126]. Authentication is based on detecting known chemical compounds obtained with instrumental signals [127]. However, these kits yield many characteristics or compounds, making it necessary to establish the chemical markers of authenticity [128]. Spectral fingerprinting can also detect and quantify adulterations using statistical data [127]. Chemometrics uses mathematical and statistical methods to create a correlation between the sample properties and chemical data obtained from analytical instruments [129]; this area is based on optimizing the experimental design and extracting useful information from large and complex data sets [122]. Therefore, analytical chemometric coupling could notably decrease the number of characteristics/compounds/signals and generate the markers responsible for different authenticity issues (adulteration detection, variety or geographical origin, discrimination, organoleptic profile, maturation, and production method). In addition, the identified markers would help to establish databases containing complete and standardized information on the chemical profiles [128].

The following research summary is based on determining chemical compounds as authentication markers (of genuine saffron or adulterants used) using different analytical techniques to determine the spectral fingerprints and/or even using chemometrics to obtain the amount of the adulterant or even the detection limits of the adulterant. Saffron adulteration determination by the inclusion of tepals and/or stamens was carried out by Senizza et al. [9]. They determined 232 compounds using UHPLC-QTO-MS. Among them, 77 chemicals were present in trace quantities including the presence of flavonoids: 11 flavanols (tepals had a high content) and 7 anthocyanins (pigments of flowers, fruits, and other plant organs), which increased in the adulterated samples. On the other hand, lignans (12 compounds) were found in low amounts in the authentic samples. Zeaxanthin and picrocrocin, which decreased in the adulterated samples, suggested a possible "dilution effect" when adding adulterants. Moras et al. [106] determined, through UHPLC-DAD-MS, the presence of iridoids as a marker for saffron adulteration, yielding positive test results when gardenia extract was added.

Investigations using analytical techniques and chemometrics to quantify the adulterant and the minimum detection to detect fraud have been presented. A method for deducing saffron authenticity using LC-MS with derivatives of kaempferol and geniposide was developed by Guijarro-Díez et al. [119]. They detected a minimum quantifiable value of adulteration (0.2%) regardless of the adulterant (linear regression lineal and ANOVA), the specific method, and saffron quality control. Sabatino et al. [85] used HPLC-PDA-ESI-MS to identify unusual concentrations of adulterants in saffron (10–67% safflower, calendula, and turmeric). Their results showed that the ISO did not detect the addition of 10% of adulterants. Moreover, marker molecules such as picrocrocin, trans-5-nG, trans-4-GG, trans-4-ng, cis-3-Gg, cis-4-GG, and cis-2-gg were not found in the adulterated spices. They determined the addition of 5% of safflower or calendula and 2% addition of turmeric in the analyzed samples.

Saffron stigma adulteration with up to 20% of plant derivatives (saffron stamens, calendula, safflower, turmeric, buddleja, and gardenia) was determined by Petrakis and Polissiou [123] using a DRIFTS method and chemometric techniques. PLS-DA was applied to perform saffron authentication based on infrared fingerprints (4000–600 cm). Identification was carried out with data from the 2000–600 cm^{-1} region to develop the mathematical models and detection limits ranging from 1.0 to 3.1% (p/p). Another (NIR) spectroscopy investigation combined with multivariate data analysis was performed by Shawky et al. [130]. They performed saffron stigma authentication with other plants (safflower, pomegranate peel, calendula flower, paprika, turmeric, hibiscus, saffron stamens, and re-extracted saffron stigma), modeling them with data at the spectral region (9000–4000 cm^{-1}). The use of PLS-DA allowed them to differentiate between authentic, adulterated, and mixed adulterant samples, with a detection limit of up to 10 mg/g of the adulterant. In addition, they quantified other added adulterants.

Saffron stigma authentication using artificial intelligence (simulating senses: sight, smell) was reported by Heidarbeigi et al. [7]. They determined plant adulterants (safflower and dyed corn using beetroot as a colorant, in addition to their mixtures) through signals obtained by the e-nose (managing to differentiate adulterated and unadulterated saffron). They also applied PCA and artificial neural networks (ANN) to determine fraud in saffron stigmas, determining adulteration levels higher than 10%. Kiani et al. [83] used CVS (camera, lighting system, and software) and an e-nose in combination with multivariate methods (PCA, HCA, and SVMs) to detect saffron stigma adulterants (colored safflower and saffron style) based on color and aroma profiles. The test demonstrated the ability to identify the adulterated samples and this was achieved using ANN-MLP models, concluding that neural networks allowed color (89%) and aroma-intensity (100%) prediction. CVS was used by Minaei et al. [91] to characterize saffron color by sample image analysis. The use of PCA to group color characteristics and the use of PLS, MLR, and MLP neural networks (color characteristics used: R, Y, I, and Cr) related color and dye force (ISO 3632), with a correlation coefficient of 0.89 and a success rate of 96.67%.

Another interesting application is the use of an e-nose (non-conventional technique), compared to IR-MS and GC-MS (conventional techniques) to discriminate among saffron samples with different origins, ages, and types of drying. The e-nose, in conjunction with PLS-DA, was able to discriminate between samples of saffron with different origins; this unconventional methodology was proposed to detect adulterates [131]. Recently, molecular techniques for detecting fraud by adulterations have gained interest. Safflower adulteration stamens as saffron adulterants were also studied by Babaei et al. [124], using a multiplex PCR technique. Khilare et al. [116] described three methods to achieve saffron authentication (microscopic examination, ISO3632 standard, and DNA barcode). They evaluated 36 saffron samples and showed that the ISO only determines the color and aroma, while the microscopic method indicates color purity and uniformity (possible adulterants).

Finally, DNA codes (gene code used: rbcL) have allowed researchers to authenticate saffron's origin and quality. Torelli et al. [115] used SCAR to detect adulteration or contamination. SCAR markers can represent a rapid, reliable, and inexpensive method for saffron authentication. Other rapid techniques for determining saffron adulteration were proposed by Zhao et al. [132] via DNA extraction. They used a recombinase polymerase amplification (RPA-LFD), which allowed them to perform the rapid visual detection of the saffron and adulterated samples. Finally, when saffron was immersed in water, it expanded immediately; when a diphenylamine and sulfuric acid solution was added, the saffron was colored with a blue tone and quickly became reddish brown. Saffron phenylethanol varies according to the spice preparation and is related to the stamen pollen [93]. Table 2 shows a summary of the various research works and techniques for the determination of the different types of adulterants. As regards the adulteration of saffron by its origin or PDO products, saffron has a high value on the market so some saffron producers falsify the product's origin [15,54]. In Europe, a PDO label carries a regional valuation that identifies the products produced, processed, and prepared in a specific geographic area [103]. There are five brands recognized with this label: "Krokos Kozanis" (Greece), "Azafrán de la Mancha" (Spain), "Zafferano dell 'Aquila", "Zafferano di San Gimignano", and "Zafferano di Sardegna" (Italy) [15]. There have been a considerable number of studies on origin adulteration [31,54,101,103,131,133–135]. La Mancha in Spain and Kashmir in India are two regions where saffron maintains higher prices [134]. Therefore, labeling saffron samples with a PDO implies that the product is of high quality [54]. Moreover, Senizza et al. [9] determined the chemical markers capable of discriminating PDO saffron samples from non-PDO. Chemical fingerprints were obtained using UHPLC-ESI-QTOF-MS and multivariate statistics, obtaining the flavonoids belonging to the flavonols and flavones (pelargonidin 3-O-6-succinyl-glucoside, isoxanthohumol, nobiletin, jaceosidin, 6-hydroxyluteolin, 3-methoxysinenset, 7-dimethylquercetin, quercetin 6-O-malonylglycitin), phenolic acids (protocatechuic aldehyde, 4-hydroxybenzaldehyde, vanillin, 2/3/4-hydroxybenzoic acids, benzoic acid, sinapine, p-coumaroyl malic acid, p-coumaric acid, cinnamoyl glucose, 4-hydroxyphenylacetic acid), lignans, and other polyphenols.

Table 2. Different techniques for saffron adulterant determination.

Type	Adulterant	Adulterant Concentration	Adulterant Minimal Detection		Adulterants Indicators or Markers		Technique	Ref.
1. A	Calendula flower Curcuma rhizome Hibiscus flower Paprika fruit Pomegranate fruit Safflower	10–400 mg/g	10	mg/g	6000–5800 5400–5000 4600–4200	cm^{-1}	FT-NIR/PCA: SIMCA PLS-DA	[130]
1. A	Gardenia	0–100% *w*/*w*	5	% *w*/*w*	Geniposide Deacetyl-asperuloside acid methyl ester Gardenoside Genipin-1-β-D-gentibioside 6″-O-trans-coumaroylgenipin gentibioside Scandoside methyl ester Absence of picrocrocin derivatives		UHPLC-DAD-MS	[106]
1. A	Gardenia extract	ND	41.7	g/g	Geniposide		LC–MS	[119]
1. A	Gardenia extract	0–100%	0.8 0.2 1.8 2.5 2.2	%	Kaempferol 3,7,40-O-triglucoside Kaempferol 3-O-sophoroside 7-O-glucoside Kaempferol 3,7-O-diglucoside Kaempferol 3-O-sophoroside Kaempferol 3-O-glucoside		LC-MS	[51]
1. A	Curcuma rhizome	0.5–20% *w*/*w*	0.5	% *w*/*w*	ND		DNA isolation/ Bar-HRM	[15]

Table 2. *Cont.*

Type	Adulterant	Adulterant Concentration	Adulterant Minimal Detection		Adulterants Indicators or Markers		Technique	Ref.
1. A	Calendula Rubia Safflower	5–35% *w/w*	5	% *w/w*	4200 4750 5170 6000–5400 7100–6000 8300 cm	cm^{-1}	NIR/PLS-DA	[117]
1. A	Turmeric, Onion peels Pomegranate peels Calendula petals	0–30% *w/w*	3.7 6.2 3.6 3.5	% *w/w*	4961–4016 6388–5389 9975–7472	cm^{-1}	FT-NIR/MCR-ALS	[136]
1. A	Tumeric Safflower G. jasminoides fruit extract	20% *w/w*	20	% *w/w*	7.541, 6.751, 6.059, 7.318, 7.147, 6.819 5.205, 5.138, 5.066 7.569, 7.466, 5.679, 5.121	^{1}H ppm	^{1}NMR/OPLS-DA/O2PLS-DA	[125]
1. A	Buddleja Officinalis flower Calendula petals Gardenia fruit extract Safflower Turmeric	0–20% *w/w*	1.1–1.6 1.9–2.6 1.1–1.5 2.1–2.8 1–1.6	% *w/w*	1624–1456 and 941–771 1508–1396 and 1167–1055 1794–1626 and 1113–943 1539–1456 and 858–773 1624–1286 and 941–771	cm^{-1}	DRIFTS/PCA PLS-DA	[123]
1. B	Saffron style	5–35% *w/w*	5	% *w/w*	4200 4750 5170 6000–5400 7100–6000 8300	cm^{-1}	NIR and MIR/PLS-DA	[117]
1. B	Saffron stamens	20% *w/w*	20	% *w/w*	5.181	^{1}H ppm	^{1}NMR/OPLSDA/O2PLS-DA	[125]
1. B	Saffron stamens	0–20% *w/w*	2.2–3.1	% *w/w*	4000–600 1963–1626 and 941–771	cm^{-1}	DRIFTS/PCA PLS-DA	[123]
1. B	Saffron stamens	10–400 mg/g	10	mg/g	6000–5800 5400–5000 4600–4200	cm^{-1}	FT-NIR/PLS-DA	[130]
3	Carminic acid	0.5–20% *w/w*	10	% *w/w*	1564–1576 1445–1456 1211–1231 810–816	cm^{-1}	FT-IR/PCA/PLS-DA	[137]
3	Carminic acid	0.2–2% *w/w*	0.2	% *w/w*	Carminic acid at 4.7 min, 495 nm	min, nm	RT-HPLC-DAD	[137]

Table 2. *Cont.*

Type	Adulterant	Adulterant Concentration	Adulterant Minimal Detection		Adulterants Indicators or Markers		Technique	Ref.
3	Synthetic dyes	ND	Magenta III Rhodamine B		330.1964 (HRMS) 300.14 (EI-MS) 223.11 (EI-MS) 2.5 (HPLC) 443.2320 (HRMS) 399.17 (EI-MS) 316.21 (EI-MS) 3.4 (HPLC)	m/z min	TLC/EI-MS/HRMS HPLC	[138]
3	Sudan III Sudan I Sudan II Sudan IV	0.14–7.1 g/Kg	0.14	g/Kg	8.014 6.87 8.618 8.181	^{1}H ppm	^{1}H NMR	[139]
4	Exhausted saffron	10–400 mg/g	10	mg/g	6000–5800 5400–5000 4600–4200	cm^{-1}	FT-NIR/SIMCA PLS-DA	[130]

A suitable method is the use of NMR in conjunction with multivariate statistical analysis. Principal component analysis allowed the discrimination between the samples of Italian PDO and commercial saffron, despite the year of harvest, date of purchase, and storage time [101]. Bosmali et al. [15] proposed a molecular approach for the authentication of the "Krokos Kozanis" brand using specific ISSR (inter-simple sequence repeat) markers to evaluate the variability within the *C. sativus* L. species (differences in bands produced by other *Crocus* species). The species-specific markers such as HRM analysis were developed in conjunction with the DNA barcode regions.

8. Saffron By-Products

The preparation of saffron is expensive due to the intense harvesting work and postharvest processes (dehydration and storage) required [32]. It is known that in order to produce 1 kg of stigma, around 1000 kg of flowers are treated by weight, which represents 220,000–260,000 flowers [42,98]. Therefore, saffron cultivation is not highly profitable in terms of biomass, which increases the interest in minimizing losses and ensuring efficient waste management [140]. Several reports have focused on the stigma, which is the plant's biologically active part [141]; its bioactivity is attributed to the composition, containing the main chemical components and their synergy with other compounds [60].

However, the by-products are also important since their use could increase the *C. sativus* flower's economic value, considering that other parts of the plant contain compounds with sensorial properties or biological activity [98,140]. *C. sativus* tepals are the main by-product of saffron production [142] but the flowers have low safranal content so they cannot be consumed or sold as saffron on the spice market [42]; only the leaves are used as forage [143]. Using HPLC-DAD, Serrano-Díaz et al. [144] determined kaempferol 3-Osophoroside and delphinidin 3,5-di-O-glucoside as the main components of the aqueous by-products of saffron flowers. Tepal and stamen biomarkers were determined by Mottaghipisheh et al. [145] using HPLC-DAD; they reported crocin, crocetin, picrocrocin, safranal, kaempferol-3-O-sophoroside, kaempferol-3-O-glucoside, and quercetin-3-O-soforoside. Tepal's main component was kaempferol-3-O-sophoroside with crocin, crocetin, and picrocrocin; safranal was not detected in any of the analyzed samples. Table 3 shows the principal agro-industrial by-products of saffron that have been investigated and their possible uses. Lahmass et al. (2017) determined that the corms, leaves, and spasms of *C. sativus* may possess anti-aging or anticancer properties.

These investigations generate interest in valorizing the various parts of saffron flowers and improving small-scale farmers' incomes. These results could contribute to the development of innovative products from saffron flowers and more effective biological waste management and exploitation [146]. It is important to emphasize knowledge of the components' depth (majority or minority) within each potentially valuable plant part of the saffron plant, which could help in determining the most suitable application [10].

Table 3. Saffron by-products of different geographical origins, major components, and applications.

By-Product (Origin and Type of Extract)	Major Components	Concentration		Application	Ref.
Sepals (Fiesole, Italy; ethanolic)	Trans-4-GG Trans-3-Gg Cis-2-G Kaempferol-3-sophoroside Quercetin diglucoside Kaempferol glucoside Kaempferol sinapoyl glucoside	3.1 0.8 0.2 6.4 0.4 0.4 0.3	mg/g	Phytochemicals	[98]
Stamens (Fiesole, Italy; ethanolic)	Trans-4-GG Trans-3-Gg Cis-4-GG Trans-2-G Kaempferol-3-sophoroside Quercetin diglucoside Methyl quercetin derivative Methyl quercetin diglucoside Kaempferol-3-sophoroside-7-glucoside	112.2 33.4 22.0 20.7 1.7 1.0 0.7 0.6 0.5	mg/g	Phytochemicals	[98]
Sepals (Perugia, Italy; ethanolic)	Traces of crocin Kaempferol-3-sophoroside Quercetin diglucoside Kaempferol glucoside Kaempferol sinapoyl glucoside	nd 8.3 0.7 0.4 0.3	mg/g	Phytochemicals	[98]
Stamens. (Perugia, Italy; ethanolic)	Trans-4-GG Trans-2-G Methyl quercetin diglucoside Quercetin diglucoside Methyl quercetin derivative Kaempferol-3-sophoroside-7-glucoside Kaempferol diglucoside	4 1.3 2.1 1.2 1.2 0.9 0.8	mg/g	Phytochemicals	[98]
Petals (Srinagar, Jammu & Kashmir, India; aqueous)	Not detected			Kashmir dye green and yellow tones	[66]
Petals (Kerman, Iran; aqueous)	Methanol Biogenix aldehyde fragment Acetic acid Isobutanal Furanone 2,3-butanedione	355 303 492 694 6397 524	ppb	Volatile compounds in the pharmaceutical industry	[42]
Petals (Sardinia, Italy; aqueous)	Kaempferol-3-O-sophoroside Phenylalanine Delphinidin 3,5-di-O-glucose Tyrosine Kaempferol-3,7-di-O-glucoside Isorhamnetin-3-O-rutinoside Quercetin 3-O-sophoroside	2790 1072 822 619 368 268 207	mg/L	Antioxidant and colon anticancer activities.	[147]

Table 3. *Cont.*

By-Product (Origin and Type of Extract)	Major Components	Concentration		Application	Ref.
Petals (Northeast, Iran; ethanolic and aqueous)	Pelargonidin 3,5-glycosides 3,5 cyanidin-diglycosides Petunidin Delphinidin 3-glycosides Pelargonidin 3-glycosides	56.1 20.9 15.5 4.1 3.4	%	Antioxidant and colorant activities.	[148]
Petals and anthers (Navelli, Italy; ethanolic, oil, and aqueous)	Crocin Catechin Rutin Epicatechin p-OH benzoic acid Safranal Vanillic acid Galic acid Safranal Quercetin	0.6 0.2 0.1 0.08 0.04 0.02 0.02 0.09 0.05 0.01	%	Antioxidant and anti-inflammatory (in vivo; in vitro).	[149]
Petals (Torbat Heydariyeh region, Iran; ohmic extraction)	Crocin Safranal Catechin Epicatechin Delphinidin 3,5-di-O-glucose Petunidin 3-O-glucoside Petunidin 2,5-di-O-glucoside Quercetin 3-O-glucoside Kaempferol-3-O-sophoroside Kaempferol-3-O-glucoside Quercetin 3-O-sophoroside Kaempferol	81.2 5.5 1.4 1.2 74.2 10.3 8.6 59.5 8.2 6.1 5.5 5.4	%	Source of natural flavoring, coloring, and antioxidants.	[41]
Petals. (Torbat Heydariyeh region, Iran; ultrasound extraction)	Crocin Safranal Delphinidin 3,5-di-O-glucose Petunidin 3-O-glucoside Petunidin 3,5-di-O-glucoside Quercetin 3-O-glucoside Kaempferol-3-O-sophoroside Kaempferol-3-O-glucoside Quercetin 3-O-sophoroside	79.02 4.03 67.88 10.74 7.39 54.32 8.16 5.27 5.12	%	Source of natural flavoring, coloring, and antioxidants.	[41]
Petals. (Torbat Heydariyeh region, Iran; microwave extraction)	Crocin Safranal Epicatechin Vanillic acid Delphinidin 3,5-di-O-glucose Petunidin 3-O-glucoside Malvidin O-glucoside Quercetin 3-O-glucoside Kaempferol-3-O-sophoroside Kaempferol-3-O-glucoside Quercetin 3-O-sophoroside Kaempferol	77.42 5.03 1.02 1.03 56.36 11.44 7.94 59.49 8.16 6.13 5.51 5.42	%	Source of natural flavoring, coloring, and antioxidants.	[41]

9. Conclusions

The high commercial value of saffron is a result of the production (harvesting, drying, and storage) and low biomass yield, a critical characteristic of market fraud. Saffron is used in the food industry as an aromatic species to give flavor, color, and odor to various foods, but its extracts or extractive compounds are also used as functional ingredients in a large number of products (desserts, beverages, oils, pastes, etc.). The ISO 3632 standard proposes a spectrophotometric technique for the determination of the commercial quality of saffron. This methodology has the great advantage of being easy to prepare, accessible, and low-cost in terms of equipment. The quality of *C. sativus* is based on the quantity of the main apocarotenoids (crocin, picrocrocin, and safranal). However, for the quantification of saffron apocarotenoids, more rigorous, sensitive, selective, and related analytical techniques (UHPLC/QTO/MS, DRIFTS, NIR, SCAR, PCR, etc.), which provide more accurate concentrations, are preferred. Moreover, the results obtained by spectrophotometry yield inaccurate results (overlapping of chemical compounds, poor solubility of safranal, erroneous quantification of compounds, and non-identification of adulterants). Therefore, the ISO standard is only proposed as a preliminary methodology to rule out low-quality saffron and is not suitable for authentication and/or the detection of adulterants. The determination of the chemical profiles or fingerprints of the sample or aromatic plant is used to obtain the markers of the saffron or adulterants. These signals or fingerprints obtained by analytical techniques coupled to chemometric methods (principal component analysis, linear discriminant analysis (LDA, etc.) favor the discrimination of adulterated samples, possible adulterant plants or compounds, the detection limits of the equipment, and even the concentrations at which they are obtained. Finally, it was determined that not only the stigmas contained bioactive compounds since this work describes some research on saffron flower by-products that contain a large number of phytochemical compounds (catechin, quercetin, delphinidin, etc.). For these reasons, saffron is an interesting and aromatic spice as a colorant, antioxidant, and source of phytochemicals.

Author Contributions: Conceptualization, T.S.C.-P. and R.A.-S.; methodology, G.V.N.-M.; investigation, T.S.C.-P., R.A.-S., A.R.N.-C., C.E.O.-V., P.H.-C. and G.V.N.-M.; writing—original draft preparation, T.S.C.-P. and R.A.-S.; writing—review and editing, G.V.N.-M.; visualization, T.S.C.-P., R.A.-S., A.R.N.-C., C.E.O.-V., P.H.-C. and G.V.N.-M.; funding acquisition R.A.-S. All authors have read and agreed to the published version of the manuscript.

Funding: This paper was funded by VIEP-BUAP, grant number 100377166-VIEP2022.

Conflicts of Interest: The authors declare no conflict of interest.

Abbreviations

ANN: Artificial Neural Network; ANN–MLP: Multi-Layer Perceptron–Artificial Neural Network; ANOVA: Analysis of Variance; Bar–HRM: Barcode-DNA–High-Resolution Melting; CVS: Computer Vision System; DNA: Deoxyribo Nucleic Acid; DRIFTS: Diffuse Reflectance Infrared Fourier Transform Spectroscopy; EI–MS: Electrospray Ionization–Mass Spectrometry; E-nose: Electronic nose; FT–IR: Fourier Transform–Infrared Spectroscopy; FT–NIR: Fourier Transform–Near-Infrared; GC–MS: Gas Chromatography–Mass Spectrometry; HCA: Hierarchical Cluster Analysis; HPLC: High-Performance Liquid Chromatography; HPLC–DAD: High-Performance Liquid Chromatography coupled with Diode Array Detection; HPLC–DAD–MS: High-Performance Liquid Chromatography coupled with Diode Array Detection–Mass Spectrometry; HPLC–PDA–ESI–MS: High-Performance Liquid Chromatography coupled with Photo Diode Array–Electrospray Ionization–Mass Spectrometry; HTCC: 4-hydroxy-2,6,6-trimethyl-1-cyclohexene-1-carboxaldehyde or hydroxy-β-cyclocitral; HRM: High-Resolution Melting; HRMS: High-Resolution Mass Spectrometry; ISSR: Inter-Simple Sequence Repeat; ISO: International Organization for Standardization; IR–MS: Isotope Ratio–Mass Spectrometry; LC: Liquid Chromatography; LC–MS: Liquid Chromatography–Mass Spectrometry; MCR–ALS: Multivariate Curve Resolution–Alternating Least Squares; MIR: Mid Infrared; MLP: Multi-Layer Perceptron; MLR: Multiple Linear Regression; NIR: Near Infrared; NMR: Nuclear Magnetic Resonance; OPLS–DA: Orthogonal Projection to Latent Structures–Discriminant Analysis; O2PLS–DA: Orthogonal Projection to Latent Structures–Discriminant Analysis with bidirectional modifications; PCA: Principal Component Analysis; PDO: Protected Designations of Origin; PLS: Partial Least Squares; PLS–DA: Partial Least Squares–Discriminant Analysis; PCR: Polymerase Chain Reaction; RPA–LFD: Recombinase Polymerase Amplification in combination with-Lateral Flow dipstick; RT–HPLC–DAD: Reverse Phase–High-Performance Liquid Chromatography coupled with

Diode Array Detection; SCAR: Sequence-Characterized Amplified Regions; SIMCA: Soft Independent Modeling of Class Analogies; SVMs: Support Vector Machines; Trans-1-g: crocin-1or trans-crocetin mono-(β-D-glucosyl) ester; Trans-2-gg: crocin-2II, crocin-2′ or trans-crocetin di-(β-D-glucosyl) ester; Trans-2-G: crocin-2 or trans-crocetin (β-D-gentiobiosyl) ester; Trans-3-Gg: crocin-3 or trans-crocetin (β-D-glucosyl)-(β-D-gentiobiosyl) ester; Trans-4-GG: crocin-4 or trans-crocetin di-(β-D-gentiobiosyl) ester); Trans-5-tG: crocin-5 or trans-crocetin (β-D-triglucosyl)-(β-D-gentiobiosyl) ester; TLC: Thin Layer Chromatography; UDP-glucuronosyl transferase: Uridine-diphosphate-glucuronosyl transferase; UHPLC–QTO–MS: Ultra-High-Performance Liquid Chromatography coupled to Quadrupole Time-of-Flight–Mass Spectrometry; UHPLC–ESI–QTOF–MS: Ultra-High-Performance Liquid Chromatography with Electrospray Ionization coupled to Quadrupole Time-of-Flight–Mass Spectrometry; UHPLC–DAD–MS: Ultra-High-Performance Liquid Chromatography with Diode Array Detection–Mass Spectrometry; UHPLC–MS/MS: Ultra-High-Performance Liquid Chromatography coupled to Tandem Mass Spectrometry; UV-vis: Ultraviolet-visible spectroscopy.

References

1. Teng, X.; Zhang, M.; Devahastin, S. New Developments on Ultrasound-Assisted Processing and Flavor Detection of Spices: A Review. *Ultrason. Sonochem.* **2019**, *55*, 297–307. [CrossRef] [PubMed]
2. Mirani, A.; Goli, M. Production of the Eggplant-fiber Incorporated Cupcake and Evaluating Its Chemical, Textural and Colorimetric Properties over a Ten-day Storage Time. *J. Food Process. Preserv.* **2021**, *45*. [CrossRef]
3. Cardone, L.; Castronuovo, D.; Perniola, M.; Cicco, N.; Candido, V. Saffron (*Crocus sativus* L.), the King of Spices: An Overview. *Sci. Hortic.* **2020**, *272*. [CrossRef]
4. Farag, M.A.; Hegazi, N.; Dokhalahy, E.; Khattab, A.R. Chemometrics Based GC-MS Aroma Profiling for Revealing Freshness, Origin and Roasting Indices in Saffron Spice and Its Adulteration. *Food Chem.* **2020**, *331*, 127358. [CrossRef] [PubMed]
5. Lee, F.Y.; Htar, T.T.; Akowuah, G.A. ATR-FTIR and Spectrometric Methods for the Assay of Crocin in Commercial Saffron Spices (*Crocus savitus* L.). *Int. J. Food Prop.* **2015**, *18*, 1773–1783. [CrossRef]
6. Mykhailenko, O.; Desenko, V.; Ivanauskas, L.; Georgiyants, V. Standard Operating Procedure of Ukrainian Saffron Cultivation According with Good Agricultural and Collection Practices to Assure Quality and Traceability. *Ind. Crops Prod.* **2020**, *151*, 112376. [CrossRef]
7. Heidarbeigi, K.; Mohtasebi, S.S.; Foroughirad, A.; Ghasemi-Varnamkhasti, M.; Rafiee, S.; Rezaei, K. Detection of Adulteration in Saffron Samples Using Electronic Nose. *Int. J. Food Prop.* **2015**, *18*, 1391–1401. [CrossRef]
8. Hadavi, R.; Jafari, S.M.; Katouzian, I. Nanoliposomal Encapsulation of Saffron Bioactive Compounds; Characterization and Optimization. *Int. J. Biol. Macromol.* **2020**, *164*, 4046–4053. [CrossRef] [PubMed]
9. Senizza, B.; Rocchetti, G.; Ghisoni, S.; Busconi, M.; De Los Mozos Pascual, M.; Fernandez, J.A.; Lucini, L.; Trevisan, M. Identification of Phenolic Markers for Saffron Authenticity and Origin: An Untargeted Metabolomics Approach. *Food Res. Int.* **2019**, *126*, 108584. [CrossRef] [PubMed]
10. de Castro, M.D.; Quiles-Zafra, R. Appropriate Use of Analytical Terminology–Examples Drawn From Research on Saffron. *Talanta Open* **2020**, *2*, 100005. [CrossRef]
11. Biancolillo, A.; Maggi, M.A.; De Martino, A.; Marini, F.; Ruggieri, F.; D'Archivio, A.A. Authentication of PDO Saffron of L'Aquila (*Crocus sativus* L.) by HPLC-DAD Coupled with a Discriminant Multi-Way Approach. *Food Control* **2020**, *110*, 107022. [CrossRef]
12. Dias, C.; Mendes, L. Protected Designation of Origin (PDO), Protected Geographical Indication (PGI) and Traditional Speciality Guaranteed (TSG): A Bibiliometric Analysis. *Food Res. Int.* **2018**, *103*, 492–508. [CrossRef] [PubMed]
13. Lu, X.; Xia, Z.; Qu, F.; Zhu, Z.; Li, S. Identification of Authenticity, Quality and Origin of Saffron Using Hyperspectral Imaging and Multivariate Spectral Analysis. *Spectrosc. Lett.* **2020**, *53*, 76–85. [CrossRef]
14. Bakshi, R.A.; Sodhi, N.S.; Wani, I.A.; Khan, Z.S.; Dhillon, B.; Gani, A. Bioactive Constituents of Saffron Plant: Extraction, Encapsulation and Their Food and Pharmaceutical Applications. *Appl. Food Res.* **2022**, *2*, 100076. [CrossRef]
15. Bosmali, I.; Ordoudi, S.A.; Tsimidou, M.Z.; Madesis, P. Greek PDO Saffron Authentication Studies Using Species Specific Molecular Markers. *Food Res. Int.* **2017**, *100*, 899–907. [CrossRef] [PubMed]
16. Halvorson, S. Saflron Cultivation and Culture in Central Spain. *Focus Geogr.* **2008**, *51*, 17–24. [CrossRef]
17. Ordoudi, S.A.; Tsimidou, M. Saffron Quality: Effect of Agricultural Practices, Processing and Storage. In *Production Practices and Quality Assessment of Food Crops*; Dris, R., Jain, S.M., Eds.; Kluwer Academic: Amsterdam, The Netherland, 2004; Volume 1, pp. 209–260. ISBN 1402025335.
18. Winterhalter, P.; Straubinger, M. Saffron-Renewed Interest in an Ancient Spice. *Food Rev. Int.* **2000**, *16*, 39–59. [CrossRef]
19. Bathaie, S.Z.; Farajzade, A.; Hoshyar, R. A Review of the Chemistry and Uses of Crocins and Crocetin, the Carotenoid Natural Dyes in Saffron, with Particular Emphasis on Applications as Colorants Including Their Use as Biological Stains. *Biotech. Histochem.* **2014**, *89*, 401–411. [CrossRef] [PubMed]
20. Ghanbari, J.; Khajoei-Nejad, G.; van Ruth, S.M. Effect of Saffron (*Crocus sativus* L.) Corm Provenance on Its Agro-Morphological Traits and Bioactive Compounds. *Sci. Hortic. (Amst.)* **2019**, *256*, 108605. [CrossRef]

21. Lage, M.; Cantrell, C.L. Quantification of Saffron (*Crocus sativus* L.) Metabolites Crocins, Picrocrocin and Safranal for Quality Determination of the Spice Grown under Different Environmental Moroccan Conditions. *Sci. Hortic. (Amst.)* **2009**, *121*, 366–373. [CrossRef]
22. Ahrazem, O.; Rubio-Moraga, A.; Nebauer, S.G.; Molina, R.V.; Gómez-Gómez, L. Saffron: Its Phytochemistry, Developmental Processes, and Biotechnological Prospects. *J. Agric. Food Chem.* **2015**, *63*, 8751–8764. [CrossRef] [PubMed]
23. Condurso, C.; Cincotta, F.; Tripodi, G.; Verzera, A. Bioactive Volatiles in Sicilian (South Italy) Saffron: Safranal and Its Related Compounds. *J. Essent. Oil Res.* **2017**, *29*, 221–227. [CrossRef]
24. Douglas, M.H.; Smallfield, B.M.; Wallace, A.R.; McGimpsey, J.A. Saffron (*Crocus sativus* L.): The Effect of Mother Corm Size on Progeny Multiplication, Flower and Stigma Production. *Sci. Hortic. (Amst.)* **2014**, *166*, 50–58. [CrossRef]
25. Mzabri, I.; Addi, M.; Berrichi, A. Traditional and Modern Uses of Saffron (*Crocus sativus*). *Cosmetics* **2019**, *6*, 2–11. [CrossRef]
26. Serrano-Díaz, J.; Sánchez, A.M.; Martínez-Tomé, M.; Winterhalter, P.; Alonso, G.L. A Contribution to Nutritional Studies on *Crocus sativus* Flowers and Their Value as Food. *J. Food Compos. Anal.* **2013**, *31*, 101–108. [CrossRef]
27. Parizad, S.; Dizadji, A.; Habibi, M.K.; Winter, S.; Kalantari, S.; Movi, S.; Lorenzo Tendero, C.; Alonso, G.L.; Moratalla-Lopez, N. The Effects of Geographical Origin and Virus Infection on the Saffron (*Crocus sativus* L.) Quality. *Food Chem.* **2019**, *295*, 387–394. [CrossRef] [PubMed]
28. Aghhavani, M.S.; Rezvani, P.M. Are the Apocarotenoids Content and Colorimetric Traits of Saffron (*Crocus sativus* L.) Affected by Some Post Harvesting Operations? *J. Stored Prod. Res.* **2022**, *97*, 101967. [CrossRef]
29. Kumar, R.; Singh, V.; Devi, K.; Sharma, M.; Singh, M.K.; Ahuja, P.S. State of Art of Saffron (*Crocus sativus* L.) Agronomy: A Comprehensive Review. *Food Rev. Int.* **2009**, *25*, 44–85. [CrossRef]
30. Gracia, L.; Perez-Vidal, C.; Gracia-López, C. Automated Cutting System to Obtain the Stigmas of the Saffron Flower. *Biosyst. Eng.* **2009**, *104*, 8–17. [CrossRef]
31. Rubert, J.; Lacina, O.; Zachariasova, M.; Hajslova, J. Saffron Authentication Based on Liquid Chromatography High Resolution Tandem Mass Spectrometry and Multivariate Data Analysis. *Food Chem.* **2016**, *204*, 201–209. [CrossRef] [PubMed]
32. Acar, B.; Sadikoglu, H.; Ozkaymak, M. Freeze Drying of Saffron (*Crocus sativus* L.). *Dry. Technol.* **2011**, *29*, 1622–1627. [CrossRef]
33. Anastasaki, E.; Kanakis, C.; Pappas, C.; Maggi, L.; del Campo, C.P.; Carmona, M.; Alonso, G.L.; Polissiou, M.G. Geographical Differentiation of Saffron by GC-MS/FID and Chemometrics. *Eur. Food Res. Technol.* **2009**, *229*, 899–905. [CrossRef]
34. Catinella, G.; Borgonovo, G.; Dallavalle, S.; Contente, M.L.; Pinto, A. From Saffron Residues to Natural Safranal: Valorization of Waste through a β-Glucosidase. *Food Bioprod. Process.* **2022**, *131*, 144–148. [CrossRef]
35. Chen, D.; Xing, B.; Yi, H.; Li, Y.; Zheng, B.; Wang, Y.; Shao, Q. Effects of Different Drying Methods on Appearance, Microstructure, Bioactive Compounds and Aroma Compounds of Saffron (*Crocus sativus* L.). *LWT Food Sci. Technol.* **2020**, *120*, 108913. [CrossRef]
36. Cid-Pérez, T.S.; Nevárez-Moorillón, G.V.; Ochoa-Velasco, C.E.; Navarro-Cruz, A.R.; Hernández-Carranza, P.; Avila-Sosa, R. The Relation between Drying Conditions and the Development of Volatile Compounds in Saffron (*Crocus sativus*). *Molecules* **2021**, *26*, 6954. [CrossRef]
37. Eslami, M.; Bayat, M.; Mozaffari Nejad, A.S.; Sabokbar, A.; Anvar, A.A. Effect of Polymer/Nanosilver Composite Packaging on Long-Term Microbiological Status of Iranian Saffron (*Crocus sativus* L.). *Saudi J. Biol. Sci.* **2016**, *23*, 341–347. [CrossRef]
38. Giaccio, M. Crocetin from Saffron: An Active Component of an Ancient Spice. *Crit. Rev. Food Sci. Nutr.* **2004**, *44*, 155–172. [CrossRef]
39. Mykhailenko, O.; Ivanauskas, L.; Bezruk, I.; Marksa, M.; Borodina, O.; Georgiyants, V. Effective and Simple Approach for Colchicine Determination in Saffron Parts. *Food Chem.* **2022**, *368*, 130862. [CrossRef]
40. Wang, D.; Cui, L.; Ren, H.; Wang, Y.; Long, D.; Niu, Y. Anti-Fungal Activity and Preliminary Active Components Separation from Ethanol Extracts in Saffron (*Crocus sativus* L.) Lateral Buds. *Ind. Crops Prod.* **2021**, *173*, 114081. [CrossRef]
41. Hashemi Gahruie, H.; Parastouei, K.; Mokhtarian, M.; Rostami, H.; Niakousari, M.; Mohsenpour, Z. Application of Innovative Processing Methods for the Extraction of Bioactive Compounds from Saffron (*Crocus sativus*) Petals. *J. Appl. Res. Med. Aromat. Plants* **2020**, 100264. [CrossRef]
42. Ghanbari, J.; Khajoei-Nejad, G.; Erasmus, S.W.; van Ruth, S.M. Identification and Characterisation of Volatile Fingerprints of Saffron Stigmas and Petals Using PTR-TOF-MS: Influence of Nutritional Treatments and Corm Provenance. *Ind. Crops Prod.* **2019**, *141*, 111803. [CrossRef]
43. Armellini, R.; Peinado, I.; Asensio-Grau, A.; Pittia, P.; Scampicchio, M.; Heredia, A.; Andres, A. In Vitro Starch Digestibility and Fate of Crocins in Pasta Enriched with Saffron Extract. *Food Chem.* **2019**, *283*, 155–163. [CrossRef]
44. Gani, A.; Jan, R.; Ashwar, B.A.; Ashraf, Z.U.; Shah, A.; Gani, A. Encapsulation of Saffron and Sea Buckthorn Bioactives: Its Utilization for Development of Low Glycemic Baked Product for Growing Diabetic Population of the World. *LWT* **2021**, *142*, 111035. [CrossRef]
45. Bhat, N.A.; Hamdani, A.M.; Masoodi, F.A. Development of Functional Cookies Using Saffron Extract. *J. Food Sci. Technol.* **2018**, *55*, 4918–4927. [CrossRef]
46. Bhat, N.A.; Wani, I.A.; Hamdani, A.M.; Gani, A. Development of Functional Cakes Rich in Bioactive Compounds Extracted from Saffron and Tomatoes. *J. Food Sci. Technol.* **2022**, *59*, 2479–2491. [CrossRef] [PubMed]
47. Armellini, R.; Peinado, I.; Pittia, P.; Scampicchio, M.; Heredia, A.; Andres, A. Effect of Saffron (*Crocus sativus* L.) Enrichment on Antioxidant and Sensorial Properties of Wheat Flour Pasta. *Food Chem.* **2018**, *254*, 55–63. [CrossRef] [PubMed]

48. Sena-Moreno, E.; Alvarez-Ortí, M.; Serrano-Díaz, J.; Pardo, J.E.; Carmona, M.; Alonso, G.L. Olive Oil Aromatization with Saffron by Liquid–Liquid Extraction. *J. Food Sci. Technol.* **2018**, *55*, 1093–1103. [CrossRef] [PubMed]
49. Almodóvar, P.; Prodanov, M.; Arruñada, O.; Inarejos-García, A.M. Affron® Eye, a Natural Extract of Saffron (*Crocus sativus* L.) with Colorant Properties as Novel Replacer of Saffron Stigmas in Culinary and Food Applications. *Int. J. Gastron. Food Sci.* **2018**, *12*, 1–5. [CrossRef]
50. Dabbagh Moghaddam, A.; Garavand, F.; Razavi, S.H.; Dini Talatappe, H. Production of Saffron-Based Probiotic Beverage by Lactic Acid Bacteria. *J. Food Meas. Charact.* **2018**, *12*, 2708–2717. [CrossRef]
51. Guijarro-Díez, M.; Castro-Puyana, M.; Crego, A.L.; Marina, M.L. A Novel Method for the Quality Control of Saffron through the Simultaneous Analysis of Authenticity and Adulteration Markers by Liquid Chromatography-(Quadrupole-Time of Flight)-Mass Spectrometry. *Food Chem.* **2017**, *228*, 403–410. [CrossRef] [PubMed]
52. Garavand, F.; Rahaee, S.; Vahedikia, N.; Jafari, S.M. Different Techniques for Extraction and Micro/Nanoencapsulation of Saffron Bioactive Ingredients. *Trends Food Sci. Technol.* **2019**, *89*, 26–44. [CrossRef]
53. Caballero-Ortega, H.; Pereda-Miranda, R.; Abdullaev, F.I. HPLC Quantification of Major Active Components from 11 Different Saffron (*Crocus sativus* L.) Sources. *Food Chem.* **2007**, *100*, 1126–1131. [CrossRef]
54. Liu, J.; Chen, N.; Yang, J.; Yang, B.; Ouyang, Z.; Wu, C.; Yuan, Y.; Wang, W.; Chen, M. An Integrated Approach Combining HPLC, GC/MS, NIRS, and Chemometrics for the Geographical Discrimination and Commercial Categorization of Saffron. *Food Chem.* **2018**, *253*, 284–292. [CrossRef] [PubMed]
55. Ghanbari, J.; Khajoei-Nejad, G.; van Ruth, S.M.; Aghighi, S. The Possibility for Improvement of Flowering, Corm Properties, Bioactive Compounds, and Antioxidant Activity in Saffron (*Crocus sativus* L.) by Different Nutritional Regimes. *Ind. Crops Prod.* **2019**, *135*, 301–310. [CrossRef]
56. Moratalla-López, N.; José Bagur, M.; Lorenzo, C.; Martínez-Navarro, M.E.; Mora-talla-López, N.; José Bagur, M. Alonso Bioactivity and Bioavailability of the Major Metabolites of *Crocus sativus* L. Flower. *Molecules* **2019**, *24*, 2827. [CrossRef] [PubMed]
57. Verma, R.S.; Middha, D. Analysis of Saffron (*Crocus sativus* L. Stigma) Components by LC-MS-MS. *Chromatographia* **2010**, *71*, 117–123. [CrossRef]
58. Sarfarazi, M.; Rajabzadeh, Q.; Tavakoli, R.; Ibrahim, S.A.; Jafari, S.M. Ultrasound-Assisted Extraction of Saffron Bioactive Compounds; Separation of Crocins, Picrocrocin, and Safranal Optimized by Artificial Bee Colony. *Ultrason. Sonochem.* **2022**, *86*, 105971. [CrossRef] [PubMed]
59. Rajabi, H.; Jafari, S.M.; Feizi, J.; Ghorbani, M.; Mohajeri, S.A. Surface-Decorated Graphene Oxide Sheets with Nanoparticles of Chitosan-Arabic Gum for the Separation of Bioactive Compounds: A Case Study for Adsorption of Crocin from Saffron Extract. *Int. J. Biol. Macromol.* **2021**, *186*, 1–12. [CrossRef] [PubMed]
60. Bagur, M.J.; Alonso Salinas, G.L.; Jiménez-Monreal, A.M.; Chaouqi, S.; Llorens, S.; Martínez-Tomé, M.; Alonso, G.L. Saffron: An Old Medicinal Plant and a Potential Novel Functional Food. *Molecules* **2017**, *23*, 30. [CrossRef]
61. Azarabadi, N.; Özdemir, F. Determination of Crocin Content and Volatile Components in Different Qualities of Iranian Saffron. *Gida J. Food* **2018**, 476–489. [CrossRef]
62. Gismondi, A.; Serio, M.; Canuti, L.; Canini, A. Biochemical, Antioxidant and Antineoplastic Properties of Italian Saffron (*Crocus sativus* L.) . *Am. J. Plant Sci.* **2012**, *03*, 1573–1580. [CrossRef]
63. Kosar, M.; Demirci, B.; Goger, F.; Kara, I.; Baser, K.H.C. Volatile Composition, Antioxidant Activity, and Antioxidant Components in Saffron Cultivated in Turkey. *Int. J. Food Prop.* **2017**, *20*, S746–S754. [CrossRef]
64. Masi, E.; Taiti, C.; Heimler, D.; Vignolini, P.; Romani, A.; Mancuso, S. PTR-TOF-MS and HPLC Analysis in the Characterization of Saffron (*Crocus sativus* L.) from Italy and Iran. *Food Chem.* **2016**, *192*, 75–81. [CrossRef]
65. Moratalla-López, N.; Sánchez, A.M.; Lorenzo, C.; López-Córcoles, H.; Alonso, G.L. Quality Determination of *Crocus sativus* L. Flower by High-Performance Liquid Chromatography. *J. Food Compos. Anal.* **2020**, *93*, 103613. [CrossRef]
66. Rahaiee, S.; Moini, S.; Hashemi, M.; Shojaosadati, S.A. Evaluation of Antioxidant Activities of Bioactive Compounds and Various Extracts Obtained from Saffron (*Crocus sativus* L.): A Review. *J. Food Sci. Technol.* **2015**, *52*, 1881–1888. [CrossRef] [PubMed]
67. Reddy, C.N.; Bharate, S.B.; Vishwakarma, R.A.; Bharate, S.S. Chemical Analysis of Saffron by HPLC Based Crocetin Estimation. *J. Pharm. Biomed. Anal.* **2020**, *181*, 113094. [CrossRef] [PubMed]
68. Rocchi, R.; Mascini, M.; Sergi, M.; Compagnone, D.; Mastrocola, D.; Pittia, P. Crocins Pattern in Saffron Detected by UHPLC-MS/MS as Marker of Quality, Process and Traceability. *Food Chem.* **2018**, *264*, 241–249. [CrossRef] [PubMed]
69. Shahi, T.; Assadpour, E.; Jafari, S.M. Main Chemical Compounds and Pharmacological Activities of Stigmas and Tepals of 'Red Gold'; Saffron. *Trends Food Sci. Technol.* **2016**, *58*, 69–78. [CrossRef]
70. Tabtabaei, S.; D'Archivio, A.A.; Maggi, M.A.; Brutus, M.; Bajracharya, D.H.; Konakbayeva, D.; Soleimani, A.; Brim, H.; Ashktorab, H. Geographical Classification of Iranian and Italian Saffron Sources Based on HPLC Analysis and UV–Vis Spectra of Aqueous Extracts. *Eur. Food Res. Technol.* **2019**, *245*, 2435–2446. [CrossRef]
71. Urbani, E.; Blasi, F.; Simonetti, M.S.; Chiesi, C.; Cossignani, L. Investigation on Secondary Metabolite Content and Antioxidant Activity of Commercial Saffron Powder. *Eur. Food Res. Technol.* **2016**, *242*, 987–993. [CrossRef]
72. Vahedi, M.; Kabiri, M.; Salami, S.A.; Rezadoost, H.; Mirzaie, M.; Kanani, M.R. Quantitative HPLC-Based Metabolomics of Some Iranian Saffron (*Crocus sativus* L.) Accessions. *Ind. Crops Prod.* **2018**, *118*, 26–29. [CrossRef]

73. Bhat, Z.S.; Jaladi, N.; Khajuria, R.K.; Shah, Z.H.; Arumugam, N. Comparative Analysis of Bioactive N-Alkylamides Produced by Tissue Culture Raised versus Field Plantlets of Spilanthes Ciliata Using LC-Q-TOF (HRMS). *J. Chromatogr. B Anal. Technol. Biomed. Life Sci.* **2016**, *1017–1018*, 195–203. [CrossRef]
74. Guclu, G.; Kelebek, H.; Selli, S. Saffron (*Crocus sativus* L.): Its Aroma and Key Odorants. In *Saffron The Age-Old Panacea in a New Light*; Sarwat, M., Sumaiya, S., Eds.; Academic Press: London, UK, 2020; pp. 69–82.
75. Melnyk, J.P.; Wang, S.; Marcone, M.F. Chemical and Biological Properties of the World's Most Expensive Spice: Saffron. *Food Res. Int.* **2010**, *43*, 1981–1989. [CrossRef]
76. Hosseini, A.; Razavi, B.M.; Hosseinzadeh, H. Pharmacokinetic Properties of Saffron and Its Active Components. *Eur. J. Drug Metab. Pharmacokinet.* **2018**, *43*, 383–390. [CrossRef]
77. Jalali-Heravi, M.; Parastar, H.; Ebrahimi-Najafabadi, H. Characterization of Volatile Components of Iranian Saffron Using Factorial-Based Response Surface Modeling of Ultrasonic Extraction Combined with Gas Chromatography-Mass Spectrometry Analysis. *J. Chromatogr. A* **2009**, *1216*, 6088–6097. [CrossRef]
78. Ordoudi, S.A.; Cagliani, L.R.; Lalou, S.; Naziri, E.; Tsimidou, M.Z.; Consonni, R. 1H NMR-Based Metabolomics of Saffron Reveals Markers for Its Quality Deterioration. *Food Res. Int.* **2015**, *70*, 1–6. [CrossRef]
79. García-Rodríguez, M.V.; Moratalla-López, N.; López-Córcoles, H.; Alonso, G.L. Saffron Quality Obtained under Different Forcing Conditions, Considering Various Vegetative Stages of Corms. *Sci. Hortic. (Amst.)* **2021**, *277*, 109811. [CrossRef]
80. Fallahi, H.-R.; Aghhavani-Shajari, M.; Sahabi, H.; Behdani, M.A.; Sayyari-Zohan, M.H.; Vatandoost, S. Influence of Some Pre and Post-Harvest Practices on Quality of Saffron Stigmata. *Sci. Hortic. (Amst.)* **2021**, *278*, 109846. [CrossRef]
81. Sereshti, H.; Ataolahi, S.; Aliakbarzadeh, G.; Zarre, S.; Poursorkh, Z. Evaluation of Storage Time Effect on Saffron Chemical Profile Using Gas Chromatography and Spectrophotometry Techniques Coupled with Chemometrics. *J. Food Sci. Technol.* **2018**, *55*, 1350–1359. [CrossRef]
82. Sarfarazi, M.; Jafari, S.M.; Rajabzadeh, G.; Feizi, J. Development of an Environmentally-Friendly Solvent-Free Extraction of Saffron Bioactives Using Subcritical Water. *LWT Food Sci. Technol.* **2019**, *114*, 108428. [CrossRef]
83. Kiani, S.; Minaei, S.; Ghasemi-Varnamkhasti, M. Integration of Computer Vision and Electronic Nose as Non-Destructive Systems for Saffron Adulteration Detection. *Comput. Electron. Agric.* **2017**, *141*, 46–53. [CrossRef]
84. Neri, L.; Giancaterino, M.; Rocchi, R.; Tylewicz, U.; Valbonetti, L.; Faieta, M.; Pittia, P. Pulsed Electric Fields (PEF) as Hot Air Drying Pre-Treatment: Effect on Quality and Functional Properties of Saffron (*Crocus sativus* L.). *Innov. Food Sci. Emerg. Technol.* **2021**, *67*, 102592. [CrossRef]
85. Sabatino, L.; Scordino, M.; Gargano, M.; Belligno, A.; Traulo, P.; Gagliano, G. HPLC/PDA/ESI-MS Evaluation of Saffron (*Crocus sativus* L.) Adulteration. *Nat. Prod. Commun.* **2011**, *6*, 1873–1876. [CrossRef] [PubMed]
86. Aiello, D.; Siciliano, C.; Mazzotti, F.; Di Donna, L.; Athanassopoulos, C.M.; Napoli, A. A Rapid MALDI MS/MS Based Method for Assessing Saffron (*Crocus sativus* L.) Adulteration. *Food Chem.* **2020**, *307*. [CrossRef]
87. Chaouqi, S.; Moratalla-López, N.; Lage, M.; Lorenzo, C.; Alonso, G.L.; Guedira, T. Effect of Drying and Storage Process on Moroccan Saffron Quality. *Food Biosci.* **2018**, *22*, 146–153. [CrossRef]
88. Cossignani, L.; Urbani, E.; Simonetti, M.S.; Maurizi, A.; Chiesi, C.; Blasi, F. Characterisation of Secondary Metabolites in Saffron from Central Italy (Cascia, Umbria). *Food Chem.* **2014**, *143*, 446–451. [CrossRef]
89. Kiani, S.; Minaei, S.; Ghasemi-Varnamkhasti, M. Instrumental Approaches and Innovative Systems for Saffron Quality Assessment. *J. Food Eng.* **2018**, *216*, 1–10. [CrossRef]
90. Kabiri, M.; Rezadoost, H.; Ghassempour, A. A Comparative Quality Study of Saffron Constituents through HPLC and HPTLC Methods Followed by Isolation of Crocins and Picrocrocin. *LWT Food Sci. Technol.* **2017**, *84*, 1–9. [CrossRef]
91. Minaei, S.; Kiani, S.; Ayyari, M.; Ghasemi-Varnamkhasti, M. A Portable Computer-Vision-Based Expert System for Saffron Color Quality Characterization. *J. Appl. Res. Med. Aromat. Plants* **2017**, *7*, 124–130. [CrossRef]
92. Pei, Y.; Li, Z.; Xu, W.; Song, C.; Li, J.; Song, F. Effects of Ultrasound Pretreatment Followed by Far-Infrared Drying on Physicochemical Properties, Antioxidant Activity and Aroma Compounds of Saffron (*Crocus sativus* L.). *Food Biosci.* **2021**, *42*, 101186. [CrossRef]
93. Sánchez, A.M.; Winterhalter, P. Carotenoid Cleavage Products in Saffron (*Crocus sativus* L.). *ACS Symp. Ser.* **2013**, *1134*, 45–63. [CrossRef]
94. del Campo, C.P.; Carmona, M.; Maggi, L.; Kanakis, C.D.; Anastasaki, E.G.; Tarantilis, P.A.; Polissiou, M.G.; Alonso, G.L. Effects of Mild Temperature Conditions during Dehydration Procedures on Saffron Quality Parameters. *J. Sci. Food Agric.* **2010**, *90*, 719–725. [CrossRef] [PubMed]
95. Christodoulou, E.; Kadoglou, N.P.; Kostomitsopoulos, N.; Valsami, G. Saffron: A Natural Product with Potential Pharmaceutical Applications. *J. Pharm. Pharmacol.* **2015**, *67*, 1634–1649. [CrossRef]
96. García-Rodríguez, M.V.; López-Córcoles, H.; Alonso, G.L.; Pappas, C.S.; Polissiou, M.G.; Tarantilis, P.A. Comparative Evaluation of an ISO 3632 Method and an HPLC-DAD Method for Safranal Quantity Determination in Saffron. *Food Chem.* **2017**, *221*, 838–843. [CrossRef] [PubMed]
97. Urbani, E.; Blasi, F.; Chiesi, C.; Maurizi, A.; Cossignani, L. Characterization of Volatile Fraction of Saffron from Central Italy (Cascia, Umbria). *Int. J. Food Prop.* **2015**, *18*, 2223–2230. [CrossRef]
98. Vignolini, P.; Heimler, D.; Pinelli, P.; Ieri, F.; Sciullo, A.; Romani, A. Characterization of By-Products of Saffron (*Crocus sativus* L.) Production. *Nat. Prod. Commun.* **2008**, *3*, 1934578X0800301. [CrossRef]

99. Ordoudi, S.A.; De Los Mozos Pascual, M.; Tsimidou, M.Z. On the Quality Control of Traded Saffron by Means of Transmission Fourier-Transform Mid-Infrared (FT-MIR) Spectroscopy and Chemometrics. *Food Chem.* **2014**, *150*, 414–421. [CrossRef]
100. Bononi, M.; Milella, P.; Tateo, F. Gas Chromatography of Safranal as Preferable Method for the Commercial Grading of Saffron (*Crocus sativus* L.). *Food Chem.* **2015**, *176*, 17–21. [CrossRef]
101. Cagliani, L.R.; Culeddu, N.; Chessa, M.; Consonni, R. NMR Investigations for a Quality Assessment of Italian PDO Saffron (*Crocus sativus* L.). *Food Control* **2015**, *50*, 342–348. [CrossRef]
102. Suchareau, M.; Bordes, A.; Lemée, L. Improved Quantification Method of Crocins in Saffron Extract Using HPLC-DAD after Qualification by HPLC-DAD-MS. *Food Chem.* **2021**, *362*, 130199. [CrossRef]
103. D'Archivio, A.A.; Maggi, M.A. Geographical Identification of Saffron (*Crocus sativus* L.) by Linear Discriminant Analysis Applied to the UV–Visible Spectra of Aqueous Extracts. *Food Chem.* **2017**, *219*, 408–413. [CrossRef] [PubMed]
104. Maggi, L.; Sánchez, A.M.; Carmona, M.; Kanakis, C.D.; Anastasaki, E.; Tarantilis, P.A.; Polissiou, M.G.; Alonso, G.L. Rapid Determination of Safranal in the Quality Control of Saffron Spice (*Crocus sativus* L.). *Food Chem.* **2011**, *127*, 369–373. [CrossRef]
105. Carmona, M.; Zalacain, A.; Sánchez, A.M.; Novella, J.L.; Alonso, G.L. Crocetin Esters, Picrocrocin and Its Related Compounds Present in *Crocus Sativus* Stigmas and *Gardenia Jasminoides* Fruits. Tentative Identification of Seven New Compounds by LC-ESI-MS. *J. Agric. Food Chem.* **2006**, *54*, 973–979. [CrossRef] [PubMed]
106. Moras, B.; Loffredo, L.; Rey, S. Quality Assessment of Saffron (*Crocus sativus* L.) Extracts via UHPLC-DAD-MS Analysis and Detection of Adulteration Using Gardenia Fruit Extract (*Gardenia jasminoides* Ellis). *Food Chem.* **2018**, *257*, 325–332. [CrossRef]
107. Guijarro-Díez, M.; Nozal, L.; Marina, M.L.; Crego, A.L. Metabolomic Fingerprinting of Saffron by LC/MS: Novel Authenticity Markers. *Anal. Bioanal. Chem.* **2015**, *407*, 7197–7213. [CrossRef] [PubMed]
108. Amanpour, A.; Kelebek, H.; Selli, S. *GLC/HPLC Methods for Saffron (Crocus sativus L.)*; Springer: Cham, Switzerland, 2018; pp. 1–49.
109. Moratalla-López, N.; Parizad, S.; Habibi, M.K.; Winter, S.; Kalantari, S.; Bera, S.; Lorenzo, C.; García-Rodríguez, M.V.; Dizadji, A.; Alonso, G.L. Impact of Two Different Dehydration Methods on Saffron Quality, Concerning the Prevalence of Saffron Latent Virus (SaLV) in Iran. *Food Chem.* **2021**, *337*, 127786. [CrossRef]
110. Rahaiee, S.; Hashemi, M.; Moini, S.; Shojaosadati, S.A.; Razavi, S.H. Comparison of Phytochemical Constituents and Antioxidant Activities of Aqueous and Alcoholic Extracts of Saffron. *Qual. Assur. Saf. Crop. Foods* **2015**, *7*, 521–529. [CrossRef]
111. Yousuf, B.; Qadri, O.S.; Srivastava, A.K. Recent Developments in Shelf-Life Extension of Fresh-Cut Fruits and Vegetables by Application of Different Edible Coatings: A Review. *LWT Food Sci. Technol.* **2018**, *89*, 198–209. [CrossRef]
112. Kyriakoudi, A.; Chrysanthou, A.; Mantzouridou, F.; Tsimidou, M.Z. Revisiting Extraction of Bioactive Apocarotenoids from *Crocus sativus* L. Dry Stigmas (Saffron). *Anal. Chim. Acta* **2012**, *755*, 77–85. [CrossRef]
113. Zhang, H.; Zeng, Y.; Yan, F.; Chen, F.; Zhang, X.; Liu, M.; Liu, W. Semi-Preparative Isolation of Crocins from Saffron (*Crocus sativus* L.). *Chromatographia* **2004**, *59*. [CrossRef]
114. Carmona, M.; Zalacain, A.; Salinas, M.R.; Alonso, G.L. A New Approach to Saffron Aroma. *Crit. Rev. Food Sci. Nutr.* **2007**, *47*, 145–159. [CrossRef] [PubMed]
115. Torelli, A.; Marieschi, M.; Bruni, R. Authentication of Saffron (*Crocus sativus* L.) in Different Processed, Retail Products by Means of SCAR Markers. *Food Control* **2014**, *36*, 126–131. [CrossRef]
116. Khilare, V.; Tiknaik, A.; Prakash, B.; Ughade, B.; Korhale, G.; Nalage, D.; Ahmed, N.; Khedkar, C.; Khedkar, G. Multiple Tests on Saffron Find New Adulterant Materials and Reveal That Ist Grade Saffron Is Rare in the Market. *Food Chem.* **2019**, *272*, 635–642. [CrossRef] [PubMed]
117. Amirvaresi, A.; Nikounezhad, N.; Amirahmadi, M.; Daraei, B.; Parastar, H. Comparison of Near-Infrared (NIR) and Mid-Infrared (MIR) Spectroscopy Based on Chemometrics for Saffron Authentication and Adulteration Detection. *Food Chem.* **2021**, *344*, 128647. [CrossRef] [PubMed]
118. Masoum, S.; Gholami, A.; Hemmesi, M.; Abbasi, S. Quality Assessment of the Saffron Samples Using Second-Order Spectrophotometric Data Assisted by Three-Way Chemometric Methods via Quantitative Analysis of Synthetic Colorants in Adulterated Saffron. *Spectrochim. Acta Part A Mol. Biomol. Spectrosc.* **2015**, *148*, 389–395. [CrossRef]
119. Guijarro-Díez, M.; Castro-Puyana, M.; Crego, A.L.; Marina, M.L. Detection of Saffron Adulteration with Gardenia Extracts through the Determination of Geniposide by Liquid Chromatography–Mass Spectrometry. *J. Food Compos. Anal.* **2017**, *55*, 30–37. [CrossRef]
120. Nenadis, N.; Heenan, S.; Tsimidou, M.Z.; Van Ruth, S. Applicability of PTR-MS in the Quality Control of Saffron. *Food Chem.* **2016**, *196*, 961–967. [CrossRef]
121. Paul, A.; de Boves Harrington, P. Chemometric Applications in Metabolomic Studies Using Chromatography-Mass Spectrometry. *TrAC Trends Anal. Chem.* **2021**, *135*, 116165. [CrossRef]
122. Medina, S.; Perestrelo, R.; Silva, P.; Pereira, J.A.M.; Câmara, J.S. Current Trends and Recent Advances on Food Authenticity Technologies and Chemometric Approaches. *Trends Food Sci. Technol.* **2019**, *85*, 163–176. [CrossRef]
123. Petrakis, E.A.; Polissiou, M.G. Assessing Saffron (*Crocus sativus* L.) Adulteration with Plant-Derived Adulterants by Diffuse Reflectance Infrared Fourier Transform Spectroscopy Coupled with Chemometrics. *Talanta* **2017**, *162*, 558–566. [CrossRef] [PubMed]
124. Babaei, S.; Talebi, M.; Bahar, M. Developing an SCAR and ITS Reliable Multiplex PCR-Based Assay Forsafflower Adulterant Detection in Saffron Samples. *Food Control* **2014**, *35*, 323–328. [CrossRef]

125. Petrakis, E.A.; Cagliani, L.R.; Polissiou, M.G.; Consonni, R. Evaluation of Saffron (*Crocus sativus* L.) Adulteration with Plant Adulterants by 1H NMR Metabolite Fingerprinting. *Food Chem.* **2015**, *173*, 890–896. [CrossRef] [PubMed]
126. Li, Y.; Shen, Y.; Yao, C.; Guo, D. Quality Assessment of Herbal Medicines Based on Chemical Fingerprints Combined with Chemometrics Approach: A Review. *J. Pharm. Biomed. Anal.* **2020**, *185*, 113215. [CrossRef] [PubMed]
127. Campmajó, G.; Saez-Vigo, R.; Saurina, J.; Núñez, O. High-Performance Liquid Chromatography with Fluorescence Detection Fingerprinting Combined with Chemometrics for Nut Classification and the Detection and Quantitation of Almond-Based Product Adulterations. *Food Control* **2020**, *114*, 107265. [CrossRef]
128. Kalogiouri, N.P.; Aalizadeh, R.; Dasenaki, M.E.; Thomaidis, N.S. Application of High Resolution Mass Spectrometric Methods Coupled with Chemometric Techniques in Olive Oil Authenticity Studies—A Review. *Anal. Chim. Acta* **2020**, *1134*, 150–173. [CrossRef] [PubMed]
129. Monago-Maraña, O.; Durán-Merás, I.; Muñoz de la Peña, A.; Galeano-Díaz, T. Analytical Techniques and Chemometrics Approaches in Authenticating and Identifying Adulteration of Paprika Powder Using Fingerprints: A Review. *Microchem. J.* **2022**, *178*, 107382. [CrossRef]
130. Shawky, E.; Abu El-Khair, R.A.; Selim, D.A. NIR Spectroscopy-Multivariate Analysis for Rapid Authentication, Detection and Quantification of Common Plant Adulterants in Saffron (*Crocus sativus* L.) Stigmas. *Lwt* **2020**, *122*, 109032. [CrossRef]
131. Rocchi, R.; Mascini, M.; Faberi, A.; Sergi, M.; Compagnone, D.; Di Martino, V.; Carradori, S.; Pittia, P. Comparison of IRMS, GC-MS and E-Nose Data for the Discrimination of Saffron Samples with Different Origin, Process and Age. *Food Control* **2019**, *106*, 106736. [CrossRef]
132. Zhao, M.; Wang, B.; Xiang, L.; Xiong, C.; Shi, Y.; Wu, L.; Meng, X.; Dong, G.; Xie, Y.; Sun, W. A Novel Onsite and Visual Molecular Technique to Authenticate Saffron (*Crocus sativus*) and Its Adulterants Based on Recombinase Polymerase Amplification. *Food Control* **2019**, *100*, 117–121. [CrossRef]
133. Karabagias, I.K.; Koutsoumpou, M.; Liakou, V.; Kontakos, S.; Kontominas, M.G. Characterization and Geographical Discrimination of Saffron from Greece, Spain, Iran, and Morocco Based on Volatile and Bioactivity Markers, Using Chemometrics. *Eur. Food Res. Technol.* **2017**, *243*, 1577–1591. [CrossRef]
134. Wakefield, J.; McComb, K.; Ehtesham, E.; Van Hale, R.; Barr, D.; Hoogewerff, J.; Frew, R. Chemical Profiling of Saffron for Authentication of Origin. *Food Control* **2019**, *106*. [CrossRef]
135. Wang, P.; Yu, Z. Species Authentication and Geographical Origin Discrimination of Herbal Medicines by near Infrared Spectroscopy: A Review. *J. Pharm. Anal.* **2015**, *5*, 277–284. [CrossRef] [PubMed]
136. Castro, R.C.; Ribeiro, D.S.M.; Santos, J.L.M.; Páscoa, R.N.M.J. Near Infrared Spectroscopy Coupled to MCR-ALS for the Identification and Quantification of Saffron Adulterants: Application to Complex Mixtures. *Food Control* **2021**, *123*, 107776. [CrossRef]
137. Ordoudi, S.A.; Staikidou, C.; Kyriakoudi, A.; Tsimidou, M.Z. A Stepwise Approach for the Detection of Carminic Acid in Saffron with Regard to Religious Food Certification. *Food Chem.* **2018**, *267*, 410–419. [CrossRef]
138. Bhooma, V.; Nagasathiya, K.; Vairamani, M.; Parani, M. Identification of Synthetic Dyes Magenta III (New Fuchsin) and Rhodamine B as Common Adulterants in Commercial Saffron. *Food Chem.* **2020**, *309*, 125793. [CrossRef]
139. Petrakis, E.A.; Cagliani, L.R.; Tarantilis, P.A.; Polissiou, M.G.; Consonni, R. Sudan Dyes in Adulterated Saffron (*Crocus sativus* L.): Identification and Quantification by 1H NMR. *Food Chem.* **2017**, *217*, 418–424. [CrossRef]
140. Lahmass, I.; Lamkami, T.; Delporte, C.; Sikdar, S.; Van Antwerpen, P.; Saalaoui, E.; Megalizzi, V. The Waste of Saffron Crop, a Cheap Source of Bioactive Compounds. *J. Funct. Foods* **2017**, *35*, 341–351. [CrossRef]
141. Li, J.; Wu, G.; Qin, C.; Chen, W.; Chen, G.; Wen, L. Structure Characterization and Otoprotective Effects of a New Endophytic Exopolysaccharide from Saffron. *Molecules* **2019**, *24*, 749. [CrossRef] [PubMed]
142. Righi, V.; Parenti, F.; Tugnoli, V.; Schenetti, L.; Mucci, A. *Crocus sativus* Petals: Waste or Valuable Resource? The Answer of High-Resolution and High-Resolution Magic Angle Spinning Nuclear Magnetic Resonance. *J. Agric. Food Chem.* **2015**, *63*, 8439–8444. [CrossRef]
143. Tirillini, B.; Pagiotti, R.; Menghini, L.; Miniati, E. The Volatile Organic Compounds from Tepals and Anthers of Saffron Flowers (*Crocus sativus* L.). *J. Essent. Oil Res.* **2006**, *18*, 298–300. [CrossRef]
144. Serrano-Díaz, J.; Sánchez, A.M.; Martínez-Tomé, M.; Winterhalter, P.; Alonso, G.L. Flavonoid Determination in the Quality Control of Floral Bioresidues from *Crocus sativus* L. *J. Agric. Food Chem.* **2014**, *62*, 3125–3133. [CrossRef]
145. Mottaghipisheh, J.; Mahmoodi Sourestani, M.; Kiss, T.; Horváth, A.; Tóth, B.; Ayanmanesh, M.; Khamushi, A.; Csupor, D. Comprehensive Chemotaxonomic Analysis of Saffron Crocus Tepal and Stamen Samples, as Raw Materials with Potential Antidepressant Activity. *J. Pharm. Biomed. Anal.* **2020**, *184*. [CrossRef]
146. Jadouali, S.M.; Atifi, H.; Mamouni, R.; Majourhat, K.; Bouzoubaâ, Z.; Laknifli, A.; Faouzi, A. Chemical Characterization and Antioxidant Compounds of Flower Parts of Moroccan *Crocus sativus* L. *J. Saudi Soc. Agric. Sci.* **2019**, *18*, 476–480. [CrossRef]
147. Tuberoso, C.I.G.; Rosa, A.; Montoro, P.; Fenu, M.A.; Pizza, C. Antioxidant Activity, Cytotoxic Activity and Metabolic Profiling of Juices Obtained from Saffron (*Crocus sativus* L.) Floral by-Products. *Food Chem.* **2016**, *199*, 18–27. [CrossRef]

148. Lotfi, L.; Kalbasi-Ashtari, A.; Hamedi, M.; Ghorbani, F. Effects of Enzymatic Extraction on Anthocyanins Yield of Saffron Tepals (*Crocus sativus*) along with Its Color Properties and Structural Stability. *J. Food Drug Anal.* **2015**, *23*, 210–218. [CrossRef]
149. Menghini, L.; Leporini, L.; Vecchiotti, G.; Locatelli, M.; Carradori, S.; Ferrante, C.; Zengin, G.; Recinella, L.; Chiavaroli, A.; Leone, S.; et al. *Crocus sativus* L. Stigmas and Byproducts: Qualitative Fingerprint, Antioxidant Potentials and Enzyme Inhibitory Activities. *Food Res. Int.* **2018**, *109*, 91–98. [CrossRef]

foods

MDPI

Article

Proposing Two Local Modeling Approaches for Discriminating PGI Sunite Lamb from Other Origins Using Stable Isotopes and Machine Learning

Ruting Zhao [1,2], Xiaoxia Liu [1,2], Jishi Wang [1,2], Yanyun Wang [1,2], Ai-Liang Chen [1,2], Yan Zhao [1,2] and Shuming Yang [1,2,*]

1 Institute of Quality Standard and Testing Technology for Agro-Products, Chinese Academy of Agricultural Sciences, Beijing 100081, China; zhaoruting1201@163.com (R.Z.); lxxiagood@163.com (X.L.); wjs2018@163.com (J.W.); 82101182143@caas.cn (Y.W.); ailiang.chen@gmail.com (A.-L.C.); zhaoyan01@caas.cn (Y.Z.)

2 Key Laboratory of Agro-Product Quality and Safety, Ministry of Agriculture, Beijing 100081, China

* Correspondence: yangshuming@caas.cn

Abstract: For the protection of Protected Geographical Indication (PGI) Sunite lamb, PGI Sunite lamb samples and lamb samples from two other banners in the Inner Mongolia autonomous region were distinguished by stable isotopes ($\delta^{13}C$, $\delta^{15}N$, $\delta^{2}H$, and $\delta^{18}O$) and two local modeling approaches. In terms of the main characteristics and predictive performance, local modeling was better than global modeling. The accuracies of five local models (LDA, RF, SVM, BPNN, and KNN) obtained by the Adaptive Kennard–Stone algorithm were 91.30%, 95.65%, 91.30%, 100%, and 91.30%, respectively. The accuracies of the five local models obtained by an approach of PCA–Full distance based on DD–SIMCA were 91.30%, 91.30%, 91.30%, 100%, and 95.65%, respectively. The accuracies of the five global models were 91.30%, 91.30%, 91.30%, 100%, and 91.30%, respectively. Stable isotope ratio analysis combined with local modeling can be used as an effective indicator for protecting PGI Sunite lamb.

Keywords: local modeling; protected geographical indication; Sunite lamb; stable isotopes; machine learning

Citation: Zhao, R.; Liu, X.; Wang, J.; Wang, Y.; Chen, A.-L.; Zhao, Y.; Yang, S. Proposing Two Local Modeling Approaches for Discriminating PGI Sunite Lamb from Other Origins Using Stable Isotopes and Machine Learning. *Foods* **2022**, *11*, 846. https://doi.org/10.3390/foods11060846

Academic Editor: Theodoros Varzakas

Received: 13 February 2022
Accepted: 14 March 2022
Published: 16 March 2022

1. Introduction

Sunite sheep were formed in the special ecological environment of Sunite grassland through long-term natural selection and artificial selection. They enjoy natural herbage and pure water in the ecological environment of natural grassland without pollution, and feed on more than 400 kinds of natural herbage, such as *Allium mongolicum regel*, *Allium polyrhizum turcz*, and *Stipa capillata*. It is this good ecological environment and primitive and extensive feeding mode that afford Sunite lamb with excellent quality and flavor [1,2]. Sunite lamb was awarded protected geographical indication (PGI) status in China in 2008. PGI Sunite lamb originates from the Sunite Right Banner and Sunite Left Banner, and it recognizes its high-quality reputation and characteristic flavors. Therefore, PGI labeling guarantees the origin and quality of food products, minimizing food safety risks, and ensures consumer confidence for the declaration of origin on this commodity [3].

In order to protect the PGI products, researchers have put forward a fingerprint tracing method, that is, using chemical parameters to build the fingerprint of geographical indication products [4] and comparing it with the fingerprint of the testing sample to determine whether the testing sample is the geographical indication product. At present, the chemical parameters used in the traceability of animal-origin food include stable isotopes [5], mineral elements [6], fatty acid content [7,8], amino acid content [9], and metabolites [10]. Stable isotopes are commonly used to characterize geographical origin information and to describe agricultural products' origin information, where δ^2H and $\delta^{18}O$ can be used to distinguish altitude, $\delta^{15}N$ can be used to determine the type of grazing vegetation, and $\delta^{13}C$ can determine the type of animal feed [11]. In addition, $\delta^{34}S$ is related to rainfall in the geological environment and traditional industrial emissions, so it indicates the geographical characteristics of animal food. Sr is obtained from the decay of ^{87}Rb, and its stable isotope abundance is mainly affected by geological conditions and rock ages. Sr has good applications in plant-derived-food tracing [12], but it is limited in tracing the origin of animal food due to its low content in animal bodies [13].

Thus, the stable isotope ratios can be used to distinguish PGI Sunite lamb from different origins. Stable isotopes have been applied to determine the origin of different animal-origin foods, such as beef [14], lamb [15,16], milk and dairy products [17], and marine products [18]. In 2007, Camin et al. [19] measured $\delta^{13}C$, $\delta^{15}N$, δ^2H, $\delta^{18}O$, and $\delta^{34}S$ in crude lamb protein from 13 European regions, and achieved correct classification rate of original grouping and cross validation of 78.7% and 77.6%, respectively. This indicated the feasibility of using stable isotopes to distinguish the geographical origin of lambs. However, the information of the samples Camin et al. [19] tested was complex, such as samples collected in different years from the same region, and samples collected in the same year for different feeding methods from the same region, meaning that the sample set covered a wide range of variations, which led to the model's lower predictive performance. On the other hand, the wide range of variations in the sample set may cover samples that will appear in the future; that is to say, it is conducive to improve the prediction ability of the model for unknown samples. Additionally, smaller sample difference coverage also leads to lower predictive performance. In the study by Sun et al. [5], the similarities of feed types, agricultural practice, and environment in two regions accounted for the overlapping of lamb samples from these two regions in the Inner Mongolia autonomous region. In subsequent studies, in order to improve the prediction ability of the geographical origin model, not only was the increase of chemical parameters considered, but also the coverage of sample differences.

In previous research on food traceability, the global modeling method was used to establish the discriminant model, that is, to create a model from all data sets that cover the whole space [20]. However, a good traceability model requires that the sample set should cover as wide a range as possible and avoid the appearance of samples with as similar chemical information as possible. In addition, the number of samples in the model should not be too large, so as to avoid the increase in interference information along with the increase in information, which will reduce the prediction performance of the model [21]. In fact, in the field of the near-infrared spectrum, scholars have focused on the coverage and representativeness of the sample set [22–24]. In a large sample set, there is a nonlinear relationship between response Y and all predictors X to varying degrees, and a sample set with a linear relationship can be obtained based on distance similarity. This is local modeling, where a set of local model data is created from all data sets according to certain rules, each covering a subspace [20]. Local modeling includes two rules; one is selecting the local model data set based on spatial similarity, and the other is selecting the most representative data subset based on the uniform design principle. Abhinav et al. [25] used the small spectral library obtained by a local modeling scheme based on spatial similarity to predict the soil property parameters of samples, which improved the prediction accuracy of soil properties compared with global modeling. This local modeling scheme referred to predicting the response of the samples by finding the most similar samples from existing databases. The similarity here was based on distance measures, such as the Euclidean distance, the covariance distance, the correlation distance, the surface difference spectrum, the information distance, optimized principal component Mahalanobis distance, and local linear embedding. Additionally, sampling representative samples can ensure that the chemical parameter characteristics and property range of the sample set can better cover the chemical parameter properties of unknown samples and improve the prediction ability of the unknown samples. In 2017, Palou et al. [26] proposed a strategy for calibration set selection of biodiesel/diesel samples based on principal component analysis (PCA) and the Kennard–Stones algorithm, and the results showed that, by using this methodology, the models could keep their robustness over time. In the future, local modeling should be more applied in the discrimination of the geographical origin of agricultural products.

In order to better discriminate PGI Sunite lamb from other origins using stable isotopes ($\delta^{13}C$, $\delta^{15}N$, $\delta^{2}H$, and $\delta^{18}O$) and machine learning, we proposed two local modeling approaches to optimize the sample set. It is worth mentioning that this is the first exploration of the protection of PGI Sunite lamb, and also a new application of local modeling in origin identification. The two local modeling approaches were (a) the Adaptive Kennard–Stone (AKS) algorithm and (b) an approach of PCA–Full Distance (FD) based on Data-Driven Soft Independent Modeling of Class Analogy (DD–SIMCA). The AKS algorithm was used to select the most representative data subset based on a uniform design principle, and the approach of PCA–FD based on DD–SIMCA was used to select the local model data set based on spatial similarity. It should be emphasized here that global modeling and local modeling in this study refer to the selection of data set coverage space, and the establishment of the discriminant model still depends on machine learning. The machine learning methods used in this work are linear discriminant analysis (LDA), random forests (RF), support vector machine (SVM), back-propagation neural network (BPNN), and k-nearest neighbor (KNN) classification. Based on the confusion matrix, we compared the predictive performance of the traceability models established by the five machine learning methods.

2. Proposed Two Local Modeling Approaches

2.1. Adaptive Kennard–Stone (AKS)

AKS is an adaptive sample selection method based on the Kennard–Stone algorithm, and its advantage is that it can determine the optimal sample set. The idea of AKS is to provide a uniform spatial design for the selection of the most representative samples from the known sample set. It ensures that the chemical parameters and property range of the sample set can better cover that of the unknown samples, and improves the prediction ability of the unknown samples. To our knowledge, there was only one report related to AKS application in the near-infrared spectrum [23]. At present, AKS has not been reported in the discrimination of the geographical origin of agricultural products, but Kennard–Stone (KS) has been reported [27].

The D-optimal criterion [28] was used as the criterion to select the samples. The minimum variance in the model could be achieved by selecting the right number of samples included in S that maximize $\log[\mathrm{Det}(\boldsymbol{M}_N)]$. The $\log[\mathrm{Det}(\boldsymbol{M}_N)]$ can be represented as the information of the selected sample set. The one we chose was the subset with the most information per sample, which was given by $\log[\mathrm{Det}(\boldsymbol{M}_N)]$, where $\boldsymbol{M}_N = \boldsymbol{S}^{\mathrm{T}}\boldsymbol{S}/N$, Det is the determinant of the matrix, and $\boldsymbol{S}$ is the principal component score matrix of the selected sample set. The number of principal components was *pc*. Figure 1 shows the steps for obtaining the optimal sample set [23].

2.2. An Approach of PCA–Full Distance (FD) Based on Data-Driven Soft Independent Modeling of Class Analogy (DD–SIMCA)

The measurement of similarity used to be based on Euclidean distance, Mahalanobis distance, principal component analysis Euclidean distance (PCA–ED), and principal component analysis Mahalanobis distance (PCA–MD). Both distances can be calculated in the original variable space and in the principal component space. In the principal component space, the correlation between variables is eliminated, simplifying the data information and making it superior to the original variable space. The Euclidean distance is the straight line distance between two points, and it is affected by the data distribution, noise, and characteristic metrics. Unlike the Euclidean distance, the Mahalanobis distance introduces a covariance matrix, and implements coordinate rotation and data compression, which makes Mahalanobis distance not affected by data distribution and feature dimensions [29]. However, these distance threshold choices were hard, often selected several times, and then compared the predictive ability of the model of the selected multiple sample sets. It took time and effort to achieve this, but the best sample set might not be found. Moreover, the whole process could not be visualized, making it harder to understand. Based on the understanding of DD–SIMCA, we found that PCA–FD integrated the advantages of MD and ED, the operation process was simple, and the results were visible. Therefore, we proposed an approach of PCA–FD based on DD–SIMCA. As far as we know, an approach of PCA–FD based on DD–SIMCA to screen the samples has not been reported.

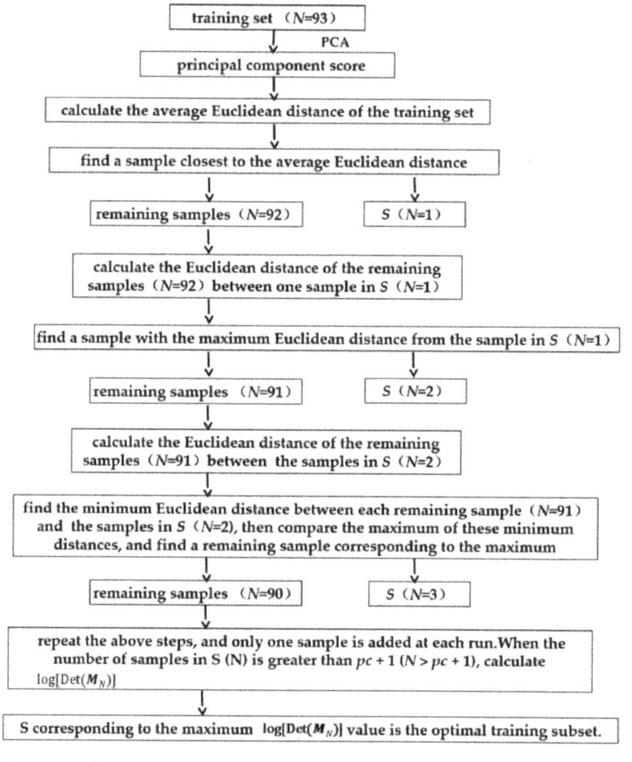

Figure 1. The steps of the AKS algorithm.

Each element of the data cloud can be presented as a sum of two vectors: a vector that lies in the subspace (a projection) and a vector transversal to the hyperplane (a residual). The lengths of these vectors are important indicators that characterize a sample position with respect to the subspace (model). These statistics are often referred to as the leverage and the residual variance. In DD–SIMCA, they are termed as the score distance (SD) and the orthogonal distance (OD) correspondingly, which are used to define the critical limits of the classification model [30]. SD is equal to the squared Mahalanobis distance from the model center to one sample within the score subspace, and OD is the squared Euclidean distance from one sample to the model subspace. FD is affected by parameters related to SD and OD (See Formula (1)), as shown below.

$$FD = \frac{Nh \times SD}{SD0} + \frac{Nq \times OD}{OD0} \tag{1}$$

where Nh and Nq are the degrees of freedom (DOF) for SD and OD, and SD0 and OD0 are the means of SD and OD of all the samples [31], respectively. PCA–FD integrated the advantages of MD and ED and eliminated the effects of data distribution and feature dimensions. Moreover, the approach of PCA–FD based on DD–SIMCA simplified the operation and could be visualized. As shown in Figure S1, the abscissa and ordinate of the acceptance plot were the parameters associated with SD and OD, respectively, and the boundary lines of regulars and outliers are given. The red line and the green line were available on the figure, and the yellow line was added later. The red line is the boundary of the outliers, and samples above the bounds are outliers; the green line is the boundary of regulars, and samples below the bounds are regulars; the points in the middle area of the green line and the red line are extremes; the yellow line is FD. The FD on the same line is the same, and the larger the FD, the farther the yellow line is away from the base point. You can obtain all samples inside the strip centered on the testing sample (a black triangular), such as a custom bound of

$$FD_1 < FD_{\text{one testing sample}} < FD_2 \tag{2}$$

where the border values, FD_1 and FD_2, have certain rules. Samples between the two yellow lines are the screened samples.

As a measure of similarity, FD was used to select samples similar to one testing sample from the original sample set to solve the nonlinear problem of large-sample data modeling. This approach greatly simplifies the threshold selection process, and part of the process is visualized.

3. Predictive Performance of the Model

The predictive performance of the model in our work based on the confusion matrix includes the sensitivity, specificity, accuracy, and kappa coefficient (these measures were calculated for each method based on the test data set). The confusion matrix summarizes the results of a classification method. For a binary classification, when we determine that class 1 is positive, the schematic table of the confusion matrix is shown in Table 1.

Table 1. The schematic table of the confusion matrix.

Confusion Matrix		**Predicted Class** **Class 1**	**Class 2**
Actual class	Class 1	True positive (TP)	False negative (FN)
	Class 2	False positive (FP)	True negative (TN)

In this study, instead of negative and positive, the classes were "non-PGI lamb" and "PGI Sunite lamb", respectively. For example, TN is the number of non-PGI lamb in the test data set correctly classified as non-PGI, and FN is the number of PGI Sunite lamb incorrectly classified as non-PGI. The sensitivity, specificity, accuracy, and kappa coefficient are defined as follows:

$$sensitivity = \frac{TP}{TP + FN} \tag{3}$$

$$specificity = \frac{TN}{TN + FP} \tag{4}$$

$$accuracy = \frac{TP + TN}{TP + TN + FP + FN} \tag{5}$$

$$kappa = \frac{P0 - Pe}{1 - Pe} \tag{6}$$

where P0 (P0 = accuracy) indicates the accuracy of the model, and Pe ($Pe = \frac{(TP+FN)\times(TP+FP)+(TN+FN)\times(TN+FP)}{(TP+TN+FP+FN)^2}$) is the expected proportion of lamb correctly classified by chance.

Sensitivity is the proportion of actual PGI Sunite lamb that is correctly classified as PGI Sunite lamb. Specificity is the proportion of actual non-PGI lamb that is correctly classified as non-PGI lamb. Accuracy is the ratio of true positive and true negative samples to the total number of testing samples, which reflects the overall accuracy. If the proportion of one class of samples is not dominant in all classes of samples, then its high error rate has little influence on the accuracy. In this case, the accuracy does not carry much meaning, and the Kappa coefficient can better reflect the discrimination effect. The Kappa coefficient can better reflect the consistency of actual classification and predict classification. The evaluation result of the Kappa coefficient is divided into three grades [32]: excellent (Kappa > 0.75), good (0.40 < Kappa ≤ 0.75), and poor (Kappa ≤ 0.40). As long as the accuracy of one class is low, the Kappa coefficient will decrease.

4. Materials and Methods

4.1. Materials

Lamb samples (n = 116) were collected from 4 banners in two cities of China's Inner Mongolia autonomous region (Table S1), where the Sunite Right Banner and Sunite Left Banner are the specified regions of PGI Sunite lamb, located in Xilin Gol League; Abaga Banner also belongs to Xilin Gol League, east of Sunite Left Banner; Siziwang Banner belongs to Ulanqab City, west of Sunite Right Banner (Figure 2). The lamb samples from each banner came from the same abattoir and were collected from the right hind leg. The samples were from 5–8-month-old grazing sheep. The fresh mutton (50 g) was dried to a constant weight and then pulverized through a 100 mesh. The sample was mixed with a chloroform/methanol (2:1, v/v) solution at 1:5, vortexed for 10 min, and centrifuged at 5000 rpm for 5 min, and the supernatant was discarded [33]. Then, the previous degreasing step was repeated twice, the supernatant was discarded, and the solid was retained and lyophilized to obtain a defatted dry matter (DDM) for the determination of stable isotopes. These samples were stored at −20 °C for subsequent analysis.

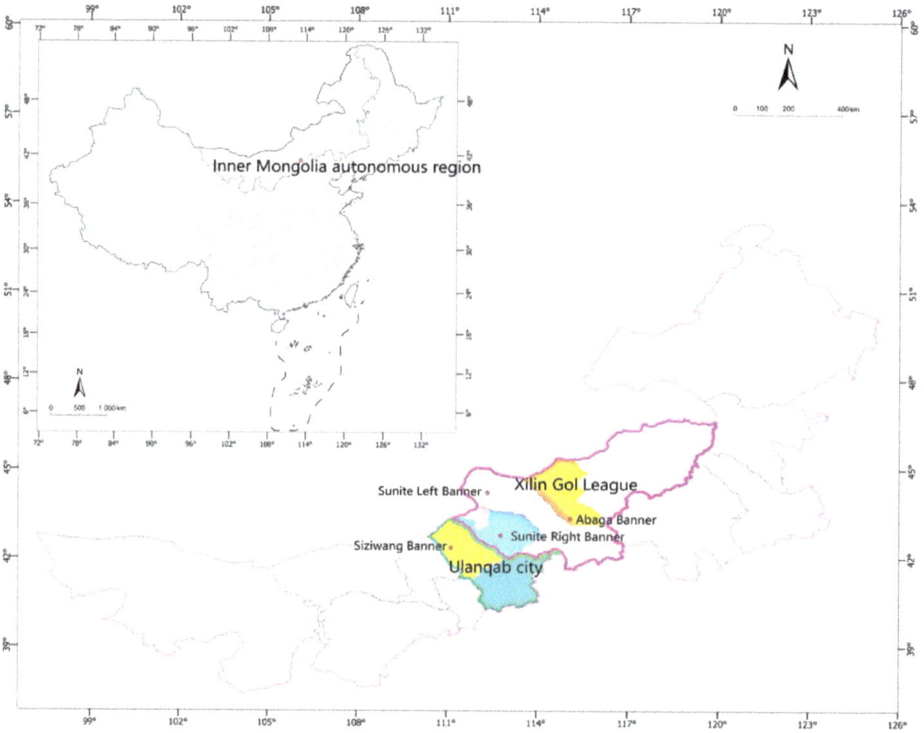

Figure 2. Regional location information for the lamb samples.

4.2. Stable Isotope Analysis

For the stable isotope analysis of $\delta^{13}C$ and $\delta^{15}N$, DDM and other international reference materials (USGS40, USGS43, and USGS62) were weighed into tin capsules (5 × 8 mm) and then introduced into an elemental analyzer (Flash 2000, Thermo, Waltham, MA, USA), converting the entire material into carbon dioxide and nitrogen gas analyzed by an isotope ratio mass spectrometer (Delta V Advantage of Thermo, Waltham, MA, USA). The calibration of $\delta^{13}C$ and $\delta^{15}N$ was analyzed with USGS40 ($\delta^{13}C = -26.39‰$, $\delta^{15}N = -4.5‰$ air N2), USGS43 (Indian Hair, $\delta^{13}C = -21.28‰$, $\delta^{15}N = 8.44‰$), and USGS62 (caffeine, $\delta^{13}C = -14.79‰$, $\delta^{15}N = 20.17‰$).

For the stable isotope ratio analysis of δ^2H and $\delta^{18}O$, DDM and international reference materials (Caribou Hoof, Kudu Horn, and EMA P2) were weighed into silver capsules (4 × 6 mm) along with other international reference materials and introduced into the elemental analyzers (Flash 2000, Thermo, Waltham, MA, USA). The reactor packing was a glassy carbon reactor and silver wool. The elements hydrogen and oxygen in the samples were converted into H_2 and CO at 1380 °C via pyrolysis with glass carbon. The gas was transferred to an isotope ratio mass spectrometer (Delta V Advantage, Thermo, Waltham, MA, USA). The calibration of δ2H and δ18O was analyzed with CBS (Caribou Hoof Standard, $\delta^2H = -197.00‰$, $\delta^{18}O = 3.80‰$), KHS (Kudu Horn Standard, $\delta^2H = -54.10‰$, $\delta^{18}O = 20.3‰$), and B2205 (EMA P2, $\delta^2H = -87.80‰$, $\delta^{18}O = 26.90‰$).

The results of the isotope analysis were expressed as δ (‰), and the formula was

$$\delta\ (‰) = \frac{Rsample - Rstandard}{Rsample} \times 1000 \tag{7}$$

where R sample and R standard are the isotope ratios of the sample and the international reference material, respectively. The references of $\delta^{13}C$, $\delta^{15}N$, δ^2H, and $\delta^{18}O$ were Vienna–Pee Dee Belemnite (V–PDB), Air, Standard Mean Ocean Water (SMOW), and SMOW, respectively.

4.3. Statistical Analysis

All of the samples ($N = 116$) were divided into a training set and a testing set (4:1). Due to the uneven sample size in the four regions, stratified random sampling was adopted to avoid contingency, and samples in each region were divided into 4:1. The training set samples ($N = 93$) were used for modeling, and the testing set samples ($N = 23$) were used to evaluate the prediction ability of the model. The training set data were imported into R Studio and the training set subset ($N < 93$) was obtained by the AKS algorithm. After that, the training set subset used five machine learning methods (LDA, RF, SVM, BPNN, and KNN) to establish the geographical origin discriminant model, and finally used the confusion matrix of testing set samples to evaluate the predictive performance of the model.

One sample (called Pi, $i = 1, 2, 3, \ldots, 23$) in the testing set ($N = 23$) and the training set samples ($N = 93$) was imported into DD–SIMCA in Microsoft Excel (SIMCA template-xlsb) to obtain the FD of all samples ($N = 93$), and the training subset was appropriately selected centering on the FD of Pi. Then, import the training set subset into R Studio and use the 5 machine learning methods to build a one-time local model for Pi. Repeat the above operation 23 times, and obtain 5 confusion matrices. The evaluation method of the model was consistent with the evaluation method of the training set subset obtained by the AKS algorithm.

All of the training set samples ($N = 93$) were imported into R Studio, and 5 machine learning methods were used to establish the geographical origin discriminant model. The evaluation method of the model was consistent with the evaluation method of the training subset obtained by the AKS algorithm.

In order to compare the changes before and after screening the training set samples ($N = 93$), we analyzed the main characteristics of the global lamb isotope libraries and local lamb isotope libraries obtained by the AKS algorithm. SPSS was used to conduct an independent-samples T-test to analyze the significance between the two groups (PGI Sunite lamb and non-PGI lamb), produce box diagrams to intuitively see the significance, conduct exploratory analysis (mean value, standard deviation, and histogram), and produce the 3-dimensional scatter plot to observe the spatial distribution of the sample set. Furthermore, to know the difference of lamb between different regions, we performed a descriptive analysis of all data (N = 116) (Table S2). Additionally, we drew a 3D–score plot of the global lamb isotope library and local lamb isotope library according to geographical origin (Figure S2).

The statistical software packages R, SPSS 25.0 (SPSS Inc., Chicago, IL, USA), and a chemometric tool employed in Excel were used. AKS was written by our laboratory using R language.

5. Results and Discussions

5.1. Training Subset Obtained by Two Local Modeling Approaches

In this work, the training subset was obtained by two local modeling approaches: (a) AKS and (b) the approach of PCA–FD based on DD–SIMCA. Based on the AKS algorithm, a line chart (Figure S3) was drawn using the $\log[\mathrm{Det}(\boldsymbol{M}_N)]$ value as the ordinate and number of samples as the abscissa. As shown in Figure S3, when the number of samples was 40, the maximum $\log[\mathrm{Det}(\boldsymbol{M}_N)]$ value appeared, and the corresponding S subset was the best training set subset. Chen et al. [23] used AKS to screen the near-infrared spectrum library of plant alkali, and 49 samples were selected from 85 samples for constructing the PLS model. The sample size was also half of the original data. However, Chen et al. [23] continued to sample the near-infrared spectrum library of the aqueous solution and selected 37 samples out of 38 samples to construct the PLS model [23]. This shows that the capacity of the training subset was not related to the capacity of the training set, but was only related to the information contained in the training subset, namely the $\log[\mathrm{Det}(\boldsymbol{M}_N)]$ value. When the maximum $\log[\mathrm{Det}(\boldsymbol{M}_N)]$ value is not reached, the $\log[\mathrm{Det}(\boldsymbol{M}_N)]$ value increases with the increase in the number of samples. When the maximum $\log[\mathrm{Det}(\boldsymbol{M}_N)]$ value is reached, the $\log[\mathrm{Det}(\boldsymbol{M}_N)]$ value decreases as the sample size increases.

According to the approach of PCA–FD based on DD–SIMCA, 23 targeted training subsets were obtained with a sample size between 20 and 47. When screening data, we found that, when the FD of Pi (Figure S1a) deviated from the central position of FD of all data (N = 93), the discriminant effect of the model established by the data set with a small sample size ($20 \leq N < 35$) was better, and that when the FD of Pi (Figure S1b) close to the central position of FD of all data, the discriminant effect of the model established by the data set with a medium sample size ($35 \leq N < 50$) was better. This may be related to the principle of screening. This method selected samples within the linear range of Pi for modeling based on similarity [22]. The linear range was probably related to the position of the FD of Pi in the FD of all of the data. In the future, this finding will continue to be verified in order to summarize the screening rules.

To sum up, during sample screening, it is necessary to follow the screening principles and consider the data characteristics to select data with an appropriate sample size. In this work, with two local modeling methods, half or less of the original sample size could be used to obtain the same model effect as the original data.

5.2. Main Characteristics of the Lamb Isotope Libraries

Taking the training set subset (N = 40) obtained by AKS and the training set (N = 93) as an example, the main characteristics of the local and global lamb isotope libraries were compared. Table 2 lists the mean, the standard deviation, and the ranges spanned by the samples and Figure 3 shows the corresponding distribution histograms. For the mean of $\delta^{13}C$, $\delta^{2}H$, and $\delta^{18}O$, the local lamb isotope library was smaller than the global lamb isotope library. For the standard deviation of $\delta^{13}C$, $\delta^{15}N$, $\delta^{2}H$, and $\delta^{18}O$, the local lamb isotope library was larger than the global lamb isotope library. The mean reflected the overall average and the degree of data concentration, while the standard deviation reflected the degree of data dispersion. This meant that the local lamb isotope library was more centralized and more dispersed, and was the ideal training set. The histogram (Figure 3) also supported this conclusion. Through the histogram, we could see the data distribution of the local and global lamb isotope libraries more intuitively. The distribution of the global lamb isotope libraries was not uniform, and some data were abrupt in the histogram, which had a great influence on the main characteristics of the whole data set. The local lamb isotope libraries weakened the influence of prominent data and better reflected the main characteristics of the overall data.

Table 2. (**a**) Descriptive statistics of the isotope attributes of samples in the lamb isotope libraries. (**b**) $\delta^{13}C$, $\delta^{15}N$, $\delta^{2}H$, and $\delta^{18}O$ values of the local and global lamb isotopes libraries from two groups.

(a)

Parameter	Mean	Standard Deviation	Minimum	Maximum
		Global lamb isotope library (Training set, n = 93)		
$\delta^{13}C$	−19.87	2.10	−24.69	−17.16
$\delta^{15}N$	7.10	0.95	5.58	8.88
$\delta^{2}H$	−103.90	11.14	−131.87	−93.77
$\delta^{18}O$	12.55	3.08	4.92	17.98
		Local lamb isotope library (Training subset by AKS, n = 40)		
$\delta^{13}C$	−21.24	2.21	−24.69	−17.26
$\delta^{15}N$	7.16	1.01	5.58	8.88
$\delta^{2}H$	−110.85	14.11	−131.87	−95.44
$\delta^{18}O$	11.15	3.90	4.92	16.56

(b)

Parameter	$\delta^{13}C$	$\delta^{15}N$	$\delta^{2}H$	$\delta^{18}O$
		Global lamb isotope library (Training set, n = 93)		
PGI Sunite lamb	-19.31 ± 1.52^{a}	7.28 ± 0.94^{a}	-99.48 ± 5.44^{a}	13.65 ± 1.90^{a}
non-PGI lamb	-21.92 ± 2.65^{b}	6.45 ± 0.70^{b}	-120.02 ± 11.84^{b}	8.53 ± 3.28^{b}
		Local lamb isotope library (Training subset by AKS, n = 40)		
PGI Sunite lamb	-20.57 ± 1.44^{a}	7.88 ± 0.73^{a}	-101.68 ± 9.58^{a}	13.78 ± 2.43^{a}
non-PGI lamb	-21.92 ± 2.65^{b}	6.45 ± 0.70^{b}	-120.02 ± 11.84^{b}	8.53 ± 3.28^{b}

Note: The values are given as mean ± SD; the small letters represent significant differences ($p < 0.05$); the sample sizes of Sunite Right Banner, Sunite Left Banner, Siziwang Banner, and Abaga Banner in the global lamb isotope and local lamb isotope libraries were 68, 5, 15, and 5, and 15, 5, 15, and 5, respectively.

In this work, we paid more attention to the influence of screening on the two categories, rather than the overall data. The significance of isotopes has an impact on the accuracy of geographical origin discrimination and traceability feasibility [34], so we compared the significance of the $\delta^{13}C$, $\delta^{15}N$, $\delta^{2}H$, and $\delta^{18}O$ values between PGI Sunite lamb and non-PGI lamb in the local and global lamb isotope libraries. The result of the T-test showed that there were both significant differences ($p < 0.05$) in the $\delta^{13}C$, $\delta^{15}N$, $\delta^{2}H$, and $\delta^{18}O$ values between PGI Sunite lamb and non-PGI lamb before and after sample screening. This meant that PGI Sunite lamb and non-PGI lamb always had a characteristic stable isotope ratio profile. Figure 4a,b show the corresponding boxplots, and consistent conclusions could be drawn. In the local and global lamb isotope libraries, PGI Sunite lamb samples exhibited the highest $\delta^{13}C$, $\delta^{15}N$, $\delta^{2}H$, and $\delta^{18}O$ values. The regional disparity of the $\delta^{13}C$ and $\delta^{15}N$ of lamb samples was a consequence of the feeding systems [35]. The $\delta^{13}C$ value in animal products was based on C_3 and C_4 plants in the animal diet. One study showed that the $\delta^{13}C$ value of C_3 plants ranged from −20‰ to −35‰, and that the $\delta^{13}C$ value of C_4 plants ranged between −9‰ and −17‰ [36]. In this work, the lamb samples were grazing sheep. We could predict that the proportion of C_3 and C_4 plants fed to PGI Sunite lamb was higher than that fed to non-PGI lamb. In another aspect, the value of $\delta^{15}N$ reflects the nitrogen cycle in soil. Compared with other C_3 plants, leguminous plants can directly utilize atmospheric nitrogen, resulting in a lower $\delta^{15}N$ value [37]. Generally, leguminous plants, such as alfalfa hay, are abundant at high altitudes, which could be the cause of lower $\delta^{15}N$ value of lambs (non-PGI lamb) from high-altitude regions, such Siziwang Banner (Table S2). The values of $\delta^{2}H$ and $\delta^{18}O$ reflect the geographical information of lamb, such as altitude. In the atmospheric circulation process, the higher the altitude, the lower the enrichment degree of ^2H_2O, and the δ^2H and $\delta^{18}O$ values in the high-altitude region are lower than those in the low-altitude region [38]. The best examples are Sunite Left Banner and Siziwang Banner in Table S2. The altitude of Sunite Left Banner is higher than that of the four sons king flag (Table 2), and the δ^2H and $\delta^{18}O$ in the lamb of Sunite Left Banner are significantly lower than those in the lamb of Siziwang Banner. In addition, the values of $\delta^{13}C$, $\delta^{15}N$, δ^2H, and $\delta^{18}O$ are affected by objective factors, such as rainfall, temperature, and geology.

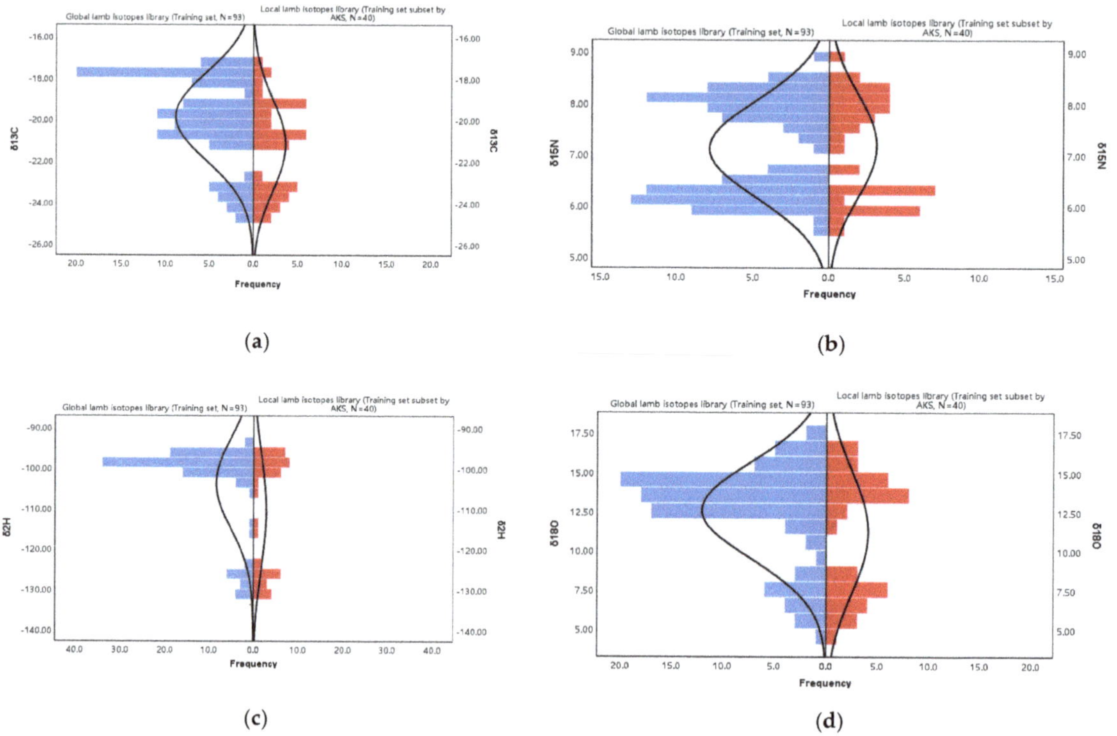

Figure 3. Histograms of the $\delta^{13}C$ (**a**), $\delta^{15}N$ (**b**), δ^2H (**c**), and $\delta^{18}O$ (**d**) values of the global lamb isotope library and local lamb isotope library.

We carried out descriptive analysis and independent-samples T-tests on the local and global lamb isotope libraries above, and concluded that sample screening can optimize the sample data. Now, we can more intuitively determine the data spatial distribution and whether the data are representative through the 3D–score plot. Additionally, we can also determine the contribution of the $\delta^{13}C$, $\delta^{15}N$, δ^2H, and $\delta^{18}O$ values to PC1 and PC2 (Figure S4). In the 3D–score plot of the global lamb isotope library (Figure 4c), PGI Sunite lamb and non-PGI lamb samples overlapped and were difficult to distinguish. From Figure S2a, the lamb samples from Abaga Banner adjacent to the PGI area had a serious overlap with PGI Sunite lamb. However, in the 3D–score plot of the local lamb isotope library (Figure 4d), PGI Sunite lamb samples were entirely separated from the non-PGI lamb samples. Additionally, the samples from the four regions were completely separated (Figure S2b). After sample screening, the spatial distribution was uniform and the samples were representative. These results provide strong evidence that the local lamb isotope libraries were superior to the global lamb isotope libraries.

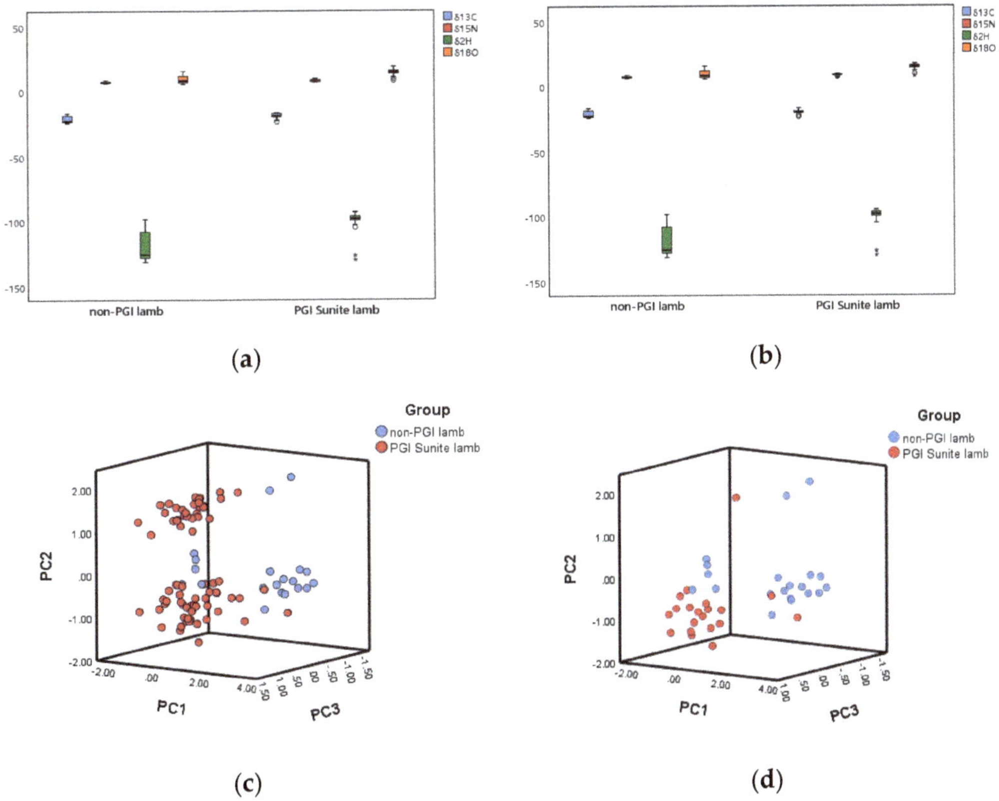

Figure 4. Boxplots of the $\delta^{13}C$, $\delta^{15}N$, $\delta^{2}H$, and $\delta^{18}O$ values of (**a**) global lamb isotope library and (**b**) local lamb isotope library according to lamb groups; 3D–score plot of (**c**) global lamb isotope library and (**d**) local lamb isotope library according to lamb groups.

5.3. Predictive Performance

According to the above data feature analysis, the local lamb isotope libraries were better than the global lamb isotope libraries. The significance analysis and 3D–score plot showed that it was feasible to use isotopes to discriminate PGI Sunite lamb from non-PGI lamb. To further compare the local and global lamb isotope libraries, machine learning was used for modeling, and some indicators of predictive performance were used to evaluate the effect of the model. For the same library, five machine learning methods were used to ensure model stability.

The origin classification results of applying the five models to the testing set lambs are shown in Table 3, together with the evaluation of prediction performance. The evaluation of each binary discriminant model was built from the confusion matrix, the records of which correctly and incorrectly recognized samples from different geographical origins. True positives were samples of PGI Sunite lamb correctly predicted, false negatives were samples of PGI Sunite lamb incorrectly predicted to be from non-PGI lamb, true negatives were samples of non-PGI lamb correctly predicted to be from non-PGI lamb, and false positives were samples of non-PGI lamb incorrectly predicted to be from PGI Sunite lamb. The evaluation of the whole model was calculated as a two-class overall classification.

Table 3. Origin classification results of applying the 5 models to the testing set lambs according to (a) the global lamb isotopes libraries, (b) the local lamb isotopes libraries screened by AKS, and (c) the local lamb isotopes libraries screened by the approach of PCA–FD based on DD–SIMCA.

(a)	Binary Discrimination Classes				
	LDA	**RF**	**SVM**	**BPNN**	**KNN**
Confusion matrix (No. of testing set samples)					
True positive (tpi)	17	17	17	18	17
False negative (fni)	1	1	1	0	1
True negative (tni)	4	4	4	5	4
False positive (fpi)	1	1	1	0	1
Performance evaluation					
Sensitivity	0.9444	0.9444	0.9444	1.0000	0.9444
Specificity	0.8000	0.8000	0.8000	1.0000	0.8000
Kappa	0.7444	0.7444	0.7444	1.0000	0.7444
Accuracy	0.9130	0.9130	0.9130	1.0000	0.9130
(b)	**Binary Discrimination Classes**				
	LDA	**RF**	**SVM**	**BPNN**	**KNN**
Confusion matrix (No. of testing set samples)					
True positive (tpi)	17	17	17	18	17
False negative (fni)	1	1	1	0	1
True negative (tni)	4	5	4	5	4
False positive (fpi)	1	0	1	0	1
Performance evaluation					
Sensitivity	0.9444	0.9444	0.9444	1.0000	0.9444
Specificity	0.8000	1.0000	0.8000	1.0000	0.8000
Kappa	0.7444	0.8808	0.7444	1.0000	0.7444
Accuracy	0.9130	0.9565	0.9130	1.0000	0.9130
(c)	**Binary Discrimination Classes**				
	LDA	**RF**	**SVM**	**BPNN**	**KNN**
Confusion matrix (No. of testing set samples)					
True positive (tpi)	17	17	17	18	17
False negative (fni)	1	1	1	0	1
True negative (tni)	4	4	4	5	5
False positive (fpi)	1	1	1	0	0
Performance evaluation					
Sensitivity	0.9444	0.9444	0.9444	1.0000	0.9444
Specificity	0.8000	0.8000	0.8000	1.0000	1.0000
Kappa	0.7444	0.7444	0.7444	1.0000	0.8808
Accuracy	0.9130	0.9130	0.9130	1.0000	0.9565

As shown in Table 3, in the global lamb isotope libraries, the five models established by machine learning achieved a good discrimination effect. Among them, the confusion matrix of models established by LDA, RF, SVM, and KNN was all one false negative and one false positive; that is to say, a sample of PGI Sunite lamb was incorrectly predicted to be from non-PGI lamb, and a sample of non-PGI lamb was incorrectly predicted to be from PGI Sunite lamb. The sensitivity and specificity of the above models were 94.44% and 80.00%, respectively. Additionally, the accuracy of the model was 91.30%, a satisfactory result for the overall classification. The Kappa coefficient was 0.7444 (0.40 < Kappa ≤ 0.75), a good consistency of actual classification and predict classification. On the other hand, the confusion matrix of the BPNN model was true positives and true negatives, and all the classes were correctly discriminated. This indicates that the five origin models established based on global lamb isotope libraries were stable, and the BPNN model had the best predictive performance.

In Table 3, using local lamb isotope libraries screened by AKS, the five models established by machine learning had a better discriminating effect, in which the order of predictive performance was LDA, SVM, and KNN < RF < BPNN. It was the same as the predictive performance of four models (LDA, SVM, BPNN, and KNN) based on the local and global lamb isotope libraries. Compared to global modeling, locally modeled RF models are superior to globally modeled RF models. In the RF model, only one PGI Sunite lamb was incorrectly predicted to be from non-PGI lamb, and all of the non-PGI lamb samples were identified as non-PGI lamb (specificity = 100.00%). The accuracy of the RF model was 95.65%, a very satisfactory overall classification result. Additionally, the Kappa coefficient was 0.8808 (Kappa > 0.75), indicating excellent consistency of actual classification and prediction classification. In other words, the local lamb isotope libraries obtained by AKS were better than the global lamb isotope libraries. The KS algorithm has also been applied to Protected Designation of Origin (PDO) cheeses recently, and good results were obtained. In 2021, Coppa et al. found that mid-infrared spectroscopy (MIR) enables the authentication of the cow feeding restrictions included in the specification of two PDO cheeses (Cantal and Laguiole). The classification result of the testing sample showed that the accuracy, sensitivity, and specificity of Cantal PDO cheeses were 90.3%, 91.1%, and 89.2% respectively; and the predictive performances of the model for Laguiole PDO cheeses were 99.5%, 100%, and 99.4%, which all outperformed the AKS modeling effect in this paper [27]. However, it must be said that Coppa et al. used the KS algorithm to select training sets and testing sets. We consider it inappropriate to select the testing sets, because it changes the true distribution of the sample, which may be the reason for the over-good classification results.

There was a better discrimination effect of five models using local lamb isotope libraries screened by the approach of PCA–FD based on DD–SIMCA (Table 3), in which the order of predictive performance was LDA, RF, and SVM < KNN < BPNN. It was the same as the predictive performance of four models (LDA, RF, SVM, and BPNN) based on local and global lamb isotope libraries. Compared to global modeling, locally modeled KNN models were superior to globally modeled RF models. In the KNN model, only one PGI Sunite lamb was incorrectly predicted to be from non-PGI lamb, and all of the non-PGI lamb samples were identified as non-PGI lamb (specificity = 100.00%). The accuracy of the KNN model was 95.65%, a very satisfactory result overall classification. Additionally, the Kappa coefficient was 0.8808 (Kappa > 0.75), indicating excellent consistency of actual classification and prediction classification. In other words, the local lamb isotope libraries obtained by the approach of PCA–FD based on DD–SIMCA were better than the global lamb isotope libraries.

The differences in the isotope profiles of the lamb's geographical origins allowed satisfactory discrimination between them, but were not sufficiently wide and systematic to be validated by adding an external set sample to the classification model. As shown in Figure 2, the lamb samples collected were PGI Sunite lambs and non-PGI lambs in their adjacent origins; that is, lambs at municipal geographical distance from the PGI Sunite lamb and lambs at banner geographical distance from the PGI Sunite lamb. This is because the geographical information difference of lambs at the provincial level and above is large and easy to distinguish [5]. Thus, this study pays more attention to the identification of lambs at municipal/banner/county geographical distances. After that, lambs from other provinces and countries would be added to enrich the sample library so that the sample library could cover as large sample differences as possible, such as geographical origin, feeding type, breed, age, and gender differences, and the external samples were verified.

To sum up, the local lamb isotope libraries obtained by AKS and the approach of PCA–FD based on DD–SIMCA were better than the global lamb isotope libraries.

6. Conclusions

In this work, stable isotope ratio ($\delta^{13}C$, $\delta^{15}N$, $\delta^{2}H$, and $\delta^{18}O$ values) analysis combined with local modeling was used to discriminate PGI Sunite lamb from other origins, and the accuracy rate reached 100%, which could be used as an effective indicator system for protecting PGI Sunite lamb. A good traceability model requires that the sample set should cover as wide a range as possible and avoid the appearance of samples with basically the same chemical information as much as possible. Therefore, local modeling is very necessary for the traceability of agricultural products, but it has not been reported. In this paper, two local modeling approaches were first proposed for the protection of PGI Sunite lamb, and the identification effect of models was better than that of global modeling, which could be used for the optimization of the training set and the traceability of agricultural products. We found that the sample set with less than or equal to half of the original sample size in this study could achieve a better predictive effect. However, the ratio of the screened sample size to the original sample size will not always be 1:2, and the screened sample size is related to the information contained. The information (geographical origin, feeding system, age, and gender) of the lamb samples collected in this paper was similar. This may account for the small changes in the data characteristics before and after screening. In the future, while increasing the sample size, we will try our best to make the sample set cover a wide range of differences, such as geographical origin, feeding system, breed, age, and gender. At that time, there will be a more obvious nonlinear relationship between the classification response and the isotope ratio, and the application of a local modeling method is more necessary.

Supplementary Materials: The following are available online at https://www.mdpi.com/article/10.3390/foods11060846/s1, Figure S1: Schematic diagram of the screening process about the approach of PCA–FD based on DD–SIMCA, Figure S2: 3D–score plot of (**a**) global lamb isotope library and (**b**) local lamb isotope library according to geographical origin, Figure S3: Line chart of the relationship between the sample number of the training subset and log[Det(*MN*)] value, Figure S4: Bi-plots of (**a**) global lamb isotope library and (**b**) local lamb isotope library according to geographical origin, Table S1: Region information of lamb samples, Table S2: $\delta^{13}C$, $\delta^{15}N$, $\delta^{2}H$ and $\delta^{18}O$ values of all lambs (n = 116) from four regions.

Author Contributions: Conceptualization, Y.Z., A.-L.C. and S.Y.; resources, S.Y.; formal analysis, R.Z.; writing—original draft preparation, R.Z.; writing—review and editing, R.Z.; investigation, X.L., J.W. and Y.W. All authors have read and agreed to the published version of the manuscript.

Funding: This work was funded by the National Key Research and Development Program (2017YFE0114400).

Institutional Review Board Statement: Not applicable.

Informed Consent Statement: Not applicable.

Data Availability Statement: Data are contained within the article or supplementary material.

Acknowledgments: The authors are very grateful to the National Key Research and Development Program for funding this study.

Conflicts of Interest: The authors declare no conflict of interest.

References

1. Luo, Y.L.; Wang, B.H.; Jin, Z.M.; Liu, X.W.; Jin, Y. Effects of two feeding conditions on nutritional quality of Sunite sheep meat. *Food Sci.* **2016**, *37*, 227–231. [CrossRef]
2. Su, R.; Luo, Y.L.; Wang, B.H.; Hou, Y.R.; Zhao, L.H.; Su, L.; Yao, D.; Qian, Y.; Jin, Y. Effects of physical exercise on meat quality characteristics of Sunite sheep. *Small Rumin. Res.* **2019**, *183*, 106023. [CrossRef]
3. Nie, J.; Shao, S.Z.; Zhang, Y.Z.; Li, C.L.; Liu, Z.; Rogers, K.M.; Wu, M.C.; Lee, C.P.; Yuan, Y.W. Discriminating protected geographical indication Chinese Jinxiang garlic from other origins using stable isotopes and chemometrics. *J. Food Compos. Anal.* **2021**, *99*, 103856. [CrossRef]
4. Zhou, G.; Feng, Y.M.; Li, Z.C.; Tao, L.Y.; Kong, W.S.; Xie, R.F.; Zhou, X. Fingerprinting and determination of hepatotoxic constituents in Polygoni Multiflori Radix Praeparata of different producing places by HPLC. *J. Chromatogr. Sci.* **2021**, 1–10. [CrossRef] [PubMed]
5. Sun, S.M.; Guo, B.L.; Wei, Y.M. Origin assignment by multi-element stable isotopes of lamb tissues. *Food Chem.* **2016**, *213*, 675–681. [CrossRef]
6. Benincasa, C.; Lewis, J.; Sindona, G.; Tagarelli, A. The use of multi element profiling to differentiate between cow and buffalo milk. *Food Chem.* **2008**, *110*, 257–262. [CrossRef]
7. Qie, M.J.; Zhang, B.; Li, Z.; Zhao, S.S.; Zhao, Y. Data fusion by ratio modulation of stable isotope, multi-element, and fatty acids to improve geographical traceability of lamb. *Food Control.* **2021**, *120*, 107549. [CrossRef]
8. Zhao, R.T.; Su, M.C.; Zhao, Y.; Chen, G.; Chen, A.L.; Yang, S.M. Chemical analysis combined with multivariate statistical methods to determine the geographical origin of milk from four regions in China. *Foods* **2021**, *10*, 1119. [CrossRef]
9. Xie, L.N.; Zhao, S.S.; Rogers, K.M.; Xia, Y.N.; Zhang, B.; Suo, R.; Zhao, Y. A case of milk traceability in small-scale districts-Inner Mongolia of China by nutritional and geographical parameters. *Food Chem.* **2020**, *316*, 126332. [CrossRef]
10. Wang, J.S.; Xu, L.; Xu, Z.Z.; Wang, Y.; Niu, C.; Yang, S.M. Liquid chromatography quadrupole time-of-flight mass spectrometry and rapid evaporative ionization mass spectrometry were used to develop a lamb authentication method: A preliminary study. *Foods* **2020**, *9*, 1723. [CrossRef]
11. Zhao, R.T.; Yang, S.M.; Zhao, Y. Research progress in traceability of agricultural products using stable isotope. *J. Nucl. Agric. Sci.* **2020**, *34* (Suppl. S1), 120–128. [CrossRef]
12. Durante, C.; Lancellotti, L.; Manzini, D.; Rossi, M.C.; Sighinolfi, S.; Marchetti, A.; Tassi, L. $^{87}Sr/^{86}Sr$ ratio as traceability marker for Modena's balsamic vinegars. *LWT* **2021**, *147*, 11571. [CrossRef]
13. Zhao, Y.; LV, J.; Yang, S.M. Research progress and application of stable isotope technology in agricultural products tracing field. *Qual. Saf. Agro-Products.* **2015**, *6*, 35–40.
14. Nie, J.; Shao, S.Z.; Xia, W.; Liu, Z.; Yu, C.C.; Li, R.; Wang, W.; Li, J.R.; Yuan, Y.W.; Rogers, K.M. Stable isotopes verify geographical origin of yak meat from Qinghai-Tibet plateau. *Meat Sci.* **2020**, *165*, 108113. [CrossRef] [PubMed]
15. Erasmus, S.W.; Muller, M.; Butler, M.; Hoffman, L.C. The truth is in the isotopes: Authenticating regionally unique South African lamb. *Food Chem.* **2018**, *239*, 926–934. [CrossRef] [PubMed]
16. Perini, M.; Camin, F.; Bontempo, L.; Rossmann, A.; Piasentier, E. Multielement (H, C, N, O, S) stable isotope characteristics of lamb meat from different Italian regions. *Rapid Commun. Mass Spectrom.* **2009**, *23*, 2573–2585. [CrossRef] [PubMed]
17. Chung, I.M.; Kim, J.K.; Yang, Y.J.; An, Y.J.; Kim, S.Y.; Kwon, C.; Kim, S.H. A case study for geographical indication of organic milk in Korea using stable isotope ratios-based chemometric analysis. *Food Control.* **2020**, *107*, 106755. [CrossRef]
18. Kang, X.M.; Zhao, Y.F.; Shang, D.R.; Zhai, Y.X.; Ning, J.S.; Ding, H.Y.; Sheng, X.F. Identification of the geographical origins of sea cucumbers in China: The application of stable isotope ratios and compositions of C., N., O and H. *Food Control.* **2020**, *111*, 107036. [CrossRef]
19. Camin, F.; Bontempo, L.; Heinrich, K.; Horacek, M.; Kelly, S.D.; Schlicht, C.; Rossmann, A. Multi-element (H, C, N, S) stable isotope characteristics of lamb meat from different European regions. *Anal. Bioanal Chem.* **2007**, *389*, 309–320. [CrossRef]
20. Kasabov, N. Global, local and personalised modeling and pattern discovery in bioinformatics: An integrated approach. *Pattern Recog Lett.* **2007**, *28*, 673–685. [CrossRef]
21. Guo, J.; Ni, L.; Zhang, L. The first Chinese Conference on Near Infrared Spectroscopy. Sampling Method for NIR Calibration Sample Based on Uniform-Design. In Proceedings of the Near Infrared Spectroscopy Technology in Modern China, Beijing, China, 18–21 October 2006.
22. Centner, V.; de Noord, O.E.; Massart, D.L. Detection of nonlinearity in multivariate calibration. *Anal. Chim. Acta* **1998**, *376*, 153–168. [CrossRef]
23. Chen, D.; Shao, X.G. The first Chinese Conference on Near Infrared Spectroscopy. Study On novel Strategies for Selecting Representative Samples of Near-Infrared Spectroscopy. In Proceedings of the Near Infrared Spectroscopy Technology in Modern China, Beijing, China, 18–21 October 2006.
24. Leonardo, R.; Thorsten, B.; Karsten, S.; Antoine, S.; Jose, A.M.; Thomas, S. The spectrum-based learner: A new local approach for modeling soil vis-NIR spectra of complex datasets. *Geoderma* **2013**, *195*, 268–279. [CrossRef]
25. Abhinav, G.; Hitesh, B.V.; Bhabani, S.D.; Aditya, K.C. Local modeling approaches for estimating soil properties in selected Indian soils using diffuse reflectance data over visible to near-infrared region. *Geoderma* **2018**, *325*, 59–71. [CrossRef]

26. Palou, A.; Miró, A.; Blanco, M.; Larraz, R.; Gómez, J.F.; Martínez, T.; González, J.M.; Alcalà, M. Calibration sets selection strategy for the construction of robust PLS models for prediction of biodiesel/diesel blends physico-chemical properties using NIR spectroscopy. *Spectrochim. Acta A* **2017**, *180*, 119–126. [CrossRef] [PubMed]
27. Coppa, M.; Martin, B.; Hulin, S.; Guillemin, J.; Gauzentes, J.V.; Pecou, A.; Andueza, D. Prediction of indicators of cow diet composition and authentication of feeding specifications of Protected Designation of Origin cheese using mid-infrared spectroscopy on milk. *J. Dairy Sci.* **2021**, *104*, 112–125. [CrossRef] [PubMed]
28. Rius, A.; Callao, M.P.; Ferré, J.; Rius, F.X. Assessing the validity of principal component regression models in different analytical conditions. *Anal. Chim. Acta* **1997**, *337*, 287–296. [CrossRef]
29. De Maesschalck, R.; Jouan-Rimbaud, D.; Massart, D.L. The Mahalanobis distance. *Chemom. Intell. Lab. Syst.* **2000**, *50*, 1–18. [CrossRef]
30. Pomerantsev, A.L. Acceptance areas for multivariate classification derived by projection methods. *J. Chemometr.* **2008**, *22*, 601–609. [CrossRef]
31. Pomerantsev, A.L.; Rodionova, O. Concept and role of extreme objects in PCA/SIMCA. *J. Chemometr.* **2013**, *28*, 429–438. [CrossRef]
32. Foody, G.M. Explaining the unsuitability of the kappa coefficient in the assessment and comparison of the accuracy of thematic maps obtained by image classification. *Remote Sens. Environ.* **2020**, *239*, 111630. [CrossRef]
33. Zhao, S.S.; Zhang, H.B.; Zhang, B.; Xu, Z.Z.; Chen, A.L.; Zhao, Y. A rapid sample preparation method for the analysis of stable isotope ratios of beef samples from different countries. *Rapid Commun. Mass Spectrom.* **2020**, *34*, 8795. [CrossRef] [PubMed]
34. Nietner, T.; Haughey, S.A.; Ogle, N.; Fauhl-Hassek, C.; Elliott, C.T. Determination of geographical origin of distillers dried grains and solubles using isotope ratio mass spectrometry. *Food Res. Inter.* **2014**, *60*, 146–153. [CrossRef]
35. Zhao, Y.; Zhang, B.; Chen, G.; Chen, A.L.; Yang, S.M.; Ye, Z.H. Recent developments in application of stable isotope analysis on agro-product authenticity and traceability. *Food Chem.* **2014**, *145*, 300–305. [CrossRef] [PubMed]
36. Yuan, H.C.; Li, C.Y.; Jian, Y.; Geng, M.M.; Xu, L.W.; Wang, J.R. Stable isotope technique in soil carbon cycling research of agricultural ecosystems. *J. Isot.* **2014**, *27*, 170–178. [CrossRef]
37. Zhao, Y.; Zhang, B.; Chen, G.; Chen, A.L.; Yang, S.M.; Ye, Z.H. Tracing the geographic origin of beef in China on the basis of the combination of stable isotopes and multielement analysis. *J. Agric. Food Chem.* **2013**, *61*, 7055–7060. [CrossRef] [PubMed]
38. Zhao, H.Y.; Wang, F.; Yang, Q.L. Origin traceability of peanut kernels based on multi-element fingerprinting combined with multivariate data analysis. *J. Sci. Food. Agric.* **2020**, *100*, 4040–4048. [CrossRef] [PubMed]

foods MDPI

Article

Discrimination of Four *Cinnamomum* Species with Physico-Functional Properties and Chemometric Techniques: Application of PCA and MDA Models

Priya Rana [1], Shu-Yi Liaw [2], Meng-Shiou Lee [3] and Shyang-Chwen Sheu [4,*]

[1] Department of Tropical Agriculture and International Cooperation, National Pingtung University of Science and Technology, Pingtung 91201, Taiwan; priyarana198933@gmail.com
[2] Department of Business Management, National Pingtung University of Science and Technology, Pingtung 91201, Taiwan; syliaw@mail.npust.edu.tw
[3] Department of Chinese Pharmaceutical Science and Chinese Medicine Resources, China Medical University, Taichung 40402, Taiwan; leemengshiou@mail.cmu.edu.tw
[4] Department of Food Science, National Pingtung University of Science and Technology, Pingtung 91201, Taiwan
* Correspondence: ssheu@mail.npust.edu.tw; Tel.: +886-8-7740375

Abstract: Discrimination of highly valued and non-hepatotoxic *Cinnamomum* species (*C. verum*) from hepatotoxic (*C. burmannii*, *C. loureiroi*, and *C. cassia*) is essential for preventing food adulteration and safety problems. In this study, we developed a new method for the discrimination of four *Cinnamomum* species using physico-functional properties and chemometric techniques. The data were analyzed through principal component analysis (PCA) and multiclass discriminant analysis (MDA). The results showed that the cumulative variability of the first three principal components was 81.70%. The PCA score plot indicated a clear separation of the different *Cinnamomum* species. The training set was used to build the discriminant MDA model. The testing set was verified by this model. The prediction rate of 100% proved that the model was valid and reliable. Therefore, physico-functional properties coupled with chemometric techniques constitute a practical approach for discrimination of *Cinnamomum* species to prevent food fraud.

Keywords: cinnamon; chemometrics; food fraud; identification model; physico-functional

Citation: Rana, P.; Liaw, S.-Y.; Lee, M.-S.; Sheu, S.-C. Discrimination of Four *Cinnamomum* Species with Physico-Functional Properties and Chemometric Techniques: Application of PCA and MDA Models. *Foods* **2021**, *10*, 2871. https://doi.org/10.3390/foods10112871

Academic Editor: Theodoros Varzakas

Received: 14 October 2021
Accepted: 17 November 2021
Published: 19 November 2021

1. Introduction

Cinnamon is one of the important spices and is obtained from the dried inner bark of the evergreen tree belonging to the genus *Cinnamomum*. There are four main economically available species of cinnamon in the spice market, including *C. verum* (Ceylon cinnamon, CV), *C. burmannii* (Indonesian cinnamon, CB), *C. loureiroi* (Vietnamese cinnamon, CL), and *C. cassia* (Chinese cinnamon, CC). Consumers' growing awareness of the health benefits of cinnamon is driving the global cinnamon market, which is expected to reach to US$1.9 billion by 2025 [1]. Because of worldwide demand and the direct relationship between food quality and commercial value, the cinnamon supply chain is susceptible to food fraud. Important types of food fraud are deliberate substitution, dilution or addition, or misrepresentation of food ingredients [2]. According to reports, cinnamon is at very high risk for adulteration involving substitution, and the increased trading of cinnamon substitutes has increased that risk [3].

Many *Cinnamomum* plants are morphologically similar. Some cheaper and hepatotoxic adulterant *Cinnamomum* species, such as *C. burmannii*, *C. loureiroi*, and *C. cassia*, are easily confused with the highly valued and non-hepatotoxic *C. verum* [4,5]. Consuming such substitutes, however, is dangerous due to the high amount of coumarin present in comparison to *C. verum*. Furthermore, the number of these lower-priced substitutes is increasing in the consumer market [6]. The task of differentiation becomes more challenging and difficult when cinnamon is converted into powder [7]. As a result, distinguishing between cinnamon species is critical for ensuring food quality and avoiding safety issues associated with fraudulent adulteration.

So far, efforts are made for evaluating the quality and safety [8] of cinnamon. Currently, there is increasing demand on governmental agencies and industries to combat the rising threat of food fraud [9]. However, most of the quality-monitoring analytical methods are expensive, have high environmental impact, require skilled analysts, and can only be employed in well-equipped laboratories [10]. To overcome these problems, more recent research trends have emphasized the evaluation of physico-functional properties of food materials, since they could serve as quality control indexes. Physico-functional properties such as pH, moisture content, and density can be determined with limited laboratory resources and are easily accessible in laboratories of less developed or developing countries [11]. Therefore, these analyses can be determined in all steps during routine quality inspections of foods at the industrial or supplier levels.

Chemometric techniques have been successfully employed as useful tools for data analysis in food-related studies [9], for example, assessing food quality, confirming food authenticity, detecting food adulteration, and distinguishing cultivars [11]. At present, there is a growing body of literature discussing the importance of principal component analysis (PCA) and multiclass discriminant analysis (MDA) in discriminating peach varieties, *Boletus edulis* [12], rice varieties [13], and vinegar varieties [14]. Although there has been some discussion of the physical or functional properties of *Cinnamomum* species [15,16], none has reported discrimination of *Cinnamomum* species based on physico-functional properties coupled with chemometric techniques.

This study was aimed at investigating the relative contribution of 13 physico-functional properties of *C. verum*, *C. burmannii*, *C. loureiroi*, and *C. cassia*. PCA was first employed for exploratory purposes and tested the suitability of the physico-functional properties for discrimination of four *Cinnamomum* species. Then, MDA was employed for classification and prediction purposes [17,18].

2. Materials and Methods

2.1. Sample Collection and Preparation

Twenty cinnamon samples (6 CV, 6 CB, 4 CL, and 4 CC) were collected from different Asian countries over the period 2018 to 2020 (Supplementary Table S1). All samples were identified at the Department of Biological Sciences, National Sun Yat-sen University, Taiwan, based on morpho-anatomical features [19,20]. Dried cinnamon bark samples were crushed manually and then pulverized to powder using a laboratory-scale stainless steel grinder. The powder samples were placed in plastic bags and stored in a vacuum desiccator until use.

2.2. Determination of Physico-Functional Properties

A total of 13 physico-functional properties were assessed on each sample.

2.2.1. Bulk Density (BD) and Tapped Density (TD)

The BD and TD were calculated by the ratio of the weight to the unsettled or tapped volume of the sample and expressed in grams per cubic centimeter (g/cm^3) [21,22].

2.2.2. True Density

The true density was measured by a gas pycnometer (AccuPyc 1340, Micromeritics, Norcross, GA, USA) and calculated using Equation (1) [15].

$$\text{True density } (\rho_t) = \left[\frac{W_s}{V_s}\right]$$
$$V_s = \left[V_{cell} - \left(\frac{V_{exp}}{\left(\frac{P_1}{P_2}\right) - 1}\right)\right] \quad (1)$$

In Equation (1), W_s = weight of the sample (g), V_s = volume of the sample (cm^3), V_{cell} = volume of the cell, V_{exp} = observed volume (experimental), and P_1 and P_2 = pressure of the multivolume pycnometer before and after nob revolution, respectively, in psi.

2.2.3. Porosity

Porosity (ε) was determined as the ratio of the difference between true density and bulk density to the true density [15]. The percentage porosity (ε%) was calculated using Equation (2).

$$\text{Porosity } (\varepsilon\%) = \left[\left(1 - \frac{\rho_b}{\rho_t}\right) \times 100\right] \quad (2)$$

where ρ_b = bulk density (g/cm^3) and ρ_t = true density (g/cm^3).

2.2.4. pH

Sample pH was determined according to the procedure of Jeong, et al. [23]. One gram of sample was mixed with 40 mL of doubly-deionized (2D) water and shaken for 3 h at 200 rpm. The mixture was then centrifuged (Himac CR 21F, Hitachi Koki Co., Ltd., Tokyo, Japan) at 1294× *g* for 10 min, and the filtrate was collected for pH measurement by a pH meter (sensION ™ + PH3, Hach Lange GmbH, Düsseldorf, Germany).

2.2.5. Moisture Content

The moisture content of the sample was measured using an automated moisture balance (MA 35, Sartorius Weighing Technology GmbH, Goettingen, Germany) and expressed as % moisture content on a dry basis.

2.2.6. Color

The sample color was determined using a colorimeter (ZE 2000, Nippon Denshoku Industries Co. Ltd., Tokyo, Japan) and evaluated by means of CIELAB coordinates [23]. The total color difference (ΔE) was determined by taking an unsieved sample as a reference and using Equation (3) [11].

$$\Delta E = \left[\Delta L^2 + \Delta a^2 + \Delta b^2\right]^{\frac{1}{2}} \quad (3)$$

where ΔL = difference in lightness, Δa = difference in red intensity, and Δb = difference in yellow intensity.

2.2.7. Aspect Ratio

The aspect ratio was determined according to the method described by Charles and Alamsjah [11]. Samples were mounted on a microscope slide without overlap of particles and observed under a microscope (Eclipse E100, Nikon Instruments Inc., Melville, NY, USA). The parameters of the aspect ratio, including the particle major axis (l) and minor axis (b), were analyzed by Image-Pro® 10 [24]. The aspect ratio (φ_{AR}) was calculated according to Equation (4).

$$\text{Aspect ratio } (\varphi_{AR}) = \left[\frac{\text{minor axis (b)}}{\text{major axis (l)}}\right] \quad (4)$$

2.2.8. Water Absorption Index (WAI) and Water Solubility Index (WSI)

The WAI and WSI were determined by the procedure described by Kraithong, et al. [25]. One gram of sample was added to 10 mL of 2D water and vortexed for 1 min. The suspension was submerged in a water bath at 30 ± 2 °C for 30 min with intermittent stirring and centrifuged at 1294× *g* for 10 min. The supernatant was transferred to a preweighed aluminum moisture dish and dried overnight at 105 °C. The weight of the sediment was recorded. The WAI and WSI were calculated and expressed as g/g of sample and %, respectively, as shown in Equations (5) and (6).

$$\text{Water absorption index (WAI, g/g)} = \left[\frac{\text{weight of wet sediment (g)}}{\text{dry weight of sample (g)}}\right] \quad (5)$$

$$\text{Water solubility index (WSI, \%)} = \left[\left(\frac{\text{weight of dried supernatant (g)}}{\text{dry weight of sample (g)}}\right) \times 100\right] \quad (6)$$

2.2.9. Oil Absorption Index (OAI)

The OAI was determined as described by Kraithong, et al. [25]. Samples (1 g) were added to commercial soybean oil (10 mL) and centrifuged at 2301× *g* for 20 min. The weight of oil absorbed was recorded. The amount of oil absorbed by the samples was calculated according to Equation (7).

$$\text{Oil absorption index (OAI, g/g)} = \left[\frac{\text{weight of oil absorbed (g)}}{\text{weight of sample (g)}}\right] \quad (7)$$

2.2.10. Swelling Power (SP)

The SP was determined according to the method described by Moutaleb, et al. [26]. One gram of sample was mixed with 10 mL of 2D water and then incubated at room temperature for 24 h. The SP was calculated using Equation (8).

$$\text{Swelling power (SP, mL/g)} = \left[\frac{\text{total volume of the swollen sample (mL)}}{\text{original dry weight of sample (g)}}\right] \quad (8)$$

2.2.11. Emulsifying Activity (EA)

The EA was performed by adapting the method by Chandra, et al. [27]. One gram of sample was mixed with 10 mL of 2D water and 10 mL of soybean oil. The mixture was vortexed thoroughly and centrifuged at 2000× *g* for 5 min. The EA was calculated according to Equation (9).

$$\text{Emulsifying activity (EA, \%)} = \left[\left(\frac{\text{height of emulsified layer}}{\text{total height of mixture}}\right) \times 100\right] \quad (9)$$

2.3. Data Processing and Analysis

A data matrix consisting of 120 observations (20 cinnamon samples × 6 replicates) and 13 physico-functional variables were used in this study. The replicates were used to enlarge the sample size. One-way ANOVA (analysis of variance) was first performed to determine the significant ($p < 0.05$) variables that could be used to discriminate among *Cinnamomum* species. Then, trials for different combinations of significant variables were conducted for the two selected groups of *Cinnamomum* species using an independent samples *t*-test ($p < 0.05$). Finally, the analysis, providing the best discriminative variables with better discrimination power for the established identification model was used for the chemometric approach.

2.4. Chemometric Techniques

IBM SPSS Statistics for Windows, Version 22.0 [28] was employed for chemometric analyses.

2.4.1. Multivariate Analysis of Variance (MANOVA)

Raw data for selected physico-functional properties were subjected to MANOVA to determine the significant interactions between the species and selected variables. Physico-functional properties were taken as the dependent variables, while species were used as the independent variables. Two multivariate tests, Wilk's lambda (Λ) and Pillai's trace, were computed to determine significant effects of selected variables on the species.

2.4.2. Principal Component Analysis (PCA)

A total of 120 observations (36 observations each for CV and CB; 24 observations each for CL and CC) were selected. Prior to PCA, we computed the Kaiser–Meyer–Olkin (KMO) measure of sampling adequacy and Bartlett's test of sphericity to assess the multicollinearity of the data for PCA suitability [29]. A factor extraction method with varimax rotation was employed. The extracted principal components (PCs) with eigenvalues equal to or higher than 1 were used to calculate the PC scores and establish a PCA model. The PC score was calculated according to Zhao, et al. [12], as shown in Equation (10).

$$PC_n = \left[FAC_n \times \sqrt{\lambda_n}\right] \tag{10}$$

where PC_n = principal component score, FAC_n = factor score obtained directly through SPSS analysis, λ = principal component eigenvalue equal to or higher than 1, and n = number of principal component extracted.

2.4.3. Multiclass Discriminant Analysis (MDA)

The PC scores of 120 observations were divided randomly into the training set (83.3%) and the testing set (16.7%) using the Microsoft Excel® 2016 Add-In function, Ablebits tools [30]. The former set included 100 observations of the four species, and the latter set contained the remaining 20 observations. The PC scores of the training set were taken as the input for stepwise discriminant analysis (DA) to build the MDA model [12]. Finally, typical discriminant functions were established for the species distinction models.

3. Results and Discussion

3.1. Descriptive Statistics of Physico-Functional Properties

The 13 physico-functional properties of *Cinnamomum* species are listed in Table 1, and box-and-whisker plots are shown in Supplementary Figure S1. The bulk density (BD) values for CB (0.45 ± 0.05 g/cm^3) showed a highly significant difference ($p < 0.05$), while nearly identical values were reported for CV (0.35 ± 0.02 g/cm^3), CL (0.34 ± 0.01 g/cm^3), and CC (0.33 ± 0.02 g/cm^3). Hermanto, et al. [21] reported BD values for CB samples between 0.43 g/cm^3 and 0.49 g/cm^3, consistent with our study. A similar trend was followed for the tapped density (TD), where no significant differences ($p > 0.05$) were found for CL (0.58 ± 0.02 g/cm^3), CC (0.58 ± 0.03 g/cm^3), and CV (0.57 ± 0.04 g/cm^3), but that of CB (0.71 ± 0.06 g/cm^3) was different. Slight variations among the species could be associated with their origin and environmental conditions. The true density varied significantly ($p < 0.05$) among *Cinnamomum* species and ranged from 1.46 ± 0.02 g/cm^3 (CC) to 1.51 ± 0.00 g/cm^3 (CV). The increase in the true density of cinnamon samples might be affected by the moisture content [15]. The porosity values were similar for CC (77.40 ± 1.01%), CL (76.71 ± 0.73%), and CV (76.49 ± 1.01%) but not for CB (69.76 ± 3.58%). The higher porosities might be due to drastic changes occurring after the grinding process [31]. The pH values of the four species were reported as moderately acidic pH values ranging between 4.73 ± 0.24 (CV) and 5.04 ± 0.17 (CC). Jeong, et al. [23] reported similar pH values (4.93 to 5.07) for cinnamon powder samples available in the Korean spice market. The observed acidic pH might be associated with the presence of organic compounds (e.g., cinnamaldehyde and cinnamyl acetate) in cinnamon. The highest moisture content was recorded for CC (11.63 ± 0.67%), while the lowest was recorded for CV (9.89 ± 0.44%). These results are in line with the results of Jeong, et al. [23], in which recorded moisture contents ranged from 7.25 to 12.73% in various cinnamon samples. Additionally, the variations in moisture content might have influenced the true densities of the samples [15], which was also evident from our findings. The color differences (ΔEs) showed a significantly ($p < 0.05$) wide range of values from 3.28 ± 0.36 (CC) to 7.89 ± 1.14 (CB). The wide variations observed for ΔE highlighted the diversity of the samples. The aspect ratios varied little between the species, ranging from 1.66 ± 0.27 (CV) to 3.63 ± 0.33 (CB). The similarities in particle aspect ratio might be related to grinding and sieving methods used for sample preparation [11].

Table 1. Descriptive statistics for physico-functional properties of cinnamon samples from four *Cinnamomum* species.

Variables	*Cinnamomum* Species															
	C. verum (n = 6)				*C. burmannii* (n = 6)				*C. loureiroi* (n = 4)				*C. cassia* (n = 4)			
	Mean	SD	Min	Max	Mean	SD	Min	Max	Mean	SD	Min	Max	Mean	SD	Min	Max
Bulk density (g/cm^3)	0.35 ^{b}	0.02	0.33	0.39	0.45 ^{a}	0.05	0.39	0.53	0.34 ^{bc}	0.01	0.32	0.36	0.33 ^{c}	0.02	0.31	0.36
Tapped density (g/cm^3)	0.57 ^{b}	0.04	0.50	0.65	0.71 ^{a}	0.06	0.63	0.82	0.58 ^{b}	0.02	0.54	0.63	0.58 ^{b}	0.03	0.54	0.63
True density (g/cm^3)	1.51 ^{a}	0.00	1.50	1.52	1.49 ^{b}	0.01	1.48	1.50	1.48 ^{c}	0.01	1.47	1.49	1.46 ^{d}	0.02	1.42	1.48
Porosity (%)	76.49 ^{a}	1.01	74.01	78.36	69.76 ^{b}	3.58	64.44	73.69	76.71 ^{a}	0.73	75.55	78.39	77.40 ^{a}	1.01	75.72	79.24
pH	4.73 ^{c}	0.24	4.42	5.11	4.90 ^{b}	0.10	4.78	5.08	4.93 ^{b}	0.09	4.83	5.08	5.04 ^{a}	0.17	4.75	5.21
Moisture content (%)	9.89 ^{d}	0.44	9.03	10.58	11.15 ^{b}	0.40	10.20	11.85	10.80 ^{c}	0.50	10.03	11.76	11.63 ^{a}	0.67	10.57	12.92
Color	3.85 ^{c}	0.78	2.94	5.22	7.89 ^{a}	1.14	5.90	9.36	5.76 ^{b}	0.58	5.06	6.66	3.28 ^{d}	0.36	2.81	3.79
Aspect ratio	1.66 ^{c}	0.27	1.11	2.17	3.63 ^{a}	0.33	3.26	4.67	2.85 ^{b}	0.14	2.44	3.09	2.78 ^{b}	0.04	2.69	2.85
Water absorption index (g/g)	3.78 ^{b}	0.20	3.27	4.07	5.18 ^{a}	0.82	4.17	6.40	3.09 ^{c}	0.17	2.90	3.33	2.93 ^{c}	0.19	2.58	3.15
Water solubility index (%)	5.61 ^{b}	1.90	2.81	8.62	7.89 ^{a}	2.66	4.14	12.68	8.63 ^{a}	0.44	7.95	9.42	7.83 ^{a}	1.12	6.11	9.29
Oil absorption index (g/g)	3.16 ^{a}	0.25	2.64	3.48	2.53 ^{b}	0.20	2.30	2.81	2.38 ^{c}	0.09	2.24	2.51	2.32 ^{c}	0.08	2.21	2.44
Swelling power (mL/g)	4.56 ^{b}	0.17	4.19	4.80	8.93 ^{a}	2.02	5.20	11.40	3.53 ^{c}	0.25	3.10	3.90	3.05 ^{c}	0.19	2.70	3.40
Emulsifying activity (%)	2.97 ^{b}	0.48	2.27	4.84	27.53 ^{a}	13.89	7.14	46.34	1.58 ^{b}	0.39	0.82	2.61	1.72 ^{b}	0.32	1.40	2.54

Data is mean of six replicates. Mean values followed by different superscripts (a–d) within the same row are significantly different ($p < 0.05$) based on Duncan's test (One-way ANOVA). n is the number of samples and SD is standard deviation.

On the other hand, similar water absorption indexes (WAIs) were documented for CC (2.93 ± 0.19 g/g) and CL (3.09 ± 0.17 g/g), whereas CV (3.78 ± 0.20 g/g) and CB (5.18 ± 0.82 g/g) showed differences. A high WAI may be associated with large hydrophilic molecules, such as polysaccharides. Other factors, including the nature, concentration and conformation of proteins and the level of protein interaction with water, might also influence the WAI [27]. The water solubility index (WSI) of CL (8.63 ± 0.44%) was high but showed no significant difference ($p > 0.05$) from those of CB (7.89 ± 2.66%) and CC (7.83 ± 1.12%) but differed from that of CV (5.61 ± 1.90%). This trend might be attributable to particle size resulting from similar grinding and sieving processes. The oil absorption index (OAI) values for CV (3.16 ± 0.25 g/g) and CB (2.53 ± 0.20 g/g) differed significantly ($p < 0.05$) from those of CL (2.38 ± 0.09 g/g) and CC (2.32 ± 0.08 g/g). However, OAI is mainly affected by the hydrophilic or hydrophobic nature of the proteins [27], which highlights the partial interdependence between WAI and OAI properties of *Cinnamomum* species. The swelling power (SP) ranged from 3.05 ± 0.19 mL/g (CC) to 8.93 ± 2.02 mL/g (CB). The SP might be affected by the species, particle sizes, and different processing methods or unit operations employed [27]. The emulsifying activity (EA) showed a wide range of values from 1.58 ± 0.39% (CL) to 27.53 ± 13.89% (CB). There are, however, possible explanations, including geographical origin and differences in packaging or storage of cinnamon powder samples [23], which could have affected the physico-functional properties of cinnamon samples. The findings from this study have made several contributions to the current literature by providing useful and practical information on the physico-functional properties of *Cinnamomum* species.

3.2. Selection of Discriminative Variables

In general, it is important to understand the major contributing variables (within 13 physico-functional variables) that could provide the maximum information for differentiation of *Cinnamomum* species. We used an independent samples *t*-test to compare the two groups [(CV and CB) ∩ (CL and CC)] and variables to enable the correct identification among different species. The analysis identified the nine most informative physico-functional variables, including BD, true density, porosity, pH, moisture content, color, WAI, WSI, and SP (Supplementary Table S2), with less crossreactivity between the samples.

3.3. Multivariate General Linear Analysis

MANOVA was employed to perform multisignificant tests with nine selected physico-functional variables (Supplementary Table S3). The *p*-value was rounded to three decimal places due to generation of very low values, indicating very high significance. A study conducted by Karabagias, et al. [32] supported Pillai's trace and Wilks' Λ as the preferred test statistics for MANOVA and suggested the appropriateness of MANOVA by considering possible multi-significant effects of dependent variables on independent variables. In this study, Pillai's trace (F (27,330) = 54.65, p = 0.000 < 0.05; Pillai's trace = 2.45) and Wilks' Λ (F (27,316.06) = 84.52, p = 0.000 < 0.05; Wilks' Λ = 0.00) tests were considered. The results showed the existence of statistically significant multivariate effects of physico-functional properties among the cinnamon samples. Hence, we further applied PCA and MDA for a clear and in-depth understanding of variations among *Cinnamomum* species.

3.4. Data Dimensional Reduction through PCA

The dimensionality of the data for nine selected physico-functional variables was reduced to principal components (PCs) using PCA. In the present study, a KMO value of 0.61 and statistically significant (p < 0.05) Bartlett's test of sphericity supported the appropriateness of the data for performing PCA. In addition, the three significant variables (true density, moisture content, and color) determined by one-way ANOVA collectively failed to yield acceptable KMO value (0.45), thus making PCA inapplicable. Therefore, it was not considered in this study. Only the first three PCs presented eigenvalues exceeding 1 (PC1–46.69%, PC2–21.49%, and PC3–13.51%) and explained 81.70% of the cumulative variability (Supplementary Table S4). A three-dimensional (3-D) score plot shows the separation of cinnamon samples into four groups (Figure 1a). The CB samples presented relatively different physico-functional properties and thus formed a distinct group to the left of the score plot. Although CL and CC samples were found close to each other due to similarities in their respective physico-functional properties, significant boundaries were observed between them. Notably, the CV samples were placed towards the bottom of the plot and distinguished from the CB, CL, and CC samples. These results agreed with those from the study by Shawky and Selim [33], which applied PCA to demonstrate a clear separation of CV samples from adulterated cinnamon samples based on near-infrared (NIR) fingerprints. Similarly, Jeong, et al. [23] employed PCA to establish clear variations among different cinnamon powders based on physico-chemical parameters. Our results implied that physico-functional information can be utilized to discriminate among different species of *Cinnamomum* samples.

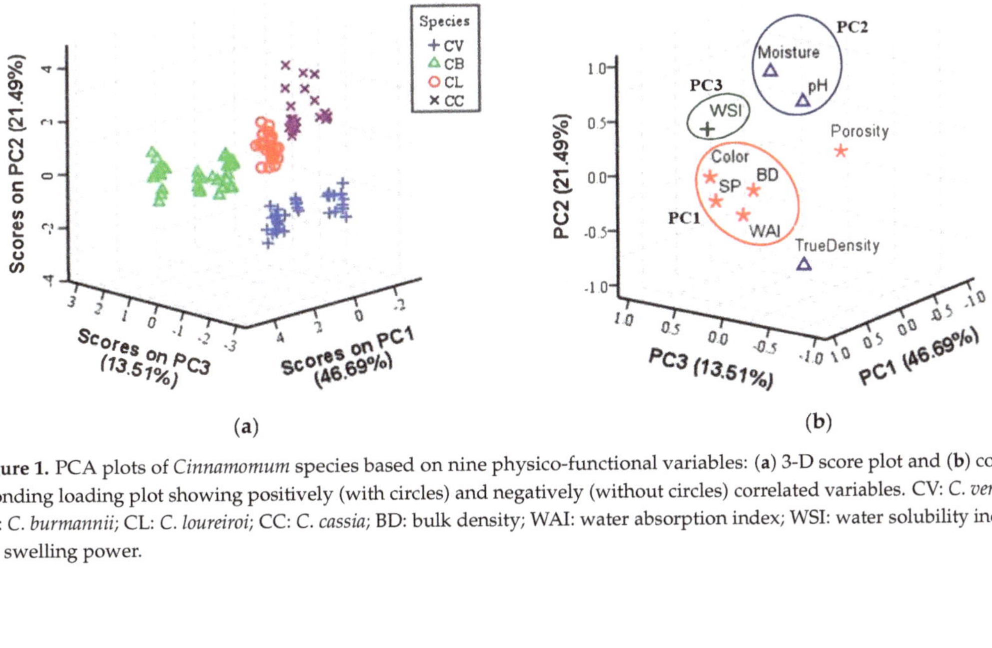

Figure 1. PCA plots of *Cinnamomum* species based on nine physico-functional variables: (**a**) 3-D score plot and (**b**) corresponding loading plot showing positively (with circles) and negatively (without circles) correlated variables. CV: *C. verum*; CB: *C. burmannii*; CL: *C. loureiroi*; CC: *C. cassia*; BD: bulk density; WAI: water absorption index; WSI: water solubility index; SP: swelling power.

Moreover, the principal component loading matrix (Supplementary Table S4) of the first three PCs extracted and the corresponding loading plot (Figure 1b) illustrate the relationships among the variables and describe the variables effecting the separation of samples. The CB samples obtained higher scores in PC1 due to high positive loadings for BD, color, WAI, and SP, whilst CV samples showed higher scores in PC1 due to high negative loading for porosity. Similarly, the CC samples reported higher scores in PC2 due to higher values of pH and moisture content, but lower values of true density. Finally, the CL samples projected towards PC3 with a strong positive weight of WSI. This showed the contribution of positively correlated (BD, pH, moisture content, color, WAI, WSI, and SP) and negatively correlated (true density and porosity) variables in explaining variations among the samples.

These findings demonstrated that the cumulative contribution rate of the first three PCs reached 81.70%, indicating that these PCs represented the original variables. From the contained information, the number of original nine selected physico-functional variables was reduced to three new variables called three PCs. Overall, the reliabilities of the three new variables demonstrated that physico-functional analysis with PCA is a promising strategy for discrimination among *Cinnamomum* species. Furthermore, MDA was applied to implement the comprehensive use of *Cinnamomum* physico-functional information from different species to predict the species of a test sample.

3.5. *Establishment of the MDA Model for Cinnamomum Species*

3.5.1. MDA Characteristics

The first three PC scores of 100 observations were used as independent variables, and *Cinnamomum* species were used as grouping variables. The highlighted MDA characteristics of nine physico-functional variables using stepwise DA are summarized (Supplementary Table S5). The results showed that three significant discriminant functions (DF1 = 60.30%, DF2 = 39.20%, and DF3 = 0.50%) accounted for 100% of the total variance. We exclude the discussion of DF3 since it represented a very small fraction of the total information. The Wilks' Λ values for DF1 ($\chi^2 = 443.66$, $p = 0.000 < 0.05$) and DF2 ($\chi^2 = 207.26$, $p = 0.000 < 0.05$) were 0.10 and 0.11, respectively. The existence of small Wilks' Λ and large chi-square (χ^2) values indicated significantly high discriminatory ability of a function and that the groups appeared to differ [11]. Moreover, DF1 exhibited a high eigenvalue (10.89) and canonical correlation of 0.96, followed by DF2 with an eigenvalue of 7.08 and canonical correlation of 0.94. These findings also revealed that a larger eigenvalue explained more variance in the grouping variable in the function test. Similarly, a higher canonical correlation indicated significant differences in physico-functional properties among *Cinnamomum* species.

3.5.2. Identification Model for Cinnamon Samples

The identification model was developed using stepwise DA to correctly identify cinnamon samples based on nine physico-functional variables. Fisher's linear discrimination functions were established for the species distinction models according to the following Equations (11)–(14):

$$\text{CV}: \ \text{Y}_1(x) = -10.97 - 3.22x_1 - 8.09x_2 - 4.97x_3 \tag{11}$$

$$\text{CB}: \ \text{Y}_2(x) = -12.71 + 7.40x_1 + 2.36x_2 + 3.55x_3 \tag{12}$$

$$\text{CL}: \ \text{Y}_3(x) = -3.82 - 2.01x_1 + 2.87x_2 + 1.21x_3 \tag{13}$$

$$\text{CC}: \ \text{Y}_4(x) = -10.81 - 4.31x_1 + 6.18x_2 + 1.13x_3 \tag{14}$$

where $\text{Y}_1(x)$, $\text{Y}_2(x)$, $\text{Y}_3(x)$, and $\text{Y}_4(x)$ are the identification values for CV, CB, CL, and CC, respectively. x_1, x_2, and x_3 are the values of the first three PC scores.

The values of the first three PC scores of 100 training observations were taken into the established identification functions to validate the functions. Out of 100 observations, two observations originating from CB and one observation originating from CC were misclassified as CL. This could be explained by a minor and unavoidable experimental handling error or the close relationship among CB, CL, and CC. As shown in Table 2, the correct identification rates were 100%, 93.30%, 100%, and 95% for CV, CB, CL, and CC, respectively. The overall correct rate of 97% showed that the established identification model was feasible. Therefore, MDA could be employed to build a distinction model for *Cinnamomum* species with a high percentage of correct identification based on physico-functional properties. In order to establish a full-scale quality evaluation and discrimination system for *Cinnamomum* species, collecting more cinnamon samples from different species should be required in the further study.

Table 2. Correct identification and prediction rates of the training and testing sets based on the MDA model.

	Actual Species	**Species Discriminated by Model**				**Total**	**Correct Identification Rate (%)**
		CV	**CB**	**CL**	**CC**		
Training set	CV	30	0	0	0	30	100
	CB	0	28	2	0	30	93.30
	CL	0	0	20	0	20	100
	CC	0	0	1	19	20	95
	Total	30	28	23	19	100	97
	Actual Species	**Species Discriminated by Model**				**Total**	**Correct Prediction Rate (%)**
		CV	**CB**	**CL**	**CC**		
Testing set	CV	6	0	0	0	6	100
	CB	0	6	0	0	6	100
	CL	0	0	4	0	4	100
	CC	0	0	0	4	4	100
	Total	6	6	4	4	20	100

CV: *C. verum*; CB: *C. burmannii*; CL: *C. loureiroi*; CC: *C. cassia*.

A two-dimensional (2-D) score plot (DF1 × DF2) represents the qualitative identification of *Cinnamomum* species (Figure 2). The results showed that CL and CC samples were spread out in the second quadrant, while CV samples were located in the third quadrant. On the other hand, CB samples were distributed in the fourth quadrant. Therefore, CV samples were completely separated from other samples. We concluded that there was a good cluster result for *Cinnamomum* species based on the first two DFs.

3.5.3. Analytical Model Prediction for Cinnamon Samples

To further test the reliability of the established identification model, the 20 testing observations were set into the four identification functions. The function with a larger value determined the predicted species (Supplementary Table S6). The results of the testing set to validate the built model showed a 100% prediction rate for the assigned samples to their respective categories (Table 2). Therefore, the established identification model was valid and reliable.

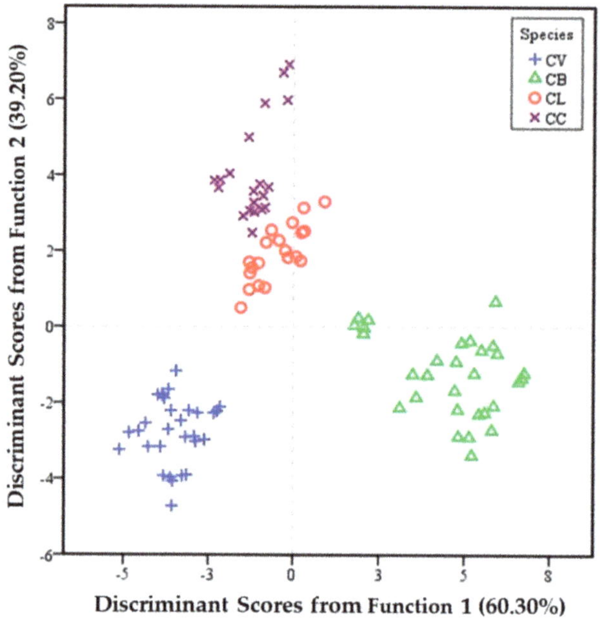

Figure 2. 2-D score plot of *Cinnamomum* species based on the two discriminant functions. CV: *C. verum*; CB: *C. burmannii*; CL: *C. loureiroi*; CC: *C. cassia*.

4. Conclusions

This study presented a valid and reliable model for *Cinnamomum* discrimination with the potential use of selected physico-functional variables coupled with chemometric techniques. By combining the PCA and MDA techniques, a relationship was established between the physico-functional properties and the *Cinnamomum* species. Additionally, by applying PCA, the training and testing of the MDA model have become feasible. The data correct identification and prediction rates realized by using the MDA model provide a checkpoint for food authorities. The combination of characteristic physico-functional variables with different *Cinnamomum* species constitutes the novelty of the present work designed to ensure future food safety. However, this method may be time-consuming way to train and test the model. Future research is recommended on the integration of feature selection and data mining approaches to decrease training time and accelerate learning from testing a large number of samples.

Supplementary Materials: The following are available online at https://www.mdpi.com/article/10.3390/foods10112871/s1, Figure S1: Box-and-whisker plots representing different physico-functional properties for the studied *Cinnamomum* species, Table S1: Detailed information on the *Cinnamomum* samples used in the study, Table S2: Independent samples *t*-test results of the nine selected discriminative physico-functional variables, Table S3: MANOVA results of *Cinnamomum* species using nine physico-functional variables, Table S4: Total variance explained and principal component loading matrix, Table S5: Stepwise DA characteristics of *Cinnamomum* species using nine physico-functional variables, Table S6: Prediction result of the testing set by the MDA model.

Author Contributions: Conceptualization, P.R., M.-S.L. and S.-Y.L.; methodology, P.R. and S.-C.S.; software, P.R. and S.-Y.L.; validation, S.-Y.L. and S.-C.S.; formal analysis, P.R.; investigation, P.R.; data curation, P.R. and S.-C.S.; writing—original draft preparation, P.R.; writing—review and editing, M.-S.L. and S.-C.S.; visualization, S.-C.S.; supervision, S.-C.S.; funding acquisition, S.-C.S. All authors have read and agreed to the published version of the manuscript.

Funding: This research received no external funding.

Institutional Review Board Statement: Not applicable.

Informed Consent Statement: Not applicable.

Data Availability Statement: The data that support the findings of this study are available from the corresponding author upon reasonable request.

Acknowledgments: The authors are grateful to the Taiwan ICDF for the scholarship. Also, special thanks to Rachael Chu from the Food Science Department, National Pingtung University of Science and Technology, Taiwan for her technical assistance.

Conflicts of Interest: The authors declare no conflict of interest.

References

1. Grand View Research. Cinnamon Market Size Worth $1.9 Billion by 2025 | CAGR: 13.8%. 2019. Available online: https://www.grandviewresearch.com/press-release/global-cinnamon-market (accessed on 4 January 2021).
2. Tibola, C.S.; Silva, S.; Dossa, A.A.; Patrício, D.I. Economically Motivated Food Fraud and Adulteration in Brazil: Incidents and Alternatives to Minimize Occurrence. *J. Food Sci.* **2018**, *83*, 2028–2038. [CrossRef] [PubMed]
3. Food Fraud Advisors. Cinnamon Fraud Warning. 2017. Available online: https://www.foodfraudadvisors.com/cinnamon-fraud/ (accessed on 6 May 2021).
4. Farag, M.A.; Labib, R.M.; Noleto, C.; Porzel, A.; Wessjohann, L.A. NMR approach for the authentication of 10 cinnamon spice accessions analyzed via chemometric tools. *LWT* **2018**, *90*, 491–498. [CrossRef]
5. Avula, B.; Smillie, T.J.; Wang, Y.-H.; Zweigenbaum, J.; Khan, I.A. Authentication of true cinnamon (*Cinnamon verum*) utilising direct analysis in real time (DART)-QToF-MS. *Food Addit. Contam. Part A* **2015**, *32*, 1–8. [CrossRef] [PubMed]
6. Ananthakrishnan, R.; Chandra, P.; Kumar, B.; Rameshkumar, K.B. Quantification of coumarin and related phenolics in cinnamon samples from south India using UHPLC-ESI-QqQ$_{LIT}$-MS/MS method. *Int. J. Food Prop.* **2018**, *21*, 50–57. [CrossRef]
7. Jeremić, K.; Kladar, N.; Vučinić, N.; Todorović, N.; Hitl, M.; Lalic-Popovic, M.; Gavarić, N. Morphological characterization of cinnamon bark and powder available in the Serbian market. *Biol. Serb.* **2019**, *41*, 89–93.
8. Bua, D.G.; Annuario, G.; Albergamo, A.; Cicero, N.; Dugo, G. Heavy metals in aromatic spices by inductively coupled plasma-mass spectrometry. *Food Addit. Contam. Part B* **2016**, *9*, 210–216. [CrossRef]
9. Granato, D.; Putnik, P.; Kovačević, D.B.; Santos, J.S.; Calado, V.; Rocha, R.S.; Da Cruz, A.G.; Jarvis, B.; Rodionova, O.Y.; Pomerantsev, A. Trends in Chemometrics: Food Authentication, Microbiology, and Effects of Processing. *Compr. Rev. Food Sci. Food Saf.* **2018**, *17*, 663–677. [CrossRef]
10. Hassoun, A.; Måge, I.; Schmidt, W.F.; Temiz, H.T.; Li, L.; Kim, H.-Y.; Nilsen, H.; Biancolillo, A.; Aït-Kaddour, A.; Sikorski, M.; et al. Fraud in Animal Origin Food Products: Advances in Emerging Spectroscopic Detection Methods over the Past Five Years. *Foods* **2020**, *9*, 1069. [CrossRef]
11. Charles, A.L.; Alamsjah, M.A. Application of chemometric techniques: An innovative approach to discriminate two seaweed cultivars by physico-functional properties. *Food Chem.* **2019**, *289*, 269–277. [CrossRef]
12. Zhao, G.; Luo, X.; Yuan, X.; Jiang, S.; Zhang, H.; Zhang, W.; Bai, P. Study on principal component analysis and multi-class discrimination against the combination to identify easy-confused fur. *J. Ind. Text.* **2021**, *50*, 794–811. [CrossRef]
13. Zi-li, Z.; Chun-Feng, J.; Di, W.; Yong, H.; Xiao-li, L.; Yong-ni, S. Discrimination of varieties of rice using near infrared spectral by PCA and MDA model. In Proceedings of the 6th International Conference on Computer Science & Education (ICCSE 2011), Singapore, 3–5 August 2011; pp. 38–40.
14. Zhao, Y.; Zhang, S.; Zhao, H.; Zhang, H.; Liu, Z. Fast discrimination of mature vinegar varieties with Visible_NIR spectroscopy. In *IFIP Advances in Information and Communication Technology, Proceedings of the Computer and Computing Technologies in Agriculture IV-4th IFIP TC 12 Conference, Nanchang, China, 22–25 October 2010*; Springer: Berlin/Heidelberg, Germany; pp. 721–728.
15. Ghodki, B.M.; Goswami, T.K. Effect of moisture on physical and mechanical properties of cassia. *Cogent Food Agric.* **2016**, *2*, 1192975. [CrossRef]
16. Muhammad, D.R.A.; Dewettinck, K. Cinnamon and its derivatives as potential ingredient in functional food—A review. *Int. J. Food Prop.* **2017**, *20*, 2237–2263. [CrossRef]
17. Mottese, A.F.; Fede, M.R.; Caridi, F.; Sabatino, G.; Marcianò, G.; Calabrese, G.; Albergamo, A.; Dugo, G. Chemometrics and innovative multidimensional data analysis (MDA) based on multi-element screening to protect the Italian porcino (*Boletus* sect. *Boletus*) from fraud. *Food Control* **2020**, *110*, 107004. [CrossRef]
18. Albergamo, A.; Mottese, A.F.; Bua, G.D.; Caridi, F.; Sabatino, G.; Barrega, L.; Costa, R.; Dugo, G. Discrimination of the Sicilian Prickly Pear (*Opuntia Ficus-Indica* L., CV. Muscaredda) According to the Provenance by Testing Unsupervised and Supervised Chemometrics. *J. Food Sci.* **2018**, *83*, 2933–2942. [CrossRef]
19. RV, B.R.; Rajesh, K.S.; Mahadevan, S.; Rosamma, M.P.; Meena, C.V. Detection of adulteration in commercial samples of *Cinnamomum verum* J. S. Presl from Kerala. *Eur. J. Biomed.* **2018**, *5*, 297–304.
20. Amalesh, N.; Nirankush, P.; Kumar, G.A.; Partha, G.; Dipankar, B.; Rahul, S.; Chandrakant, K. Identification of adulterants by pharmacognostical evaluation: Tvak (*Cinnamomum zeylanicum* blume.) & Naluka/Cassia [*Cinnamomum cassia* (Nees & T. Nees.) J. Presl]. *Int. J. Res. Pharma. Biosci.* **2015**, *2*, 1–4.

21. Hermanto, R.F.; Khasanah, L.U.; Atmaka, W.; Manuhara, G.J.; Utami, R. Physical characteristics of cinnamon oil microcapsule. In *IOP Conference Series: Materials Science and Engineering*; IOP Publishing: Bristol, UK, 2016; Volume 107, p. 012064.
22. World Health Organization. S.3.6. Bulk Density and Tapped Density of Powders. 2012. Available online: https://www.who.int/medicines/publications/pharmacopoeia/Bulk-tapped-densityQAS11_450FINAL_MODIFIEDMarch2012.pdf (accessed on 20 March 2021).
23. Jeong, M.-S.; Akram, K.; Ahn, J.-J.; Kwon, J.-H. Applicability of Irradiation Detection Techniques and Quality Characterization of Cinnamon Powders Available in the Korean Market. *Int. J. Food Prop.* **2014**, *17*, 2192–2206. [CrossRef]
24. Media Cybernetics. Image Pro-Plus 7. 2020. Available online: https://www.mediacy.com/imageproplus (accessed on 16 July 2020).
25. Kraithong, S.; Lee, S.; Rawdkuen, S. Physicochemical and functional properties of Thai organic rice flour. *J. Cereal Sci.* **2018**, *79*, 259–266. [CrossRef]
26. Moutaleb, O.H.; Amadou, I.; Amza, T.; Zhang, M. Physico-functional and sensory properties of cowpea flour based recipes (akara) and enriched with sweet potato. *J. Nutr. Health Food Eng.* **2017**, *7*, 325–330.
27. Chandra, S.; Singh, S.; Kumari, D. Evaluation of functional properties of composite flours and sensorial attributes of composite flour biscuits. *J. Food Sci. Technol.* **2015**, *52*, 3681–3688. [CrossRef]
28. IBM. IBM SPSS Statistics for Windows, Version 22.0. IBM Corp: Armonk, NY, USA, 2013. Available online: https://www.ibm.com/support/pages/how-cite-ibm-spss-statistics-or-earlier-versions-spss (accessed on 26 August 2020).
29. Sridhar, K.; Dikkala, P.K.; Yadla, A.K.; Shaik, R.; Jeepalem, Y.; Dhoolipalla Venkata, B.P.; Sodadasi, V. Preparation of microwave-processed oil-free potato snacks with assorted natural flavours: Application of advanced chemometrics. *Int. J. Food Sci. Technol.* **2021**, *56*, 112697. [CrossRef]
30. Ablebits. Ultimate Suite for Excel Comprehensive Set of Time-Saving Tools. 2021. Available online: https://www.ablebits.com/excel-suite/index-2021.php?visitfrom=blog-text (accessed on 17 January 2021).
31. Gao, W.; Chen, F.; Wang, X.; Meng, Q. Recent advances in processing food powders by using superfine grinding techniques: A review. *Compr. Rev. Food Sci. Food Saf.* **2020**, *19*, 2222–2255. [CrossRef]
32. Karabagias, I.K.; Badeka, A.; Kontakos, S.; Karabournioti, S.; Kontominas, M. Characterization and classification of *Thymus capitatus* (L.) honey according to geographical origin based on volatile compounds, physicochemical parameters and chemometrics. *Food Res. Int.* **2014**, *55*, 363–372. [CrossRef]
33. Shawky, E.; Selim, D.A. Rapid Authentication and Quality Evaluation of *Cinnamomum* verum Powder Using Near-Infrared Spectroscopy and Multivariate Analyses. *Planta Med.* **2018**, *84*, 1380–1387. [CrossRef] [PubMed]

foods

MDPI

Article

Multivariate Analysis Coupled with M-SVM Classification for Lard Adulteration Detection in Meat Mixtures of Beef, Lamb, and Chicken Using FTIR Spectroscopy

Muhammad Aadil Siddiqui [1,*], Mohd Haris Md Khir [1], Gunawan Witjaksono [2], Ali Shaan Manzoor Ghumman [3], Muhammad Junaid [4], Saeed Ahmed Magsi [1] and Abdul Saboor [5]

1 Department of Electrical and Electronic Engineering, Universiti Teknologi Petronas, Seri Iskandar 32610, Malaysia; harisk@utp.edu.my (M.H.M.K.); saeed_19001716@utp.edu.my (S.A.M.)
2 BRI Research Institute, Jl. Harsono RM No. 2, Ragunan, Passsar Minggu, Jakarta 12550, Indonesia; gunawan.witjaksono@utp.edu.my
3 Department of Chemical Engineering, Universiti Teknologi Petronas, Seri Iskandar 32610, Malaysia; ali_19001079@utp.edu.my
4 Department of Electronics Engineering, FICT, BUITEMS, Quetta 87300, Pakistan; m.junaid@buitms.edu.pk
5 High Performance Cloud Computing Centre (HPC), Universiti Teknologi Petronas, Seri Iskandar 32610, Malaysia; abdul_19001745@utp.edu.my
* Correspondence: muhammad_18003606@utp.edu.my

Citation: Siddiqui, M.A.; Khir, M.H.M.; Witjaksono, G.; Ghumman, A.S.M.; Junaid, M.; Magsi, S.A.; Saboor, A. Multivariate Analysis Coupled with M-SVM Classification for Lard Adulteration Detection in Meat Mixtures of Beef, Lamb, and Chicken Using FTIR Spectroscopy. *Foods* **2021**, *10*, 2405. https://doi.org/10.3390/foods10102405

Academic Editor: Theodoros Varzakas

Received: 16 August 2021
Accepted: 21 September 2021
Published: 11 October 2021

Abstract: Adulteration of meat products is a delicate issue for people around the globe. The mixing of lard in meat causes a significant problem for end users who are sensitive to halal meat consumption. Due to the highly similar lipid profiles of meat species, the identification of adulteration becomes more difficult. Therefore, a comprehensive spectral detailing of meat species is required, which can boost the adulteration detection process. The experiment was conducted by distributing samples labeled as "Pure (80 samples)" and "Adulterated (90 samples)". Lard was mixed with the ratio of 10–50% v/v with beef, lamb, and chicken samples to obtain adulterated samples. Functional groups were discovered for pure pork, and two regions of difference (RoD) at wavenumbers 1700–1800 cm^{-1} and 2800–3000 cm^{-1} were identified using absorbance values from the FTIR spectrum for all samples. The principal component analysis (PCA) described the studied adulteration using three principal components with an explained variance of 97.31%. The multiclass support vector machine (M-SVM) was trained to identify the sample class values as pure and adulterated clusters. The acquired overall classification accuracy for a cluster of pure samples was 81.25%, whereas when the adulteration ratio was above 10%, 71.21% overall accuracy was achieved for a group of adulterated samples. Beef and lamb samples for both adulterated and pure classes had the highest classification accuracy value of 85%, whereas chicken had the lowest value of 78% for each category. This paper introduces a comprehensive spectrum analysis for pure and adulterated samples of beef, chicken, lamb, and lard. Moreover, we present a rapid M-SVM model for an accurate classification of lard adulteration in different samples despite its low-level presence.

Keywords: food adulteration; halal authentication; Fourier transform infrared (FTIR) spectroscopy; principal component analysis (PCA); chemometric methods; multiclass support vector machine (M-SVM)

1. Introduction

The verification of authenticity and the detection of adulterants are critical aspects of food control, particularly in high-value items. As a measure of food quality and authenticity, laboratory data as well as chemical, physical, and visual pictures of foodstuffs are employed. The authenticity of the food is a major concern in the worldwide food industry; with the abundance of packaged food with a lengthy supply chain on the market, food authenticity is still an issue, as introduced by Spink and Mayor [1]. Nowadays, manual inspection, which is highly impacted by subjective variables, is nevertheless used frequently in quality evaluation. As a result, detecting pork in a variety of food items has become a major research topic in many countries, particularly in those where religious laws restrict the eating of pig products. Food adulterations may only financially impact a part of the population, but others may be more seriously affected [2–4] due to food poisoning, their religious views [5,6], etc. Some of the food tampering has been poisonous, for instance, such as the addition of sawdust to make white bread [7,8], the melamine adulteration of formula milk [7,9,10], the mixing of oil for engines with oil for human consumption in Spain [11]; some cases also involved the misrepresentation of food ingredients such as the UK horse meat issue in 2013 [12–15]. There are several ways of determining the provenance of animal species in meat products that are based on nucleic acid resources, commonly known as molecular techniques, which include DNA finger printing, PCR assays and PCR simple sequence repeat (PCR-SSR) [16,17], chromatographic techniques, isotopic techniques, vibrational and fluorescence spectroscopy, elemental techniques, nuclear magnetic resonance spectroscopy, sensory analysis, non-chromatographic mass spectrometry, immunological techniques, along with chemometrics and bioinformatics [18]. However, each methodology has its own set of drawbacks such as being costly, time-consuming, and inefficient, as well as requiring a wide range of equipment and making it difficult to understand the acquired data; moreover, most of these methods often require extensive sample preparation or are very susceptible to impurities. Unless all the protocols are strictly followed, they may lead to unpredictable outcomes. As a result, establishing a quick and reliable identification procedure to recognize meat species is critical. To address these restrictions, individuals have increasingly turned to spectroscopic methods in recent years. Fourier transform infrared spectroscopy (FTIR) has been widely used in the identification of agricultural commodities such as wine, olive oil, tea, and meat due to its quick and easy operation [19–22]. Research into food-authentication vibrational spectroscopy technologies today has been growing [22–26], partly because the sample preparation using the FTIR technique is relatively simple, results are relatively rapid, and this process is non-destructive in nature. The FTIR spectroscopic methods are thus fast becoming popular [27–33]. Some researchers have started to veer to Near Infrared (NIR) spectroscopy, mainly because its feasibility would open the possibilities of making the food authentication instrumentation set-up portable [28–30]. FTIR is quick and relatively inexpensive, with an easier sample preparation and a non-destructive process [18,19,24,34]. FTIR spectroscopy can distinguish meat and lard in meatball broth quickly and with high accuracy [19,21]; it has also been used with chocolate [24,34] and vegetable oils [22]. Table 1 presents the summary of methods and adulterants used in the literature, along with the multivariate techniques used for detecting the adulteration in different meat species. Therefore, the aim of this study was to utilize in-depth FTIR spectral analysis to improve the accuracy of lard adulteration detection by employing the classification of pure and adulterated samples combined with an M-SVM analysis for lard adulterated in mixtures of beef, lamb, and chicken.

Table 1. Summary of food analyses using multivariate techniques with infrared spectroscopy for the detection of meat species adulteration [35–40].

Method	Meat Adulterant	Analysis Technique
Fourier Transform Infrared Spectroscopy	Palm Oil with Chicken Fat	Linear Discriminant Analysis
E-Nose	Lard, Chicken, and Beef	K-Nearest Neighbors algorithm (KNN), Support Vector Machine (SVM)
Fourier Transform Infrared Spectroscopy	Beef Jerky with pork	LDA, SIMCA, and SVM
Fourier Transform Infrared Spectroscopy	Lard, Mutton, and Cow	PLS Regression
Raman Spectroscopy	Beef and Horsemeat	PCA
Fourier Transform Infrared Spectroscopy	Lard and Palm Oil	PLS
Fourier Transform Infrared Spectroscopy	Lard, Beef Meatballs	PCA and PLS
Fourier Transform Infrared Spectroscopy	Lard in Palm Oil	PCA and PLS

2. Materials and Methods

2.1. Meat Sample Collection

All meat samples were obtained from the local market at Seri Iskander in Malaysia. After that, the meat was washed with purified water and cut into small parts (1 cm × 1 cm) and held at −10 °C. Total samples were then divided into two different classes, as pure and adulterated. There were 80 pure and 90 adulterated samples produced for the spectral analysis. The sample preparation was designed to be straightforward, with no extra chemical substances used. Beef, lamb, and chicken loin cuts were used, and all pork was lean meat taken from chops.

2.2. Extraction Procedure and Sample Distribution

Lard and other animal body fats from meat such as chicken fat, beef fat, and mutton fat were extracted according to the method stated by [34], with little variation. All samples were gradually heated from 50 °C to 150 °C for 45 min until the fat was extracted from all the samples on the petri dish. The discharged fat was then filtered as the concentration contained solid minute particles. Moreover, samples were centrifuged at 3000 rpm for 20 min and filtered through Whatman filter paper. Pure fats produced by the extraction process were then used to make adulterated samples. All the chemicals used in this experiment were of analytical consistency. Pure and adulterated fats were then analyzed using FTIR spectroscopy. The instrument used was Frontier FT-IR by PerkinElmer. The optical system with KBr beam splitter was used to enable quality data collection over a range of 8300–350 cm^{-1} at a best resolution of 0.4 cm. The resulting spectrum contained 2500 continuous values for one sample, with intervals of 0.8 cm^{-1}. To guarantee that there was no major fluctuation between each spectra scanned, each spectrum was recorded at the same temperature. This procedure was required to remove any uncontrolled ambient influences on the instrument and the sample.

2.3. Spectral Data Pre-Processing

Smoothing and normal variate transformation (SNV) were used as spectrum pre-processing approaches in this investigation. The reflectance spectra were smoothed by Savitzky-Golay smoothing using a second-order polynomial and a 5-point window to eliminate the random disturbances caused by the system's internal components. SNV was used to adjust for scatter effects and reduce slope variation. The Savitzky-Golay smoothing filter was used to increase the precision of the data without distorting the signal tendency.

2.4. Preparing Mixture Samples

Lard was mixed with body fats of lamb, beef, and chicken to obtain a series of standard or trained sets of 80 pure and 90 adulterated samples containing 10–50% *v*/*v* of lard in lamb, beef, and chicken samples, as shown in Table 2. The following method is according to Rohman et al. [23]. We prepared six pieces for each combination of lard mixed with a defined percentage of lamb, chicken, and beef, with pork in the proportion of 10, 20, 30, 40, and 50%, whereas B-50%, L-50%, and C-50% represent a 50-50 ratio of pork with beef, lamb, and chicken, respectively; meanwhile, B-90%, L-90%, and C-90% indicate 10% lard with 90% of the respective species. The detailed distribution of samples is presented in Table 3.

Table 2. Distribution of adulterated and pure samples along with the number of pieces produced and spectra obtained for individual species.

Meat Specie	Number of Pieces	Number of Samples Obtained		Number of Spectra Obtained
		Pure Samples	Adulterated Samples (*v*/*v*)	
Beef	20	10 × 2 = 20	15 × 2 = 30	50
Lamb	20	10 × 2 = 20	15 × 2 = 30	50
Pork	20	10 × 2 = 20	-	20
Chicken	20	10 × 2 = 20	15 × 2 = 30	50
Total	80	80	90	170

Table 3. Composition of adulterated samples with the ratio of lard mixed with samples of beef, lamb, and chicken, represented by their initials (Lamb: L-90% to L50%, Beef: B-90% to B-50%, Chicken: C-90% to C-50%).

Mixture Samples Label	Pork (*v*/*v*)	Lamb (*v*/*v*)	Beef (*v*/*v*)	Chicken (*v*/*v*)	Number of Samples
L-90%	10%	90%	-	-	6
L-80%	20%	80%	-	-	6
L-70%	30%	70%	-	-	6
L-60%	40%	60%	-	-	6
L-50%	50%	50%	-	-	6
B-90%	10%	-	90%	-	6
B-80%	20%	-	80%	-	6
B-70%	30%	-	70%	-	6
B-60%	40%	-	60%	-	6
B-50%	50%	-	50%	-	6
C-90%	10%	-	-	90%	6
C-80%	20%	-	-	80%	6
C-70%	30%	-	-	70%	6
C-60%	40%	-	-	60%	6
C-50%	50%	-	-	50%	6
		Total Mixture Samples			90

3. Results and Discussion

After a careful process of sample-making and data pre-processing, the obtained spectrum for both pure and adulterated samples was analyzed separately. The developed workflow for further investigating the lard adulteration was carried out using a three-stage process. In the first stage, identification of functional groups in lard samples without any contamination was made. Secondly, pure spectral samples of beef, lamb, chicken, and lard were analyzed by overlapping the spectrums and identifying the region of difference (RoD) for highly significant regions. Moreover, the profiling of adulterated samples with the percentage difference for beef, lamb, and chicken was also carried out. After spectral analysis, the third and final stage combined the multivariate analysis with M-SVM classification for both pure and adulterated samples separately. Samples were divided into two classes, 'Haram (lard)' and 'Halal (chicken, lamb, and beef)', for M-SVM classification.

3.1. FTIR Spectra Analysis of Pure Samples

Amid the four different meat fats, the pure lard used in this study was evaluated and analyzed separately using FTIR spectroscopy. The peak is shown in Figure 1 approximately at wavenumber 2921 cm^{-1}, which was due to the tensile vibration of C-H (Sp^3) in = C-H cis. The functional group-CH_2 provided peaks at wavenumber 2853 cm^{-1} consecutively as result of asymmetrical and symmetrical vibration. The peak showed the triglyceride ester carbonyl (C=O) group at wavenumber 1750 cm^{-1}.

In the fingerprint region, vibrations of the stretching mode from the C-O group in esters were detected at wavenumber 1155 cm^{-1}, while at wavenumber 1467 cm^{-1} the bending vibrations of the CH_2 and CH_3 aliphatic groups were detected, as shown in Figure 1. Table 4 shows the details of wavenumber and the associated vibration of functional groups for the pure lard sample.

Figure 2 below shows the FTIR spectra of pure samples overlapped for the identification of wavenumbers, with associated peaks identified as the region of difference (RoD) along with the fingerprint region. This spectrum can be divided into three regions to make the analysis convenient: the first region range is at wavenumber 3000–2500 cm^{-1}, the second region range is 2000–2500 cm^{-1}, the third region range is 1500–2000 cm^{-1}, and to conclude, the fingerprint region range is at wavenumber 1500–500 cm^{-1}. Two separate regions are highlighted with dotted lines (a and b), with the overlapping of pure samples for all species, as indicated in Figure 2, where the change in absorbance values is highly prominent; wavenumbers associated with these two regions are in the spectrum ranges of 1700–1800 cm^{-1} for RoD(a) and 2800–3000 cm^{-1} for RoD(b) respectively as shown in Figure 3. The FTIR spectra of all the lipids obtained from different species were combined and overlapped.

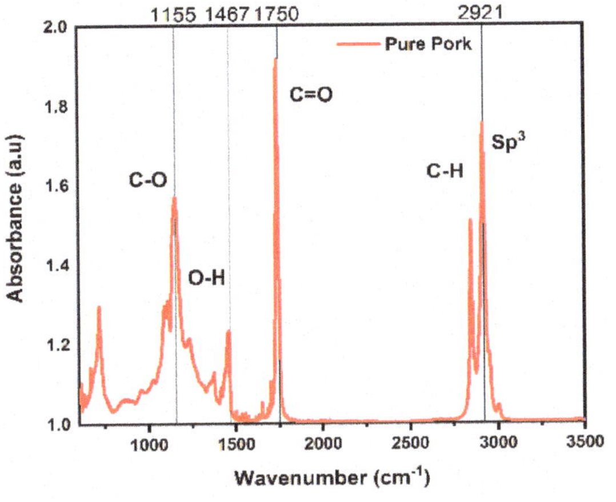

Figure 1. Spectrum analysis of pure pork identifying the frequencies for functional group vibrations.

Table 4. Functional group and associated mode of vibration for pure lard.

Frequency (cm^{-1})	Functional Group Vibration
1155	Vibrations of stretching mode from the C-O group in esters
1467	Bending vibrations of the CH_2 and CH_3 aliphatic groups
1750	Carbonyl (C=O) functional group of the ester linkage of triacylglycerol
2921	Asymmetrical or symmetrical stretching methylene (-CH_2) band vibration

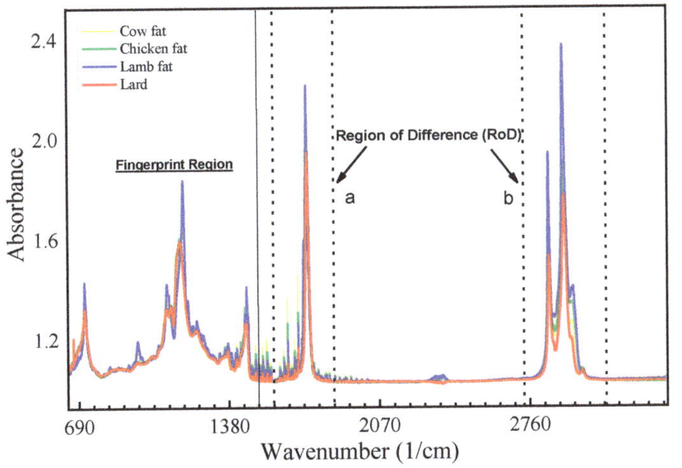

Figure 2. Overlapped spectrum from FTIR covering 3500–650 cm^{-1}, representing the fingerprint and functional group regions for pure samples of beef, lamb, lard, and chicken, with identification of potential regions of difference (RoD).

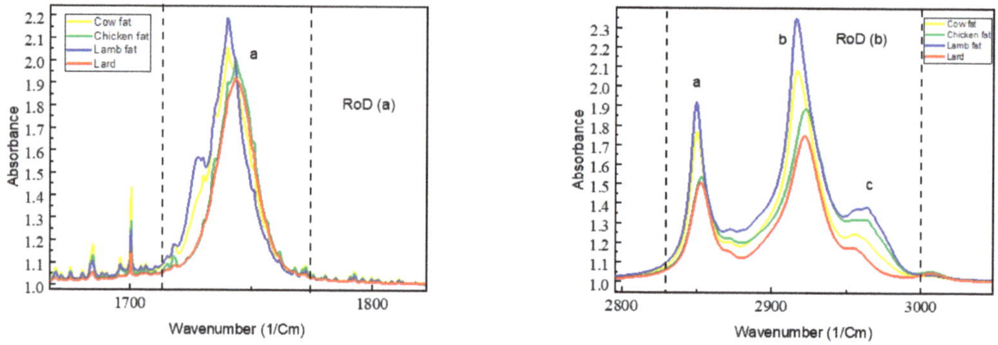

Figure 3. RoD(a) Region of difference peaks zoomed in at wavenumber 1700–1800 cm^{-1} showing the absorbance value for pure samples denoted by a; lard has the lowest value among all samples. RoD(b) Zoomed-in peaks at wavenumber 2800–3000 cm^{-1} where peaks denoted as a, b, and c represent potential regions with difference in absorbance values for all samples.

As the value for the adulteration of lard increases for both beef and chicken, the absorbance values merge with the lard, showing high contrast compared to lamb samples, which indicates negligible change when lard is mixed. This is clearly visible in the spectral analysis shown in Figure 4 for all the adulterated samples. The absorbance values in the region of RoD(b) are carefully analyzed, where the adulteration of lard can potentially be detected. This is shown in Table 5. On the other hand, beef samples are highly prone, and lard is detectable because of the significant change in absorbance value at the region of 2800–3000 cm^{-1} in the spectrum, specifically at RoD(b) a and b, which represent regions at 2840–2860 and 2900–2940 cm^{-1}, respectively. Table 5 lists all the absorbance values at the peaks of RoD(b) in Figure 2; the percentage difference is calculated with respect to lard for peak absorbance in regions with high significance.

The highest proximity of absorbance values to pure lard can be seen in the samples of B-50%, C-90%, C-80%, and C-50%, for both regions RoD(b)-a and RoD(b)-b. At the same time, adulterated beef shows a pattern of variation according to the adulteration percentage of lard. Beef samples with 10% adulteration (B-90%) have an approximate percentage difference of 7–14%, while beef with 50% adulteration (B-50%) shows approximately 3–8% change for both regions. All samples containing adulterated chicken from C-50% to C-90% show the lowest percentage difference as compared to lamb and beef. This reveals the highest similarity to be between chicken and lard, which could present some difficulty in detecting the adulteration of lard in chicken irrespective of the percentage mixing. Moreover, adulterated lamb samples depict minor variation in absorbance values throughout the mixing samples (L-50% to L-90%) and have the highest percentage difference as compared to pure lard.

3.2. Results of Principal Component Analysis

Pure lard, along with other samples of beef, chicken, and lamb, was classified using the chemometric of PCA. PCA is used to reduce the dimension of the spectral signal. The wavenumber regions for PCA were also optimized. To confirm the separation based on adulterant type, the raw data (eigenvectors of the covariance matrix) was subjected to principal component analysis (PCA). Further explanation on PCA is at Appendix A.1

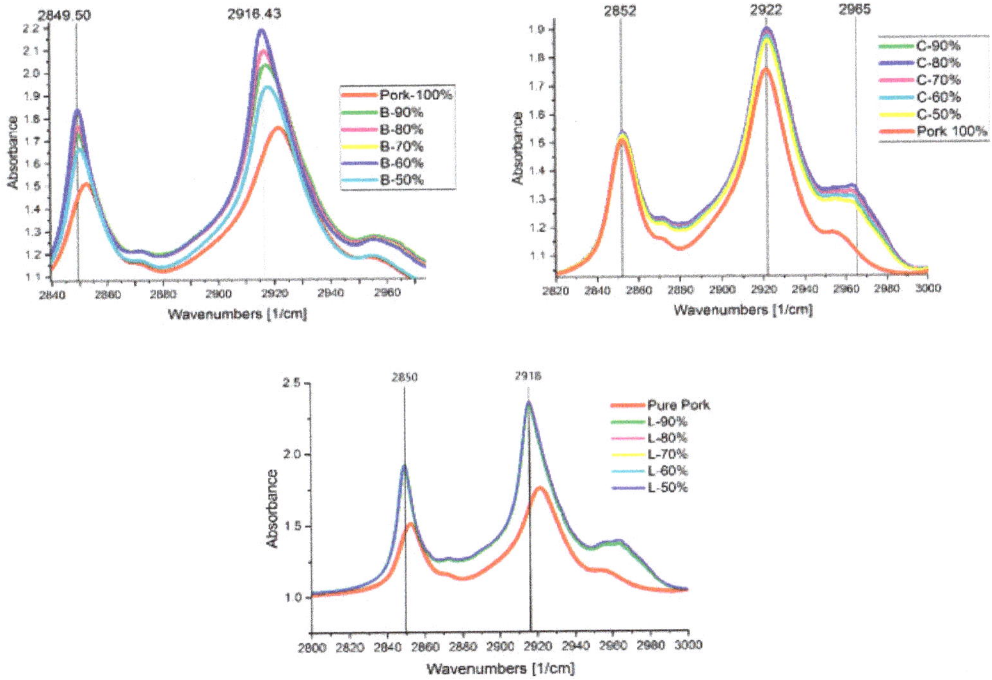

Figure 4. Peaks zoomed-in for adulterated samples overlapped with pure pork, for beef (B-50% to B-90%), lamb (L-50% to L-90%), and chicken (C-50% to C-90%), with associated adulteration ratio and associated peaks for RoD(a) and RoD(b).

Table 5. Absorbance values and percentage difference with respect to lard for adulterated samples of beef, lamb, and chicken in the region of RoD(b) at the highly significant region of 2800–3000 cm^{-1}.

Species Type	Sample	Absorbance Value at RoD(b)-a	Absorbance Value at RoD(b)-b	Percentage Difference w.r.t Pork	
				RoD(b)-a	RoD(b)-b
Pure Lard	Pork-100%	1.5963	1.75306		
Adulterated Beef	B-50%	1.6580	1.9154	3.79%	8.85%
	B-60%	1.8357	2.1793	13.95%	21.67%
	B-70%	1.8310	2.1784	13.69%	21.63%
	B-80%	1.7611	2.0906	9.81%	17.56%
	B-90%	1.7262	2.0227	7.81%	14.28%
Adulterated Chicken	C-50%	1.5256	1.8577	4.52%	5.79%
	C-60%	1.5289	1.8737	4.31%	6.65%
	C-70%	1.5312	1.8868	4.16%	7.34%
		1.5358	1.8995	3.86%	8.01%
	C-90%	1.5358	1.8995	3.86%	8.01%
Adulterated Lamb	L-50%	1.8739	2.2576	15.99%	25.15%
	L-60%	1.8739	2.2576	15.99%	25.15%
	L-70%	1.8739	2.2576	15.99%	25.15%
	L-80%	1.8739	2.2576	15.99%	25.15%
	L-90%	1.8710	2.2396	15.84%	24.37%

It is possible to observe a distinct split depending on the level of adulteration by showing the scores of the first two main components (Figure 5), which represent 99.36 percent of data variance. Only a little amount of overlap exists between the chicken samples that have been tainted with pork. The selection of wavenumbers was based on their ability to provide a useful classification between samples, as seen in Figure 5. The PCA plot showed clusters of samples based on their similarity with the first main component (PC1) and the second main component (PC2), which provided a good separation between the lamb, beef, and pork groups but was unable to separate pork and chicken. The percentage (%) variability of PC1 and PC2 was 97.31% and 2.05%, respectively. PC1 comprised the most variation of the data, as shown in Table 6.

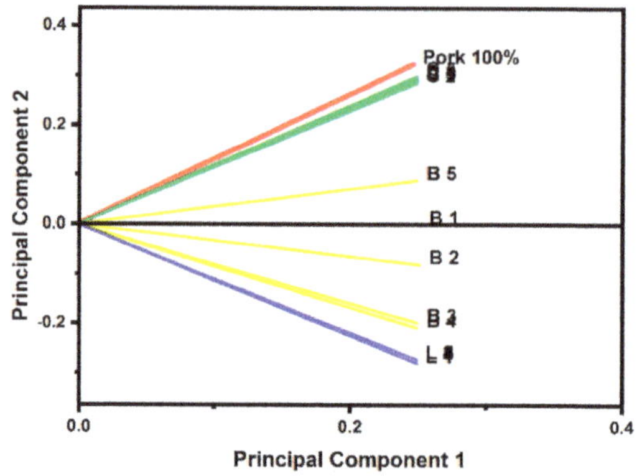

Figure 5. Principal component analysis plot showing the similarity between pork, chicken, lamb, and beef samples with adulterated mixtures. C1–C5 (10–50% Pork), B1–B5 (10–50% Pork), L1–L5 (10–50% Pork).

Table 6. Percentage of variance for each PCA component contributing to the variation of the classification.

Principal Component	Variance Contribution
PC1	97.31%
PC2	2.05%
PC3	0.64%

The FTIR spectra of the pure pork sample were compared with those of adulterated beef, chicken, and lamb. Three dimensional plots are shown in Figure 6. The PCA analysis shows the PCA projection divided into three dimensions for better analysis.

Figure 6a shows the distribution of samples across the first principal component using 1D spectra of the pure samples for beef, lamb, chicken, and pork, where chicken and pork samples overlap and correlate highly coupled values of absorbance with similar wavenumbers. At the same time, Figure 6b depicts the samples at PC1 and PC2 using 2D representation for all the adulterated species. Figure 6c combines all the three principal components using 3D for all the adulterated samples. The regions in these figures are separated based on the adulteration quantity, starting with slightly mixed, i.e., 10%, to highly adulterated, i.e., 50%. In the first projection, the plotted points representing the samples of chicken, beef, and lamb are scattered, and they are far from the pork group. The closer the dots of chicken, beef, and lamb are to the pork samples, the more significant the quantity of lard is in pure samples.

3.3. Multiclass Support Vector Machine Classification

The data obtained from the previous processes were divided into testing data (30%) and training data (70%), and subsequently evaluated with the classification model. The data acquired from the FTIR spectroscope was analyzed using the scikit-learn machine learning library in Python. The radial basis function (RBF) was used as the kernel function of SVM using the grid search method. To add an extra validation step to our model, we used the confusion matrix for both multiclass datasets, as shown in Tables 7 and 8. The confusion matrix projects the true data against predicted data. In our study, we divided the problem into two different sections: one identified pure samples correctly, and the other predicted the adulterated samples. The learning rate was 0.0001, and the regularization parameter λ was set to 1/epochs. Table 7 illustrates the user, producer, and overall accuracy of the pure samples data set. Details of the SVM is explained at Appendix A.2. Pure samples of beef and lamb using optimal parameters produced the highest accuracy (85%) among all the samples. Furthermore, pure samples of chicken had the lowest accuracy of 75%, whereas pure pork was significantly better than chicken, with 80% accuracy. Moreover, Figure 7 shows a confusion matrix using a 10-fold cross-validation for the pure samples where the a, b, and c rows represent the true label; meanwhile, according to the model prediction, the a, b, and c columns represent the number of predicted sets for each respective class.

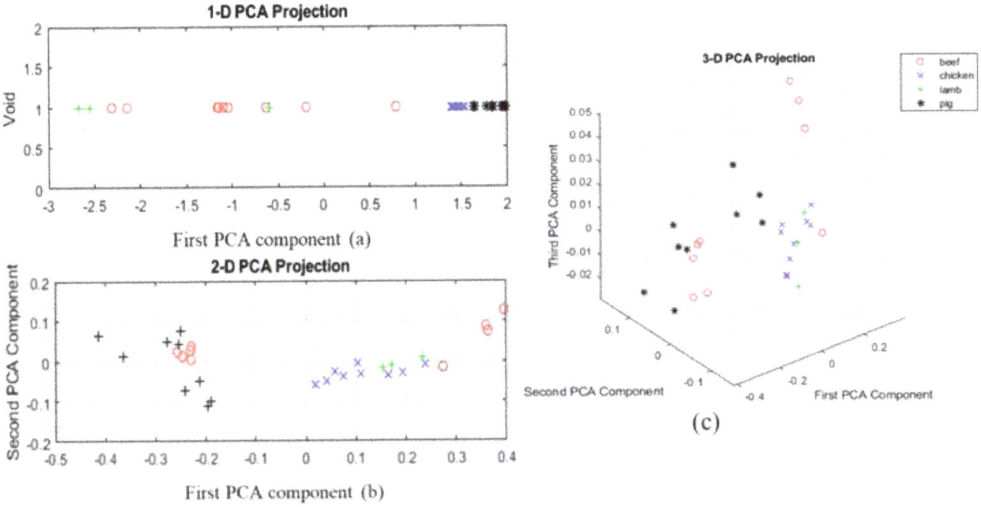

Figure 6. (**a**) Pure samples of all the species using a one-dimensional projection for first principal component. (**b**,**c**) Principal component analysis of the two and three-dimensional projections of adulterated samples of pork (*), chicken (×), lamb (+), and beef (O), showing the clustering.

Table 7. Sensitivity, precision, and classification accuracy for pure samples of beef, lamb, chicken, and pork.

Classified as	User Accuracy (Sensitivity)	Producer Accuracy (Precision)	Overall Accuracy
Beef	85%	85.00%	81.25%
Lamb	85%	85.00%	
Chicken	78%	75.00%	
Pork	76%	80.00%	

Table 8. Sensitivity, precision, and classification accuracy for adulterated samples of beef, chicken, and lamb.

Classified as	User Accuracy (Sensitivity)	Producer Accuracy (Precision)	Overall Accuracy
a = AdulteratedBeef	68.86%	73.33%	72.2%
b = AdulteratedLamb	67.19%	76.66%	
c = AdulteratedChicken	83.20%	66.00%	

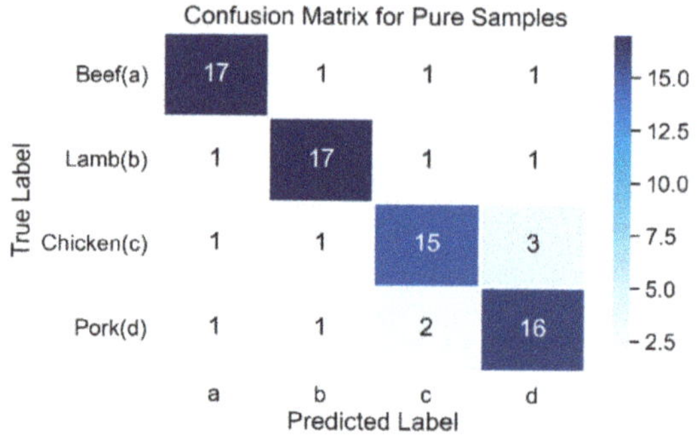

Figure 7. Heatmap confusion matrix of multiclass classification for pure samples of beef, chicken, lamb, and pork showing the predicted and true labels.

The predicted labels for pure samples shown in Figure 7 misclassified three samples of pure chicken as pure pork, while two samples of pure pork were falsely labeled as chicken. Moreover, beef and lamb both had three label misclassifications, one for each species of meat.

Table 8 shows the confusion matrix for the multiclass SVM of adulterated data samples. The adulterated data set contained all the samples that were adulterated with different proportions of lard. The AdulteratedBeef sample included samples with a v/v ratio from B-50% to B-90%. The producer accuracy was highest for AdulteratedLamb at 76.6%, whereas AdulteratedBeef had the second-highest value of 73.3%. The spectrum of lamb had no change in absorbance value when it was adulterated, irrespective of the adulteration ratio, which was also validated by the SVM classifier by getting the maximum number of correctly classified labels, as shown in Figure 8.

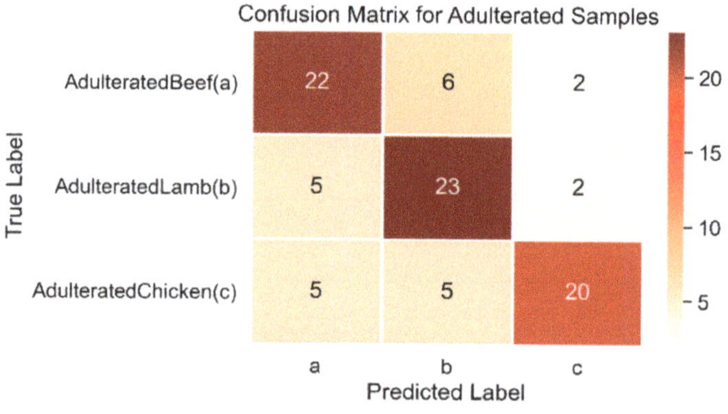

Figure 8. Heatmap confusion matrix of the multiclass SVM classifier for adulterated samples of beef, chicken, and lamb.

AdulteratedChicken samples, with 20 correctly classified samples, produced the lowest precision accuracy of 66% due to its high variation in absorbance values, as shown in Figure 8.

4. Conclusions

FTIR spectroscopy, coupled with the multivariate and M-SVM methods, seems to be an efficient and rapid technique for the discrimination of lard from other meat samples. In this paper, we demonstrated the identification and discrimination of lard from beef, chicken, and lamb fats in meat mixtures. FTIR spectral analysis in combination with Principal

Component Analysis (PCA) and M-SVM have shown that pure lard fat has unique peaks that can distinguish the pork from beef, chicken, and lamb meat at wavenumbers 1155 cm^{-1}, 1467 cm^{-1}, 1750 cm^{-1}, and 2921 cm^{-1}. The absorbance values indicate a direct correlation between lard and other species. The PCA results show that adulteration in chicken meat is positively correlated with pork meat, while lamb is negatively correlated with respect to lard. The SVM model produced an overall prediction accuracy of 81.25% for pure samples, and for adulterated samples, it showed a 72.2% prediction accuracy. The overall accuracy was computed using the sensitivity and precision values. The model accurately classified the pure samples better than the adulterated samples due to a smaller number of samples and the minimalistic difference in absorbance values of the spectrum. Thus, this study has the potential to establish as a rapid method for halal authentication and could revolutionize the in-line quality control in the meat industry. For future work, the FTIR profiles for pure and adulterated samples can be increased, and deep learning may be applied for detecting an adulteration quantity of less than 10%.

Author Contributions: M.A.S. and G.W. conceived the project; M.H.M.K. assisted and supported the experiment; A.S.M.G. supported the sample preparation; and M.J., S.A.M. and A.S. reviewed the manuscript. All authors have read and agreed to the published version of the manuscript.

Funding: This study is funded by Centre of Graduate Studies, UTP in collaboration with ITI institute for smart mobility, University Technology Petronas, Perak, Malaysia.

Institutional Review Board Statement: Not applicable.

Informed Consent Statement: Not applicable.

Data Availability Statement: Data presented in this study is available at request from the corresponding author.

Conflicts of Interest: The authors declare no conflict of interest.

Appendix A

Appendix A.1. Principal Component Analysis

Principal Component Analysis (PCA) is a statistical technique that is particularly useful in reducing observations that have many dimensions. This technique consists of transforming dimensions of a dataset into a new but smaller set of uncorrelated dimensions called principal components (PCs). An array of (q_{ij}) values can be normalized using the equation below:

$$X_{ij} = q_{ij} - q_j \tag{A1}$$

The data given to us is the array element data corresponding to the variable X_{ij}, and the mean value of the variable q_j. Then, using the new dataset array, a correlation matrix is constructed so that information about how the variables in the dataset are correlated can be obtained. To create our new correlation matrix X with the new correlation coefficients X_{ij}, the following formula is used:

$$R = X^{T} \bullet X \tag{A2}$$

Only the principal components that explain the greatest amount of data in the original are determined using the equation below:

$$S = V \bullet Q \tag{A3}$$

where S is the matrix data, known as Score; V is the eigenvectors; and Q is the original data array. The matrix S (Score) will now represent the data in a way that each column represents the projection of the initial data Q.

Appendix A.2. Support Vector Machine Classification

Most machine learning techniques have been created and statistically verified for linearly separable data. For the reduction of dimensionality, linear classifiers such as Support Vector Machines (SVMs) or the (conventional) Principal Component Analysis (PCA) are common examples. However, to efficiently accomplish tasks involving pattern analysis and discovery, most real-world data require non-linear approaches. By incorporating the kernel trick, the SVM approach has improved over time. To detect a pattern in non-linear separable data, the kernel method effectively translates the input data to higher dimensions. When the training data has many variables in comparison to the number of observations, SVMs are an excellent classification approach. In SVM, every sample x that consists of n variables is treated as an n-dimensional vector. Prediction performance can be assessed using the following three indicators: sensitivity (User Accuracy), precision (Producer Accuracy), and overall accuracy. Precision is the proportion of appropriately positive labels produced by our software to all positive labels produced. The ratio of the exactly positive labels identified by our algorithm to all positive labels is referred to as sensitivity. Accuracy is the proportion of correctly categorized topics to the total number of issues. Equations (A4)–(A6) present the formula for Precision, Accuracy, and Sensitivity.

$$Sensitivity = \frac{True\ Positive}{Predicted\ Results} \tag{A4}$$

$$Precision = \frac{True\ Positive}{Actual\ Results} \tag{A5}$$

$$Overall\ Accuracy = \frac{True\ Positive + True\ Negative}{Total} \tag{A6}$$

References

1. Spink, J.; Moyer, D.C. Defining the Public Health Threat of Food Fraud. *J. Food Sci.* **2011**, *76*, R157–R163. [CrossRef] [PubMed]
2. The Guardian. Allergic Teenager's Death after Eating Kebab Was Accidental, Rules Coroner. *The Guardian*, 2017. Available online: https://www.theguardian.com/uk-news/2017/jun/16/teenager-with-dairy-allergy-died-accidentally-rules-coroner(accessed on 16 July 2017).
3. Li, D.K. Toddler Allergic to Dairy Dies after Pre-School Serves Him Grilled Cheese. *New York Post*, 9 November 2017.
4. Barlass, T. Child Aged 10 Dies after Drinking Coconut Drink as Importer Admits Label Charges. *Sydney Morning Herald*, 6 October 2015.
5. FSA. (2013, 7/2/2018). Timeline on Horse Meat Issue. Available online: http://www.food.gov.uk/enforcement/monitoring/horse-meat/timeline-horsemeat (accessed on 16 August 2021).
6. Rohman, A.; Man, Y.B.C. Analysis of Pig Derivatives for Halal Authentication Studies. *Food Rev. Int.* **2012**, *28*, 97–112. [CrossRef]
7. Tähkäpää, S.; Maijala, R.; Korkeala, H.; Nevas, M. Patterns of food frauds and adulterations reported in the EU rapid alert system for food and feed and in Finland. *Food Control* **2015**, *47*, 175–184. [CrossRef]
8. Wood, R. Symposium on Food Identification and Authentication. Available online: https://qualityalchemist.blogspot.com/2012/01/symposium-on-food-identification-and.html (accessed on 16 July 2017).
9. Guan, N.; Fan, Q.; Ding, J.; Zhao, Y.; Lu, J.; Ai, Y.; Xu, G.; Zhu, S.; Yao, C.; Jiang, L.; et al. Melamine-Contaminated Powdered Formula and Urolithiasis in Young Children. *N. Engl. J. Med.* **2009**, *360*, 1067–1074. [CrossRef]
10. Jia, C.; Jukes, D. The national food safety control system of China e a systematic review. *Food Control* **2013**, *32*, 236–245. [CrossRef]
11. Borda, I.A.; Philen, R.M.; de la Paz, M.P.; de la Cámara, A.G.; Ruiz-Navarro, M.D.; Ribota, O.G.; Soldevilla, J.A.; Terracini, B.; Peña, S.S.; Leal, C.F.; et al. Toxic oil syndrome mortality: The first 13 years. *Int. J. Epidemiol.* **1998**, *27*, 1057–1063. [CrossRef]
12. Regenstein, J.M.; Chaudry, M.M.; Regenstein, C.E. The Kosher and Halal Food Laws. *Compr. Rev. Food Sci. Food Saf.* **2003**, *2*, 111–127. [CrossRef]
13. Chuah, L.-O.; He, X.B.; Effarizah, M.E.; Syahariza, Z.A.; Shamila-Syuhada, A.K.; Rusul, G. Mislabelling of beef and poultry products sold in Malaysia. *Food Control* **2016**, *62*, 157–164. [CrossRef]
14. Reuters, T. State of The Global Islamic Economy Report 2016/17. 2016. Available online: https://www.iedcdubai.ae/ (accessed on 16 July 2017).
15. Barnett, J.; Begen, F.; Howes, S.; Regan, A.; McConnon, A.; Marcu, A.; Rowntree, S.; Verbeke, W. Consumers' confidence, reflections and response strategies following the horsemeat incident. *Food Control* **2016**, *59*, 721–730. [CrossRef]
16. Schmutzler, M.; Beganovic, A.; Böhler, G.; Huck, C.W. Methods for detection of pork adulteration in veal product based on FT-NIR spectroscopy for laboratory, industrial and on-site analysis. *Food Control* **2015**, *57*, 258–267. [CrossRef]

17. Vlachos, A.; Arvanitoyannis, I.S.; Tserkezou, P. An Updated Review of Meat Authenticity Methods and Applications. *Crit. Rev. Food Sci. Nutr.* **2016**, *56*, 1061–1096. [CrossRef] [PubMed]
18. Danezis, G.P.; Tsagkaris, A.S.; Camin, F.; Brusic, V.; Georgiou, C.A. Food authentication: Techniques, trends & emerging approaches. *TrAC Trends Anal. Chem.* **2016**, *85*, 123–132.
19. Kurniawati, E.; Rohman, A.; Triyana, K. Analysis of lard in meatball broth using Fourier transform infrared spectroscopy and chemometrics. *Meat Sci.* **2014**, *96*, 94–98. [CrossRef] [PubMed]
20. Meza-Márquez, O.G.; Gallardo-Velázquez, T.; Osorio-Revilla, G. Application of mid-infrared spectroscopy with multivariate analysis and soft independent modeling of class analogies (SIMCA) for the detection of adulterants in minced beef. *Meat Sci.* **2010**, *86*, 511–519. [CrossRef]
21. Rahmania, H.; Sudjadi; Rohman, A. The employment of FTIR spectroscopy in combination with chemometrics for analysis of rat meat in meatball formulation. *Meat Sci.* **2015**, *100*, 301–305. [CrossRef]
22. Rohman, A.; Man, Y.B.C.; Hashim, P.; Ismail, A. FTIR spectroscopy combined with chemometrics for analysis of lard adulteration in some vegetable oils Espectroscopia FTIR combinada con quimiometría para el análisis de adulteración con grasa de cerdo de aceites vegetales. *CyTA J. Food* **2011**, *9*, 96–101. [CrossRef]
23. Rohman, A.; Che Man, Y.B. FTIR spectroscopy combined with chemometrics for analysis of lard in the mixtures with body fats of lamb, cow and chicke. *Int. Food Res. J.* **2010**, *17*, 519–527.
24. Suparman, W.S.; Sundhani, E.; Saputri, S.D. The use of Fourier transform infrared spectroscopy (FTIR) and gas chromatography mass spectroscopy (GCMS) for Halal authentication in imported chocolate with various variants. *Analysis* **2015**, *2*, 3.
25. Xu, L.; Cai, C.B.; Cui, H.F.; Ye, Z.H.; Yu, X.P. Rapid discrimination of pork in Halal and non-Halal Chinese ham sausages by Fourier transform infrared (FTIR) spectroscopy and chemometrics. *Meat Sci.* **2012**, *92*, 506–510. [CrossRef]
26. Yang, H.; Irudayaraj, J.; Paradkar, M.M. Discriminant analysis of edible oils and fats by FTIR, FT-NIR and FT-Raman spectroscopy. *Food Chem.* **2005**, *93*, 25–32. [CrossRef]
27. Alfar, I.J.; Khorshidtalab, A.; Akmeliawati, R.; Ahmad, S.; Jaswir, I. Towards authentication of beef, chicken and lard using micro near-infrared spectrometer based on support vector machine classification. *ARPN J. Eng. Appl. Sci.* **2016**, *11*, 4130–4136.
28. Alamprese, C.; Amigo, J.M.; Casiraghi, E.; Engelsen, S.B. Identification and quantification of turkey meat adulteration in fresh, frozen-thawed and cooked minced beef by FT-NIR spectroscopy and chemometrics. *Meat Sci.* **2016**, *121*, 175–181. [CrossRef] [PubMed]
29. Barbin, D.F.; Sun, D.-W.; Su, C. NIR hyperspectral imaging as non-destructive evaluation tool for the recognition of fresh and frozen–thawed porcine longissimus dorsi muscles. *Innov. Food Sci. Emerg. Technol.* **2013**, *18*, 226–236. [CrossRef]
30. Morsy, N.; Sun, D.-W. Robust linear and non-linear models of NIR spectroscopy for detection and quantification of adulterants in fresh and frozen-thawed minced beef. *Meat Sci.* **2013**, *93*, 292–302. [CrossRef] [PubMed]
31. Nizar, N.N.A.; Marikkar, J.M.N.; Hashim, D.M. Differentiation of lard, chicken fat, beef fat and mutton fat by GCMS and EA-IRMS techniques. *J. Oleo Sci.* **2013**, *62*, 459–464. [CrossRef]
32. Marikkar, J.M.N.; Ghazali, H.M.; Man, Y.B.C.; Peiris, T.S.G.; Lai, O.M. Distinguishing lard from other animal fats in admixtures of some vegetable oils using liquid chromatographic data coupled with multivariate data analysis. *Food Chem.* **2005**, *91*, 5–14. [CrossRef]
33. Wang, L.; Hang, X.; Geng, R.; Wang, L.; Hang, X.; Geng, R. Molecular detection of adulteration in commercial buffalo meat products by multiplex PCR assay. *Food Sci. Technol.* **2019**, *39*, 344–348. [CrossRef]
34. Man, Y.B.C.; Syahariza, Z.A.; Mirghani, M.E.S.; Jinap, S.; Bakar, J. Analysis of potential lard adulteration in chocolate and chocolate products using Fourier transform infrared spectroscopy. *Food Chem.* **2005**, *90*, 815–819.
35. The Harmful Effects of Counterfeit Goods–Athens State University. Available online: https://www.athens.edu/business-journal/spring-2013/asowder-couterfeit/ (accessed on 10 November 2019).
36. Food Taboos: Their Origins and Purposes. Available online: https://www.ncbi.nlm.nih.gov/pmc/articles/PMC2711054/ (accessed on 10 November 2019).
37. Alikord, M.; Momtaz, H.; keramat, J.; Kadivar, M.; Rad, A.H. Species identification and animal authentication in meat products: A review. *J. Food Meas. Charact.* **2018**, *12*, 145–155. [CrossRef]
38. Rohman, A.; Sismindari; Erwanto, Y.; Man, Y.B.C. Analysis of pork adulteration in beef meatball using Fourier transform infrared (FTIR) spectroscopy. *Meat Sci.* **2011**, *88*, 91–95. [CrossRef]
39. Manning, L.; Soon, J.M. Food Safety, Food Fraud, and Food Defense: A Fast Evolving Literature. *J. Food Sci.* **2016**, *81*, R823–R834. [CrossRef] [PubMed]
40. Yang, L.; Wu, T.; Liu, Y.; Zou, J.; Huang, Y.; Babu, V.S.; Lin, L. Rapid Identification of Pork Adulterated in the Beef and Mutton by Infrared Spectroscopy. *J. Spectrosc.* **2018**, *2018*, 2413874. [CrossRef]

foods

Article

Detecting Dye-Contaminated Vegetables Using Low-Field NMR Relaxometry

Sumaiya Shomaji *,†, Naren Vikram Raj Masna †, David Ariando, Shubhra Deb Paul, Kelsey Horace-Herron, Domenic Forte, Soumyajit Mandal and Swarup Bhunia

Department of Electrical and Computer Engineering, University of Florida, 216 Larsen Hall, P.O. Box 116200, Gainesville, FL 32611, USA; nmasna@ufl.edu (N.V.R.M.); dariando@ufl.edu (D.A.); shubhra.paul@ufl.edu (S.D.P.); khoraceherron@ufl.edu (K.H.-H.); dforte@ece.ufl.edu (D.F.); soumyajit@ece.ufl.edu (S.M.); swarup@ece.ufl.edu (S.B.)
* Correspondence: shomaji@ufl.edu
† These authors contributed equally to this work.

Abstract: Dyeing vegetables with harmful compounds has become an alarming public health issue over the past few years. Excessive consumption of these dyed vegetables can cause severe health hazards, including cancer. Copper sulfate, malachite green, and Sudan red are some of the non-food-grade dyes widely used on vegetables by untrusted entities in the food supply chain to make them look fresh and vibrant. In this study, the presence and quantity of dye-based adulteration in vegetables are determined by applying ^{1}H-nuclear magnetic resonance (NMR) relaxometry. The proposed technique was validated by treating some vegetables in-house with different dyes and then soaking them in various solvents. The resulting solutions were collected and analyzed using NMR relaxometry. Specifically, the effective transverse relaxation time constant, $T_{2,eff}$, of each solution was estimated using a Carr–Purcell–Meiboom–Gill (CPMG) pulse sequence. Finally, the estimated time constants (i.e., measured signatures) were compared with a library of existing $T_{2,eff}$ data to detect and quantify the presence of unwanted dyes. The latter consists of data-driven models of transverse decay times for various concentrations of each water-soluble dye. The time required to analyze each sample using the proposed approach is dye-dependent but typically no longer than a few minutes. The analysis results can be used to generate warning flags if the detected dye concentrations violate widely accepted standards for food dyes. The proposed low-cost detection approach can be used in various stages of a produce supply chain, including consumer household.

Keywords: food adulteration; dye additives; nuclear magnetic resonance; relaxometry

Citation: Shomaji, S.; Masna, N.V.R.; Ariando, D.; Deb Paul, S.; Horace-Herron, K.; Forte, D.; Mandal, S.; Bhunia, S. Detecting Dye-Contaminated Vegetables Using Low-Field NMR Relaxometry. *Foods* **2021**, *10*, 2232. https://doi.org/10.3390/foods10092232

Academic Editor: Theodoros Varzakas

Received: 6 August 2021
Accepted: 2 September 2021
Published: 21 September 2021

1. Introduction

Food adulteration has reportedly increased over the last few years because of the complex supply chain of food from producer to consumer. Due to urbanization, consumers rely on growth, processing, transportation, and supply of food by multiple entities in the supply chain [1]. An untrusted entity can cause adulteration of food at any of these stages. Adulteration can take numerous forms, e.g., deliberate addition of substances with adverse health outcomes, not meeting desired product quality metrics, imitating other food substances, and using false labels on food packaging [2]. Human health is sensitive to food and thus can be affected by acute or chronic exposure to adulterated products. Even major health hazards, involving liver, vision, skin, and stomach disorders, are directly associated to adulterated food intake [3]. Foods like vegetables, fruits, fish, or meat adulterated with formalin have been found to be responsible for asthma and cancer [3]. Use of chemical pesticides has been linked to severe health problems, such as nerve damage and cancer [4]. There is also evidence that dye additives are responsible for genotoxicity, hypersensitivity, and carcinogenicity [5].

Synthetic dyes are added to many foods to provide them with a fresh look and compensate for natural color variations. These dyes are often harmful for the health and may even be carcinogenic [6]. Therefore, it is very important to understand the ingredients of food items before consuming them. This information is generally available for packaged foods since genuine product labels include the names of any dyes within the list of ingredients. However, fresh fruits and vegetables are generally not labeled. Dishonest entities in the food supply chain can exploit this lack of information to add toxic dyes to fruits and vegetables that make them appear fresh and vibrant to customers. Some real-world examples of this practice are shown in Figure 1. Existing methods to detect many of these dyes have been thoroughly reviewed in [6]. For example, chromatographic, physiochemical, sensory, spectroscopy, and DNA-based detection methods have been combined with chemometrics for a wide range of adulteration-detection applications [7]. Detection approaches that are particularly suitable for dyes include capillary electrophoresis, electrochemical voltametric analysis, and amperometry [6–8]. To illustrate, carcinogenic compounds, like malachite green [9] and Sudan red [10], can be easily detected by liquid chromatography, gas chromatography, capillary electrophoresis, amperometry, and plasmon resonance light scattering [6].

Figure 1. Various instances of vegetables and other consumables being adulterated with harmful chemicals. In most cases, cheap, industrial-grade dyes are used instead of food colors to maximize profits [11–14].

Traditional methods have shown promising results in detecting food dyes with very high accuracy [6]. However, they have some limitations. Firstly, they require a labor-intensive set of tasks that ranges from sample preparation to analysis. Therefore, the experiments require a large expenditure of time and human effort, making them unsuitable for at-home and field applications. Secondly, some of these methods often require expensive instrumentation that is often unavailable in the low- and middle-income countries where dye-based adulteration is most common [7]. For example, NMR spectroscopy requires highly uniform magnets, which are bulky and expensive [15,16]. Thirdly, low-cost methods generally detect adulteration by observing anomalies in basic physical or chemical properties of the suspect substance (e.g., viscosity, pH, or electrical conductivity) [6,7]. However, modern "smart" adulteration techniques can bypass such simple detection methods [17]. Therefore, to confront the food adulteration issues, i.e., the deliberate or accidental contamination of food items with banned substances, the food industry, government bodies, and consumers need sensitive, rapid, reliable, inexpensive, widely applicable, and difficult-to-attack methods to detect adulterated foods. Spectroscopy meets many of these criteria and is promising for detecting adulteration. During spectroscopy-based analysis, the chemical composition of a food product is investigated by measuring its frequency-dependent absorption or reflection spectra. Absorbance-based spectroscopy is mostly used for liquids, whereas reflection-based spectroscopy is used to identify fillers and adulterants, such as low-cost spices and dyes used to mask ageing. A variety of spectroscopic techniques, including near-infrared (IR), mid-IR, Raman, nuclear quadrupole resonance (NQR), and nuclear magnetic resonance (NMR), have been successfully used for monitoring food quality [18]. Each technique has its own advantages and disadvantages, which makes the optimum choice strongly application dependent.

NMR is rapidly emerging as an important analytical technique for food analysis and screening [19]. NMR-based methods can be grouped into three major measurement categories: imaging, spectroscopy, and relaxometry. NMR spectroscopy has many applications in food analysis and adulterant detection. For example, it has been used to detect Sudan red in paprika powder with higher sensitivity than Raman or IR spectroscopy [10]. Nevertheless, NMR is intrinsically a bulk measurement method, so detecting adulterants at extremely low concentrations (e.g., parts per billion) remains challenging [10]. Moreover, high-resolution NMR spectroscopy requires a strong and highly uniform static magnetic field (known as B_0). Such fields are typically generated using large cryogenically cooled superconducting coils, thus resulting in very high installation and maintenance costs. A recent work proves that cryogen-free, desktop-sized permanent magnets can provide a lower-cost alternative [20]. Nevertheless, such magnets must be temperature-stabilized and manually-calibrated, so costs are still quite high (typically at least $20,000) [15,16]. As a result, complete NMR spectrometers (which combine the magnet with sample interrogation and readout electronics) cost $50,000 or more. Thus, there is a need for lower-cost alternatives for analyzing food samples.

NMR relaxometry provides such an alternative since it can be performed in a relatively weak and inhomogeneous B_0 field, which in turn allows the size, complexity, and cost of the magnet to be greatly reduced [21,22]. Relaxometry focuses on measuring the nuclear spin relaxation times of specific substances present in a sample, namely the spin- lattice (T_1) and spin-spin (T_2) time constants; the translational diffusion coefficient (D) can also be measured. In a semi-classical picture, atomic nuclei with non-zero spin can be modeled as rotating magnetic dipoles. The static B_0 field tends to align these dipoles (by convention, along the z-axis) much like compass needles in the Earth's magnetic field, thus resulting in non-zero magnetization of the sample in thermal equilibrium. A second, time-varying magnetic field (known as B_1) can be applied to perturb the magnetization away from equilibrium. Once B_1 is removed, the sample gradually returns to equilibrium; this process is known as relaxation [23]. Specifically, T_1 is the time constant for re-establishment of the equilibrium "longitudinal" magnetization, while T_2 is the time constant for decay of the non-equilibrium "transverse" magnetization.

The two parameters are generally not equal to each other (in almost all cases, $T_1 \geq T_2$) and also exhibit different dependencies on B_0 field strength and temperature [21].

Several studies have used ^{1}H-NMR and ^{13}C-NMR NMR spectroscopy to detect food dyes (e.g., azo dyes) in solution [24,25]. Azo dyes are water-soluble, organic compounds that contain a functional group of the form R−N = N−R′, where R and R′ are typically aromatic groups. These dyes are widely used in some foods and also in the textile industry; common examples include Sudan red, metanil yellow, and malachite green. However, the NMR relaxation properties of aqueous solutions of azo dyes have not been carefully studied. This paper seeks to use the T_1 and T_2 relaxation time constants to detect these dies in food samples. To the best of our knowledge, it is the first to show that NMR relaxometry can be used for rapid and low-cost detection of multiple dyes (including malachite green and Sudan red) present within common vegetables.

NMR relaxometry can be used to determine the presence and quantity of a target compound with the help of a reference sample and chemometric analysis. In this approach, relaxometry was first performed on a reference sample and its relaxation time recorded. Next, relaxometry was performed on the test sample, and the relaxation time was again recorded. Finally, the relaxation times were compared to detect the presence and quantity of the target compound. Several methods, including linear regression, comparison with internal and external standards, and comparison of relaxation spectra, were used to quantitatively analyze the resulting data [26,27]. Linear or nonlinear regression on T_1 and/or T_2 values is simple to implement and numerically stable, while finding and preparing an appropriate reference compound (i.e., internal or external standard) is sometimes troublesome. However, both regression- and standards-based methods tend to fail for complex mixtures due to overlap between the T_1 and/or T_2 values of different components. Comparison of relaxation spectra generated using Laplace inversion is well-suited for such complex samples but suffers from limited resolution due to the numerically ill-conditioned nature of the inverse Laplace transform [27]. In this study, a simple and practical approach was developed for quantification of multiple food dyes by combining an external reference with nonlinear regression.

2. Materials and Methods

To simplify sample preparation, deionized (DI) water was used as the reference sample for all dyes, which is acceptable when only a single dye is present in a given test sample. The latter is a reasonable assumption since the goal of most dye-based adulteration is to impart a single color (e.g., green, orange, or red) to the vegetable or fruit in question. Finally, a general nonlinear regression method for quantitative analysis of the acquired relaxation data was used [10]. The details of this process are described next.

2.1. Dyes and Vegetables

A large number of chemical dyes have been used to make vegetables look fresh and vibrant [6], many of which are inedible and harmful to human health. For this study, three widely-used dyes were chosen: copper sulfate, malachite green, and Sudan red [28]. The first dye, copper (II) sulfate ($CuSO_4$), is an inorganic compound that dissolves in water to produce a dark blueish-green solution. When dipped in this solution, green vegetables, like bitter gourds, peas, and cucumbers, turn dark or vibrant green. Unfortunately, $CuSO_4$ is poisonous if ingested in large quantities (>1 gm) [29], with symptoms ranging from slight nausea to severe gastrointestinal infections and other diseases [28]. For this study, three different green vegetables, namely bitter gourd, okra, and pointed gourd (also known as parwal), were purchased from a local store and dyed using copper sulfate. The second dye, malachite green, is the monochloride salt of an aromatic cation (a triarylmethane) with formula $C_{23}H_{25}N_2^+$ [30]. It is generally used to color materials like leather or silk but because of its green hue is also illegally used to color vegetables, like peas and green chilies [28]. However, it is moderately toxic (even at concentrations as low as 0.1 μg/mL) and may also be carcinogenic [29]. In this study, yellow and green peas were dyed using malachite green. The third dye, Sudan red, is a reddish-orange lysochrome azo dye with formula $C_{17}H_{14}N_2O_2$ [31]. This chemical is known to be carcinogenic and banned in food items but nevertheless continues to be illegally used to color red chilies, red chili powder, red capsicum fruits, red pepper, chili jam, and tomatoes [32,33]. In this study, red chilies were dyed using Sudan red. All the dyes were purchased from Sigma-Aldrich (St. Louis, MO, USA), while the vegetables were obtained from local grocery stores (Gainesville, FL, USA).

2.2. NMR Relaxometry Instrumentation

A block diagram of the overall experimental setup is illustrated in Figure 2a. The setup uses a benchtop permanent magnet (Spincore Technologies Inc., Gainesville, FL, USA) with a measured field strength of 0.5266 T at room temperature, resulting in a nominal ^{1}H-NMR resonance frequency of 22.6 MHz. A 3D-printed holder containing the solenoid probe coil and NMR sample tube is centered between the magnetic poles [34]. The holder is coupled to a commercial benchtop NMR spectrometer (Kea^2, Magritek Inc., Malvern, PA, USA) through a two-capacitor impedance matching network [35]. The spectrometer is powered by two 12-V, sealed lead-acid (SLA) batteries with a capacity of 18 Ah (not shown in the figure) and connected to a personal computer using a USB interface. A proprietary graphical user interface (GUI)-based software, Prospa, is used to control the spectrometer and acquire experimental data.

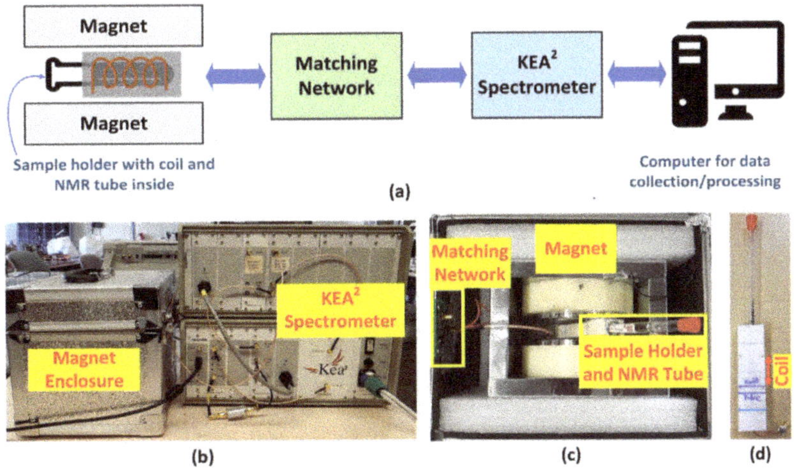

Figure 2. (**a**) A block diagram of the experimental setup; (**b**) a picture of the actual measurement setup; (**c**) an inside view of the magnet enclosure; and (**d**) a picture of the 3D-printed sample holder parts with the coil and an NMR tube inserted.

Figure 2b shows a photograph of the experimental setup. The permanent magnet, matching network, and sample holder are placed within a metallic enclosure that provides electromagnetic shielding from external radio frequency (RF) interference by acting as a Faraday cage. Figure 2c shows the internal layout of this enclosure, while Figure 2d shows a more detailed view of the sample holder with a 10-mm thin-wall precision NMR tube (Wilmad-LabGlass, Vineland, NJ, USA) inserted into it.

The probe coil was hand-wound using AWG 22 copper wire. The signal-to-noise ratio (SNR) of the NMR measurements [36] was maximized by iteratively optimizing the coil geometry to maximize its quality factor (Q) at the ^{1}H-NMR resonant frequency ($f_0 \approx 22.6$ MHz). The final design consisted of a tightly-packed solenoid with a relatively short length-to-diameter ratio ($L \approx 2$ cm and $d \approx 10$ mm, resulting in $L/d \approx 2$) but a relatively large number of turns ($N = 13$). Coil properties around f_0 were measured using a vector network analyzer (E5071C, Agilent Technologies). The results (inductance = 840 nH, series resistance = 415 mΩ) confirm adequately high quality factor (Q 287) and self-resonant frequency ($f_{SRF} \approx 130$ MHz). The estimated position of the coil within the sample holder is shown in Figure 2d.

2.3. Methodology

Instead of measuring the adulterant in situ, it was first washed out into solution. For this purpose, the sample (fruit or vegetable) was soaked in a solvent with known properties (e.g., DI water or brine) for a few minutes. The concentration of adulterant in the solvent was then measured using NMR relaxometry. This process has several advantages, including (i) eliminating the effect of sample heterogeneity from the T_1 and T_2 measurements and (ii) greatly simplifying sample preparation. The acquired relaxation data were further analyzed in two steps: (i) library creation (Figure 3a) and (ii) quantifying the concentration of adulterant (Figure 3b). Our current implementation of both steps focused on T_2 since it can be rapidly and accurately measured using the well-known Carr–Purcell–Meiboom–Gill (CPMG) pulse sequence [37,38], but the procedure can be readily extended to include T_1 data (e.g., from an inversion recovery (IR) pulse sequence).

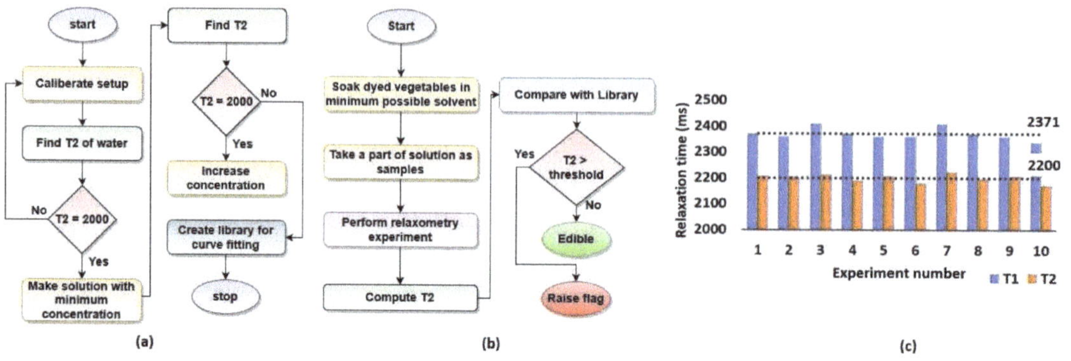

Figure 3. (**a**) Flowchart showing the process of creating a library; (**b**) flowchart showing the process of determining unknown concentrations; (**c**) measured variation of T_1 and T_2 values for the reference sample (DI water) over 10 experiments.

3. Results and Discussion

3.1. Library Creation

Calibration was carried out by using IRand CPMG pulse sequences to measure the T_1 and T_2 values of the reference sample, which is typically 12 mL of DI water. The measured values are $T_1 \approx 2370$ ms and $T_2 \approx 2200$ ms at room temperature (see Figure 3c). This value of $T_1 \approx 2400$ ms is in good agreement with earlier studies [39], while T_2 is similar to T_1, as expected for water [40].

The next goal was to confirm that aqueous solutions of all three dyes under study exhibited T_2 contrast, i.e., a reproducible dependence of T_2 on dye concentration. For this, known quantities of each dye were dissolved in a fixed amount (100 mL) of reference sample (either DI water or 0.5% NaCl solution) to create a library of solutions. For convenience, a solution containing x gram of a particular dye was referred as "x% solution". Next, 12 mL of each solution was placed in an NMR sample tube and analyzed using a CPMG pulse sequence. The measured relaxation time constant is denoted by $T_{2,eff}$ to distinguish it from that of the reference sample (DI water). In each case, the CPMG echo spacing (t_E) was kept small enough to ensure that molecular diffusion did not significantly affect the value of $T_{2,eff}$ [37].

The smallest value of x (i.e., the sample weight) used within the proposed library was experimentally adjusted for each dye to ensure that the resulting change in $T_{2,eff}$ could be accurately estimated within a few scans. For this, the measured CPMG echo decay curves were fit to mono-exponential functions of the form $Ae^{-nt_E/T_{2,eff}}$ using least-squares function minimization; here, A is the initial signal amplitude, and $n = 1, 2,...$ is the echo number. Figure 4 shows the measured dependence of $T_{2,eff}$ on concentration for all three dyes. In each case, a monotonic decrease of $T_{2,eff}$ with concentration was observed; the effect is particularly strong for $CuSO_4$. As a result, sample concentration can be unambiguously estimated from the measured value of $T_{2,eff}$.

The underlying cause for the observed decrease in $T_{2,eff}$ with concentration is increased inter-molecular dipole-dipole (D-D) relaxation of the water molecules. Inter- molecular D-D relaxation is typically the dominant relaxation mechanism in dilute aqueous solutions [21]. It arises from time-varying fluctuations in the B_0 field seen by each nucleus due to random thermal motion of other molecules or ions in the solution (which act like miniature dipole field sources). In the case of $CuSO_4$, the effect is dominated by random motion of the added Cu2+ ions, which contain unpaired electrons and are thus paramagnetic [41]. In the case of the organic dyes, the effect is likely dominated by slower motion (and thus increased D-D relaxation rates) [21] of the loosely-organized shell of water molecules that surrounds each dye molecule due to mutual electrostatic attraction. Each shell is in rapid chemical exchange with bulk water molecules, thus explaining the observed mono-exponential echo decay curves.

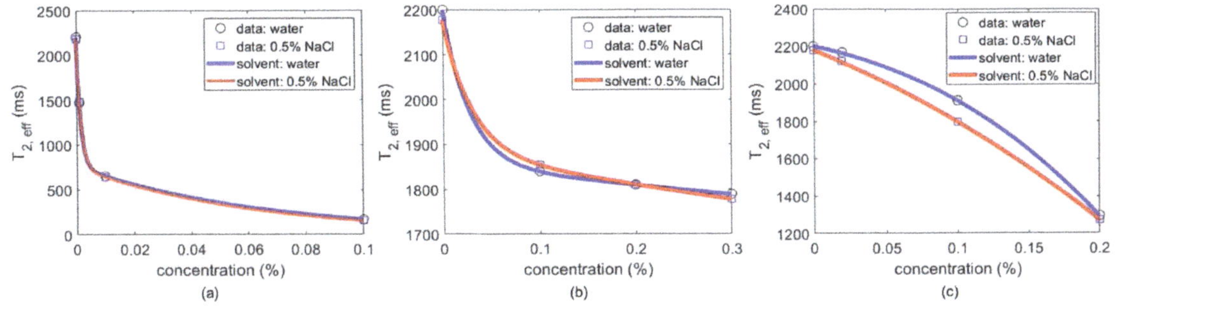

Figure 4. A library of vegetable dyes was created using the relationship between $T_{2,eff}$ and its concentration. This library can be used to quantify the amount of dye used in vegetable adulteration. The libraries exhibiting these trends are shown for 3 different dyes: (**a**) copper sulfate, (**b**) malachite green, and (**c**) Sudan red. Separate library functions are shown for two different experimental solvents, namely DI water and 0.5% NaCl solution.

The observed relationship between $T_{2,eff}$ and concentration for each dye was quantified using nonlinear regression, i.e., least-squares curve fitting. The resulting functions can be inverted to estimate unknown dye concentrations, as described in the next section.

3.2. Detection of Unknown Concentrations

The calibration curves described in the previous section were used to estimate the concentration of dye washed out from adulterated vegetables. For this purpose, non-adulterated vegetables were purchased from a local market, dyed by immersing them in the appropriate solution, and air-dried to remove extra liquid. Finally, the adulterated vegetables were soaked in the reference solvent (typically DI water) to wash out the dye. The $T_{2,eff}$ value of the solution was then analyzed using a CPMG pulse sequence.

A careful set of experiments was performed to determine the optimum sample-preparation procedure. Firstly, the optimum solution concentration for dyeing vegetables was determined. Figure 4 shows that the NMR setup can reliably detect concentrations as low as 0.1–0.3%. Thus, a higher concentration (1%) was used to dye each vegetable. Specifically, 1% $CuSO_4$ was used for pointed gourd, bitter gourd, and okra; 1% malachite green for peas; and 1% Sudan red for red dried chilies. The original (raw) and adulterated (dyed) vegetable samples are visually compared in Figure 5.

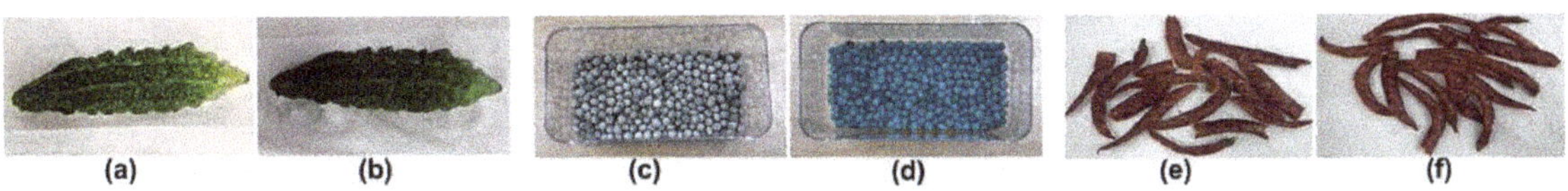

Figure 5. Comparison between the raw and dyed vegetables: (**a**) raw okra, (**b**) okra dyed with copper sulfate, (**c**) raw peas, (**d**) peas dyed with malachite green, (**e**) raw red chilies, and (**f**) red chilies dyed with Sudan red.

The vegetables were soaked in the corresponding dye solutions for 3 h and then air-dried for 12 h in room temperature. Then, it was determined the optimum combination of reference solvent, temperature, and soaking time, t_{soak}, for washing out each dye. Firstly, both DI water and 0.5% NaCl solution were studied as reference solvents; the results were similar, so DI water was chosen for convenience. Secondly, the solvent temperature and soak time were varied. For water at room temperature, t_{soak} = 5, 60, and 180 min were used. For warm water at 60 °C, t_{soak} = 1, 2, and 5 min were used since the wash-out process (which is driven by diffusion) was expected to be significantly faster. Figure 6a–c show that $T_{2,eff}$ values decreased with time as more dye (CuSO4 in this case) washed out into solution; the rate of change was significantly higher for warm water, as expected. Similarly,

Figure 6d,f confirm that (i) the estimated dye concentrations increased with time, and (ii) warm water could extract most of the dye within t_{soak} = 2 min, while much longer soak times were required at room temperature.

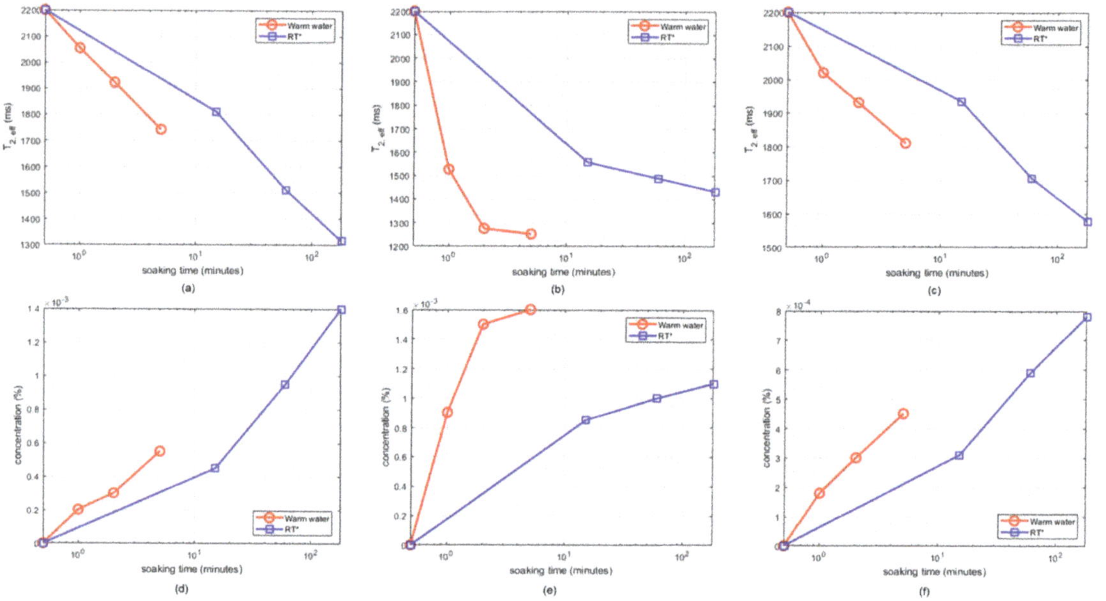

Figure 6. Estimated values of $T_{2,eff}$ (top row) and concentration of extracted dye in solution (CuSO4, bottom row) as a function of time using water at room temperature and warm water at 60 °C: (**a**,**d**) pointed gourd; (**b**,**e**) bitter gourd; and (**c**,**f**) okra.

During the experiments, the optimized procedure described above (dyeing with 1% solution, drying for 12 h, soaking in warm water for 2 min, and finally estimating dye concentration from $T_{2,eff}$ measurements) was repeated 10 times for each sample to ensure that the results are repeatable and consistent. The experiments confirm that both the presence of the chosen dyes and their extracted concentrations can be reliably estimated (with typical error < 4%) using the proposed technique.

3.3. Discussion

While the experiments in the paper were focused on three common dyes, the proposed method can be extended to any dye that exhibits NMR relaxation contrast (in T_1 and/or T_2) while in aqueous solution. Compounds containing paramagnetic ions (such as Cu^{2+} or Ni^{2+}) fall into this category since they result in increased intermolecular D-D relaxation rates. Compounds with permanent electric dipole moments, such as most azo and aryl dyes, may also exhibit a small amount of relaxation contrast due to the reduced mobility of water molecules in their associated hydration shells. Additional relaxation contrast can be obtained by performing T_1 measurements at different field strengths (e.g., by using an electromagnet to generate B_0); this process is known as field-cycling relaxometry [42].

Besides generality, additional desirable features for the proposed food-adulteration detection platform include portability and cost-effectiveness. As noted earlier, NMR spectroscopy is expensive because of the need to generate a strong and highly uniform B_0 field. While the magnet size and cost requirements can be significantly reduced by focusing on relaxometry, the large size and power consumption of the spectrometer electronics (which includes an analog front-end and a digital back-end) remains a barrier for portable and low-cost applications. Fortunately, recent work has demonstrated miniaturized and low-power versions of both the front- and back-ends. For example, a portable NMR spectrometer based on a custom front-end and a low-cost system-on-chip (SoC) back-end has been developed [43]. Such miniaturized and low-cost devices can be used to replace the benchtop spectrometer used in the current setup.

4. Conclusions

This paper has demonstrated, for the first time to our knowledge, a simple, low-cost, yet powerful technique that combines NMR relaxometry with nonlinear regression-based trend modeling to detect and quantify harmful dyes in vegetables. Our experimental results show that the proposed technique can reliably quantify the presence of three commonly used illegal dyes, namely copper sulfate, malachite green, and Sudan red, at concentrations as low as 1 g/L (0.1%). The proposed technique can be used for detecting and potentially quantifying chemical dye-based produce adulteration in various stages of a supply chain, including retail facilities and consumer households. Future work will focus on extending our approach to a wider range of chemical dyes and food items as well as further enhancing the detection sensitivity.

Author Contributions: Conceptualization, S.M. and S.B.; data curation, S.S., N.V.R.M., S.D.P. and K.H.-H.; formal analysis, N.V.R.M., S.D.P., D.F., S.M. and S.B.; funding acquisition, S.M. and S.B.; investigation, S.S., N.V.R.M., D.A., K.H.-H., D.F., S.M. and S.B.; methodology, S.S., N.V.R.M., S.M. and S.B.; project administration, S.M. and S.B.; resources, D.F., S.M. and S.B.; supervision, D.F., S.M. and S.B.; validation, D.A. and K.H.-H.; writing—original draft, S.S., N.V.R.M., D.A., S.D.P., K.H.-H., D.F., S.M. and S.B.; writing—review and editing, S.S., N.V.R.M., D.A., S.D.P., K.H.-H., D.F., S.M. and S.B. All authors have read and agreed to the published version of the manuscript.

Funding: This research was funded by the National Science Foundation (grant no. 1563924).

Institutional Review Board Statement: Not applicable.

Informed Consent Statement: Not applicable.

Data Availability Statement: Not applicable.

Conflicts of Interest: The authors declare no conflict of interest.

References

1. Schell, L.M.; Gallo, M.V.; Cook, K. What's NOT to eat—food adulteration in the context of human biology. *Am. J. Hum. Biol.* **2012**, *24*, 139–148. [CrossRef]
2. Moore, J.C.; Spink, J.; Lipp, M. Development and application of a database of food ingredient fraud and economically motivated adulteration from 1980 to 2010. *J. Food Sci.* **2012**, *77*, R118–R126. [CrossRef]
3. Bansal, S.; Singh, A.; Mangal, M.; Mangal, A.K.; Kumar, S. Food adulteration: Sources, health risks, and detection methods. *Crit. Rev. Food Sci. Nutr.* **2017**, *57*, 1174–1189. [CrossRef] [PubMed]
4. Naik, S.; Prasad, R. Pesticide residue in organic and conventional food-risk analysis. *J. Chem. Health Saf.* **2006**, *13*, 12–19.
5. Kobylewski, S.; Jacobson, M.F. Toxicology of food dyes. *Int. J. Occup. Environ. Health* **2012**, *18*, 220–246. [CrossRef]
6. Oplatowska-Stachowiak, M.; Elliott, C.T. Food colors: Existing and emerging food safety concerns. *Crit. Rev. Food Sci. Nutr.* **2017**, *57*, 524–548. [CrossRef] [PubMed]
7. Hong, E.; Lee, S.Y.; Jeong, J.Y.; Park, J.M.; Kim, B.H.; Kwon, K.; Chun, H.S. Modern analytical methods for the detection of food fraud and adulteration by food category. *J. Sci. Food Agric.* **2017**, *97*, 3877–3896. [CrossRef]
8. Uematsu, Y.; Ogimoto, M.; Kabashima, J.; Suzuki, K.; Ito, K. Fast cleanup method for the analysis of Sudan IIV and para red in various foods and paprika color (oleoresin) by high-performance liquid chromatography/diode array detection: Focus on removal of fat and oil as fatty acid methyl esters prepared by transesterification of acylglycerols. *J. AOAC Int.* **2007**, *90*, 437–445.
9. Culp, S.; Beland, F.A.; Heflich, R.H.; Benson, R.W.; Blankenship, L.R.; Webb, P.J.; Mellick, P.W.; Trotter, R.W.; Shelton, S.D.; Greenlees, K.U.J.; et al. Mutagenicity and carcinogenicity in relation to DNA adduct formation in rats fed leucomalachite green. *Mutat. Res. /Fundam. Mol. Mech. Mutagenesis* **2002**, *506*, 55–63. [CrossRef]
10. Hu, Y.; Wang, S.; Wang, S.; Lu, X. Application of nuclear magnetic resonance spectroscopy in food adulteration determination: The example of Sudan dye I in paprika powder. *Sci. Rep.* **2017**, *7*, 1–9. [CrossRef]
11. Jolene. Indian Farmers Dye and Inject Vegetables To Make Them Look 'Fresher and Bigger. 2016. Available online: https://www.worldofbuzz.com/indian-farmer-dye-inject-vegetables-make-look-fresh-big/ (accessed on 5 March 2020).
12. Julia Glotz. New Copper Sulphate Scam on The Rise in Fruit & Veg Sector. 2016. Available online: https://www.thegrocer.co.uk/food-safety/new-copper-sulphate-scam-on-the-rise-in-fruit-and-veg-sector/544132.article (accessed on 5 March 2020).

13. India Today. Common Food Adulterants in India. 2019. Available online: https://www.indiatoday.in/education-today/gk-current-affairs/story/common-food-adulterants-in-india-1370601-2018-10-19 (accessed on 5 March 2020).
14. NDTV Food. Is Your Turmeric Real or Fake? Use These Smart Tricks to Find Out. 2017. Available online: https://food.ndtv.com/food-drinks/is-your-turmeric-real-or-fake-use-these-smart-tricks-to-find-out-1709273 (accessed on 5 March 2020).
15. Anasazi Instruments, Inc. Cryogen Free NMR Spectrometers for Education and Industry. 2021. Available online: https://www.aiinmr.com/ (accessed on 8 September 2021).
16. Magritek. Discover the Spinsolve Desktop NMR Spectrometer Family. 2021. Available online: https://magritek.com/products/spinsolve/ (accessed on 8 September 2021).
17. Ellis, D.I.; Brewster, V.L.; Dunn, W.B.; Allwood, J.W.; Golovanov, A.P.; Goodacre, R. Fingerprinting food: Current technologies for the detection of food adulteration and contamination. *Chem. Soc. Rev.* **2012**, *41*, 5706–5727. [CrossRef]
18. Masna, N.V.R.; Paul, S.D.; Chen, C.; Mandal, S.; Bhunia, S. Eat, but verify: Low-cost portable devices for food safety analysis. *IEEE Consum. Electron. Mag.* **2018**, *8*, 12–18. [CrossRef]
19. Hatzakis, E. Nuclear magnetic resonance (NMR) spectroscopy in food science: A comprehensive review. *Compr. Rev. Food Sci. Food Saf.* **2019**, *18*, 189–220. [CrossRef] [PubMed]
20. Kryukov, E.; Bugoslavsky, Y.; Linde, A.J.P.; Holubar, T.; Burgess, S.; Marlow, D.; Good, J. A method for fast field settling in cryogen-free superconducting magnets for NMR. *Solid State Nucl. Magn. Reson.* **2020**, *109*, 101684. [CrossRef] [PubMed]
21. Fukushima, E.; Roeder, S.B. *Experimental Pulse NMR: A Nuts and Bolts Approach*; CRC Press: Boca Raton, FL, USA, 2018.
22. Cistola, D.P.; Robinson, M.D. Compact NMR relaxometry of human blood and blood components. *TrAC Trends Anal. Chem.* **2016**, *83*, 53–64. [CrossRef]
23. NMR Relaxation. Available online: http://chem.ch.huji.ac.il/nmr/techniques/other/t1t2/t1t2.html (accessed on 8 September 2021).
24. Fedorov, L.A. NMR spectroscopy of azo dyes. *Russ. Chem. Rev.* **1988**, *57*, 941. [CrossRef]
25. Allisy-Roberts, P.J.; Williams, J. *Farr's Physics for Medical Imaging*; Elsevier Health Sciences: New York, NY, USA, 2007.
26. Bharti, S.K.; Roy, R. Quantitative ^{1}H NMR spectroscopy. *TrAC Trends Anal. Chem.* **2012**, *35*, 5–26. [CrossRef]
27. Song, Y.Q.; Venkataramanan, L.; Hürlimann, M.; Flaum, M.; Frulla, P.; Straley, C. T1–T2 correlation spectra obtained using a fast two-dimensional Laplace inversion. *J. Magn. Reson.* **2002**, *154*, 261–268. [CrossRef]
28. Tadepalli, S. Green Veggies Made Greener with Lethal Copper Sulphate. 2015. Available online: https://times9ofindia.indiatimes.com/city/hyderabad/Green-veggies-made-greener-with-lethal-copper-sulphate/articleshow/48599444.cms (accessed on 8 September 2021).
29. Gamakaranage, C.S.S.K.; Rodrigo, C.; Weerasinghe, S.; Gnanathasan, A.; Puvanaraj, V.; Fernando, H. Complications and management of acute copper sulphate poisoning; a case discussion. *J. Occup. Med. Toxicol.* **2011**, *6*, 1–5. [CrossRef]
30. National Center for Biotechnology Information. PubChem Database. Malachite Green, CID=11294. 2020. Available online: https://pubchem.ncbi.nlm.nih.gov/compound/Malachite-green (accessed on 26 February 2020).
31. National Center for Biotechnology Information. PubChem Database. Sudan III, CID=62331. 2020. Available online: https://pubchem.ncbi.nlm.nih.gov/compound/Sudan-III (accessed on 26 February 2020).
32. Xu, J.; Zhang, Y.; Yi, J.; Meng, M.; Wan, Y.; Feng, C.; Wang, S.; Lu, X.; Xi, R. Preparation of anti-Sudan red monoclonal antibody and development of an indirect competitive enzyme-linked immunosorbent assay for detection of Sudan red in chilli jam and chilli oil. *Analyst* **2010**, *135*, 2566–2572. [CrossRef]
33. Daood, H.G.; Biacs, P.A. Simultaneous determination of Sudan dyes and carotenoids in red pepper and tomato products by HPLC. *J. Chromatogr. Sci.* **2005**, *43*, 461–465. [CrossRef] [PubMed]
34. Greer, M.; Chen, C.; Mandal, S. Automated classification of food products using 2D low-field NMR. *J. Magn. Reson.* **2018**, *294*, 44–58. [CrossRef] [PubMed]
35. Greer, M.; Chen, C.; Mandal, S. An easily reproducible, hand-held, single-sided, MRI sensor. *J. Magn. Reson.* **2019**, *308*, 106591. [CrossRef] [PubMed]
36. Hoult, D.I.; Richards, R. The signal-to-noise ratio of the nuclear magnetic resonance experiment. *J. Magn. Reson. (1969)* **1976**, *24*, 71–85. [CrossRef]
37. Carr, H.Y.; Purcell, E.M. Effects of diffusion on free precession in nuclear magnetic resonance experiments. *Phys. Rev.* **1954**, *94*, 630. [CrossRef]
38. Meiboom, S.; Gill, D. Modified spin-echo method for measuring nuclear relaxation times. *Rev. Sci. Instrum.* **1958**, *29*, 688–691. [CrossRef]
39. Tsukiashi, A.; Min, K.S.; Kitayama, H.; Terasawa, H.; Yoshinaga, S.; Takeda, M.; Lindoy, L.F.; Hayami, S. Application of spin-crossover water soluble nanoparticles for use as MRI contrast agents. *Sci. Rep.* **2018**, *8*, 1–5. [CrossRef] [PubMed]
40. T2 Relaxation. Available online: https://radiopaedia.org/articles/t2-relaxation (accessed on 8 September 2021).
41. Koenig, S.H.; Brown III, R.D. Relaxation of solvent protons by paramagnetic ions and its dependence on magnetic field and chemical environment: Implications for NMR imaging. *Magn. Reson. Med.* **1984**, *1*, 478–495. [CrossRef] [PubMed]
42. Kimmich, R. (Ed.) *Field-Cycling NMR Relaxometry: Instrumentation, Model Theories and Applications*; Royal Society of Chemistry: Cambridge, UK, 2018.
43. Ariando, D.; Chen, C.; Greer, M.; Mandal, S. An autonomous, highly portable NMR spectrometer based on a low-cost System-on-Chip (SoC). *J. Magn. Reson.* **2019**, *299*, 74–92. [CrossRef]

foods

Article

Authentication and Chemometric Discrimination of Six Greek PDO Table Olive Varieties through Morphological Characteristics of Their Stones

Sofia Agriopoulou [1,*], Maria Tarapoulouzi [2], Marie Ampères Bedine Boat [3], Catherine Rébufa [4], Nathalie Dupuy [4], Charis R. Theocharis [2], Theodoros Varzakas [1], Sevastianos Roussos [4] and Jacques Artaud [4]

1 Department of Food Science and Technology, University of the Peloponnese, Antikalamos, 24100 Kalamata, Greece; t.varzakas@uop.gr
2 Department of Chemistry, University of Cyprus, P.O. Box 20537, Nicosia CY-1678, Cyprus; tarapoulouzi.maria@ucy.ac.cy (M.T.); charis@ucy.ac.cy (C.R.T.)
3 Department of Biochemistry, University of Yaounde I, P.O. Box 812, Yaoundé, Cameroon; bedineboat@yahoo.fr
4 Aix Marseille Univ, Avignon Université, CNRS, IRD, IMBE, Campus Saint Jerome, 13007 Marseille, France; c.rebufa@univ-amu.fr (C.R.); nathalie.dupuy@univ-amu.fr (N.D.); sevastianos.roussos@imbe.fr (S.R.); jacques.artaud@univ-amu.fr (J.A.)
* Correspondence: s.agriopoulou@uop.gr

Abstract: Table olives, the number one consumed fermented food in Europe, are widely consumed as they contain many valuable ingredients for health. It is also a food which may be the subject of adulteration, as many different olive varieties with different geographical origin, exist all over the word. In the present study, the image analysis of stones of six main Greek protected designation of origin (PDO) table olive varieties was performed for the control of their authentication and discrimination, with cv. Prasines Chalkidikis, cv. Kalamata Olive, cv. Konservolia Stylidas, cv. Konservolia Amfissis, cv. Throuba Thassos and cv. Throuba Chios being the studied olive varieties. Orthogonal partial least square discriminant analysis (OPLS-DA) was used for discrimination and classification of the six Greek table olive varieties. With a 98.33% of varietal discrimination, the OPLS-DA model proved to be an efficient tool to authentify table olive varieties from their morphological characteristics.

Keywords: Greek PDO table olive varieties; chemometric analysis; OPLS-DA; discrimination; authenticity; adulteration; geographical origin; quality; safety

Citation: Agriopoulou, S.; Tarapoulouzi, M.; Bedine Boat, M.A.; Rébufa, C.; Dupuy, N.; Theocharis, C.R.; Varzakas, T.; Roussos, S.; Artaud, J. Authentication and Chemometric Discrimination of Six Greek PDO Table Olive Varieties through Morphological Characteristics of Their Stones. *Foods* **2021**, *10*, 1829. https://doi.org/10.3390/foods10081829

Academic Editors: Paolo Lucci and Simon Haughey

Received: 12 June 2021
Accepted: 4 August 2021
Published: 7 August 2021

1. Introduction

Olive growing is associated with the first steps of human existence, with a history about 5000 years [1], having acquired symbolism associated with peace and friendship as olive branches crowned the winners of the Olympics in Ancient Greece [2]. Especially in Mediterranean countries where the olive tree (*Olea europaea* L. of the family Oleaceae) is considered the most emblematic tree, there exist more than 2000 varieties [3]. From the fruit of the olive trees can produce the well-known olive oil and table olives (also called eating olives) [4,5]. According to International Olive Oil Council (IOOC), "Table olives are the product prepared from the sound fruits of varieties of the cultivated olive tree that are chosen for their production of olives whose volume, shape, flesh-to-stone ratio, fine flesh, taste, firmness and ease of detachment from the stone make them particularly suitable for processing; treated to remove its bitterness and preserved by natural fermentation, or by heat treatment with or without the addition of preservatives; packed with or without covering liquid" [6]. The main producers of table olives in Europe are Spain, Greece, and Italy, other major producers outside the Europe are Egypt, Algeria, Turkey, and Morocco, while emerging producers are Syria, Peru and USA. The word total production exceeded 2.5 million tons in the 2018/2019 season, while the precognition for the 2020/2021 season is to be exceeded 3.0 million tons [7].

2.2. Olive Stone Processing

The olive fruits were transferred to Laboratory of Environmental Biotechnology and Chemometrics, Aix Marseille University, IMBE, and the weight of fresh olives for 60 fruits was measured (30 for each set). The olive fruits were stored at −20 °C for preservation, until the beginning of the analyses. The olive stones are de-fleshed using a procedure developed by Vanloot et al. [11]. Briefly, after thawing, they were placed in hot water for ten minutes and their flesh was removed manually. The olive fruits were brushed to remove all traces of flesh and rinsed with water. The stones were then immersed in hydrogen peroxide for 24 h. They were then rinsed thoroughly to remove all traces of hydrogen peroxide, followed by drying for 48 h at room temperature to obtain a constant weight, which was then measured. This was followed by the storage of the stones in airtight glass bottles until their digital images were obtained. Images were taken from 60 olive stone and for each olive stone, two images (face and profile) were obtained with a high-resolution color camera for 103 character digital processing (Baumer TXD13C) connected on a computer for image processing (Figure 2).

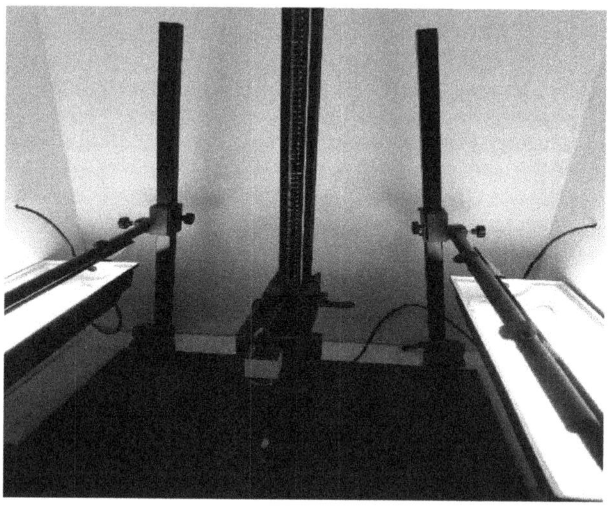

Figure 2. Olive stone image acquisition with a high-resolution color camera (Baumer TXD13C) in center and side lights.

2.3. Olive Stone Characteristics

The characterization of the stone parameters was based on the *World Catalogue of Olive Varieties*. The determination of the shape parameters was determined visually on the basis of the different shapes listed in the catalogue. In the *World Catalogue of Olive Varieties* are described with the common glossary the morphological characteristics of tree, inflorescence, leaf, fruit and endocarp (stone) of 139 olive varieties from 23 countries [12]. As it concerns the stone, according to the describing characteristics which are including in the catalogue, it will be very helpful to discriminate the varieties. According to classification there are varieties with low (<0.3 g), medium (0.3–0.45 g) and high (>0.45 g) weight of stones. The shape is characterized as spherical, ovoid, elliptic and elongated when the ratio between the length and width is <1.4, 1.4–1.8, 1.8–2.2 and >2.2, respectively. The symmetry of stone is characterized as symmetric, slightly asymmetric, and asymmetric. The base of the stone which is the part that connects the stone with the peduncle is characterized as truncate, pointed or rounded and apex which is the opposite part of stone it is characterized as pointed or rounded, with or without a mucro. The surface of stone may be smooth, rugose or scabrous [12]. The maximum transversal width can be toward the base, toward the apex or central and the fibrovascular bundles can be deep and abundant. Two positions of the stone, the face and profile, have been used for stone characterization. The first position refers to the maximum symmetry and the second is obtained after rotating 90° from the first. Images were digitized by Visilog v6.7 imaging software from Noesis (Gif sur Yvette, France). Figure 3 shows the detailed characteristics of a stone from Prasines Chalkidikis olive variety.

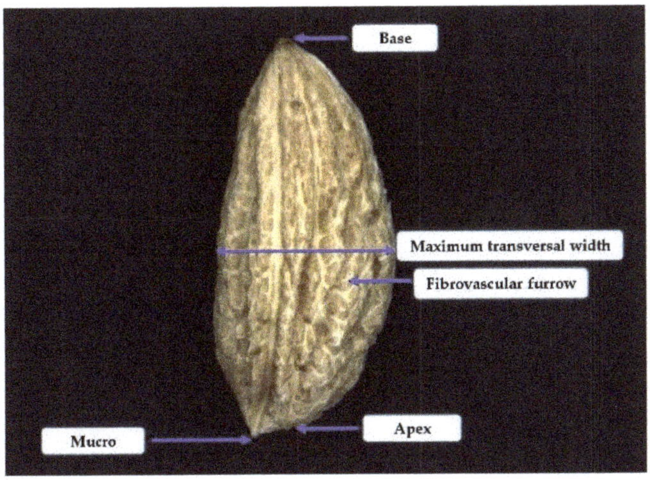

Figure 3. Morphological characteristics of a stone from cv. Prasines Chalkidikis.

2.4. Application of Chemometrics

SIMCA version 15.0.2 (Umetrics, 907 29 Umeå, Sweden) was used for chemometric analysis. The supervised OPLS-DA procedure was followed to discriminate and classify the observations (samples).

The main limitation of PLS model is its linear nature and it is not applicable for data with non-linear behavior [37,38]. OPLS-DA methods can be applied to visualize variations between sample groups and to define the discriminating performance of the variables. With the use of OPLS-DA it is possible to classify samples according to agricultural practices and predict the origin of unknown samples [39].

Thirteen parameters (variables) were used. Ten of them were related to morphological characteristics: shape, profile symmetry, front symmetry, basis, apex, mucro, MTW, surface, NFF and DFF. Three other parameters were also used related to the weight of the samples, such as average weight of olive fruits, average weight of stones and quantity of olive flesh per olive fruit.

Scaling to unit variance (UV) and mean-centering were used. The samples were discriminated into six classes, namely cv. Kalamata Olive (KO): Class 1, cv. Prasines Chalkidikis (PX): Class 2, cv. Konservolia Stylidas (KS): Class 3, cv. Konservolia Amfissis (KA): Class 4, cv. Throuba Thassos (TT): Class 5, and cv. Throuba Chios (TC): Class 6.

As described in Tarapoulouzi et al. [29] the OPLS-DA model was evaluated here by the determination coefficient, R^2, reflecting the goodness of fit and the cross-validated correlation coefficient, Q^2, reflecting the predictive ability of the model. Q^2 was obtained using the seven-fold leave out procedure (default setting in SIMCA). The ellipse in the plots defines Hotelling's T2 confidence region, which is a multivariate generalization of Student's *t* test and provides a 95% confidence interval for the observations. The number of the important components which have been chosen is given with the symbol A, therefore A = 1 + 1 components were used for all the models produced. In addition, internal validation took place with regression models which were validated using CV-ANOVA tables, via comparing $F_{statistic}$ vs. $F_{critical}$ values. F-value is a measure of the size of the effects. The larger this value, the greater the likelihood that the differences between the means are due to something other than chance alone, namely real effects. If the difference between the means is due only to chance, that is, there are no real effects, then the expected value of the F-ratio would be one (1.00). A hypothesis test takes place where the "null hypothesis" indicates that population means of the different appraisers are equal, and "alternate hypothesis" shows that one of the means is not the same. Larger values of $F_{statistic}$ than the $F_{critical}$ indicate that the difference of means of the samples is larger compared to the dispersion of the observations within each sample, and therefore, the null hypothesis should be rejected, and the alternate hypothesis is considered important. In other words, a lower $F_{statistic}$ than the $F_{critical}$ indicates that the variation within the appraisers is greater than the variation between them [40]. The misclassification table was considered important to evaluate the quality of the model, as well as permutation testing was applied (100 permutations) to check the validity and the degree of overfit for the OPLS-DA model.

Validation of the model was tested using sevenfold cross-validation. Therefore, a calibration and a validation set were set up by having 42 and 18 samples, respectively.

3. Results

3.1. Weight of Olive Stones

In Table 1, the weight of stones of the six Greek PDO table olive varieties are presented. Statistical analyses were performed with SD and these gave the same score scatter plots and classification rates as analyses which did not use them. Generally, the average weight of an olive stone is 18–22% of the olive weight [41]. The average weight of studied fresh olives varied between varieties. In this study, the maximum average weight was observed in the fruits of cv. Prasines Chalkidikis whose average weight was almost ten grams and was twice that of the cv. Kalamata Olive. The cv. Kalamata Olive had the smaller stone and plenty of flesh with the best ratio of olive flesh (90% of the weight of fresh fruit). This feature is extremely interesting to produce olive paste and other olive products from cv. Kalamata Olive. Cv. Throuba Chios had the smaller average weight of fresh olives. As it concerns the weight of olive stones the larger the olive fruit, the larger the stone. The highest average weight of the stone was observed in the cv. Prasines Chalkidikis, followed by the two varieties of Konservolia and the two varieties of Throuba while the smallest average weight of the stone was observed in the cv. Kalamata Olive.

Table 1. Weight of 60 stones of six PDO Greek table olive varieties.

Table Olive Variety	Average Weight of Stones (mg) Mean ± SD	Average Weight of Olive Fruits (mg) Mean ± SD	Quantity of Olive Flesh per Olive Fruit (mg) Mean ± SD	Percentage of Flesh (%) Mean ± SD	Percentage of Olive Stone Occupancy (%) Mean ± SD
Kalamata Olive (KO)	489 ± 6	4960 ± 11	4471 ± 7.5	90.2 ± 13	9.8 ± 2
Prasines Chalkidikis (PX)	1050 ± 12.6	9710 ± 20.7	8660 ± 45	89.2 ± 24	10.8 ± 4
Konservolia Stylidas (KS)	621 ± 5	5940 ± 9.5	5319 ± 12	89.6 ± 5.5	10.4 ± 3.5
Konservolia Amfissis (KA)	691 ± 5	5950 ± 4.9	5259 ± 9.6	88.4 ± 9	11.6 ± 1.9
Throuba Thassos (TT)	629 ± 5.4	4520 ± 7.8	3891 ± 9	86.1 ± 7	13.9 ± 3
Throuba Chios (TC)	614 ± 9	3030 ± 8.3	2416 ± 10	79.7 ± 10	20.3 ± 6

3.2. Artificial Visions of Olive Stones

Figure 4 shows the detailed characteristics of face and profile images of olive stones of six analyzed Greek PDO table olive varieties. In Table 2, they are presented the morphological characteristics of stones of six PDO Greek table olive varieties. The images prove that analyzed varieties differ quite except in the case of olive stones of varieties Konservolia. MTW, NFF, DFF and surface are also described in Table 2. All varieties have mucro except from cv. Kalamata Olive. The apexes of cv. Prasines Chalkidikis and cv. Throuba Chios are rounded and the others are pointed. The basis of the stones are pointed for cv. Konservolia Stylidas, cv. Prasines Chalkidikis, cv. Throuba Chios, cv. Throuba Thassos, and cv. Kalamata Olive and only for cv. Konservolia Amfissis is it rounded.

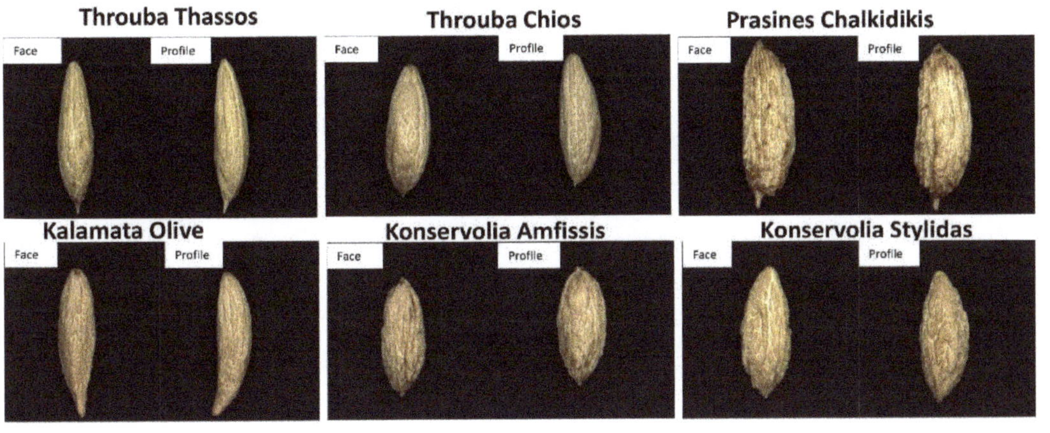

Figure 4. Morphological characteristics of face and profile images of olive stones of six analyzed Greek PDO table olive varieties.

Table 2. Morphological characteristics of 60 stones of six PDO Greek table olive varieties.

Table Olive Variety	Shape	Profile Symmetry	Front Symmetry	Basis	Apex	Mucro	MTW [a]	Surface	NFF [b]	DFF [c]
Kalamata Olive (KO)	Elongated	Asymmetrical	Slightly asymmetrical	Pointed	Pointed	Without presence	Middle	Rugged	Weak to middle	Uniform or grouped
Prasines Chalkidikis (PX)	Elongated	Slightly asymmetrical	Symmetrical	Pointed	Rounded	Presence	Middle	Rugged	Middle	Uniform
Konservolia Stylidas (KS)	Ovoid	Slightly asymmetrical	Symmetrical	Pointed	Pointed	Presence	Middle	Rough	Middle	Uniform or grouped
Konservolia Amfissis (KA)	Elliptic	Slightly asymmetrical	Symmetrical	Rounded	Pointed	Presence	Middle	Rough	Middle	Uniform or grouped
Throuba Thassos (TT)	Elongated	Very asymmetrical	Symmetrical to slightly asymmetrical	Pointed	Pointed	Presence	Middle	Smooth to rough	Middle	Uniform
Throuba Chios (TC)	Elliptic	Very asymmetrical	Symmetrical	Pointed or rounded	Rounded	Presence	Middle	Rough	Middle	Uniform

[a] Position of maximum transversal width. [b] Number of fibrovascular furrow. [c] Distribution of fibrovascular furrow.

Olive stones from varieties cv. Konservolia Amfissis were morphologically very similar to cv. Konservolia Stylidas. Regarding the shape, cv. Prasines Chalkidikis cv. Kalamata Olive and cv. Throuba Thassos have elongated shape, cv. Konservolia Amfissis, cv. Throuba Chios have elliptic shape and cv. Konservolia Stylidas has an ovoid shape. The profiles of cv. Throuba Chios and cv. Throuba Thassos stones are very asymmetrical, slightly asymmetrical for cv. Prasines Chalkidikis, cv. Konservolia Amfissis, and cv. Konservolia Stylidas stones and asymmetrical for cv. Kalamata Olive stones.

3.3. Chemometric Interpretation of the Data by Using OPLS-DA Methods

To tests the validity of the dataset, a calibration and a validation set were set up by having 42 and 18 samples, respectively, as shown in Figure 5. Both scatter plots (a) and (b) were successfully built with R2X(cum) = 0.950, R2Y(cum) = 0.946 and Q2(cum) = 0.933, and R2X(cum) = 0.995, R2Y(cum) = 0.878 and Q2(cum) = 0.789, respectively.

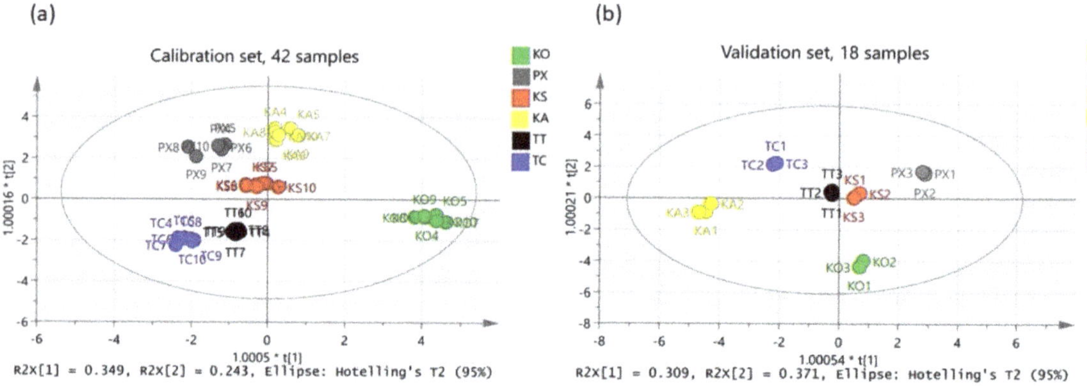

Figure 5. (**a**) OPLS-DA calibration set with R2X(cum) = 0.950, R2Y(cum) = 0.946 and Q2(cum) = 0.933, and (**b**) OPLS-DA validation set with R2X(cum) = 0.995, R2Y(cum) = 0.878 and Q2(cum) = 0.789.

After chemometric interpretation of the data, an overall OPLS-DA model was constructed, as seen in Figure 6. No outlier samples were obtained; thus, all the 60 samples of table olives were distributed in the Hotelling's T2 ellipse. KO seems to be a very special variety of table olives, as it is the only one which is located at the right part of the ellipse, while the other five varieties (i.e., PX, KS, KA, TT and TC) are located at the center to left part of the ellipse. The values of coefficients R2X(cum) = 0.991, R2Y(cum) = 0.912 and Q2(cum) = 0.855 are all good, since they are all above 0.5, and the difference between R2X(cum) and Q2(cum) is 0.136 which is satisfactory, as it is lower than 0.2–0.3.

Only one sample (PX9) was wrongly classified, as it was located away from the centre of the PX group. It seems that PX9 should belong to TC group, however, chemometric analysis and particularly misclassification table (Table 3) shows that PX9 is located closer to the center of KS group. Calculation of Euclidean distances (results not shown here) confirmed that PX9 is closer to KS group instead of TC. The incorrect classification of PX9 decreases the percentage of correct classification of the PX group to 90%, and the percentage of the overall OPLS-DA model to 98.33%.

Table 3. Misclassification table for the overall OPLS-DA model.

	Members	Correct	KO	PX	KS	KA	TT	TC
KO	10	100%	10	0	0	0	0	0
PX	10	90%	0	9	1	0	0	0
KS	10	100%	0	0	10	0	0	0
KA	10	100%	0	0	0	10	0	0
TT	10	100%	0	0	0	0	10	0
TC	10	100%	0	0	0	0	0	10
No class	0		0	0	0	0	0	0
Total	60	98.33%						
Fisher's prob.	1.1×10^{-39}							

By eliminating PX9 from the dataset, the classification rate became 95% as the other samples of PX class were not as closed between them. Thus, it was considered important to keep PX9 in the PX class and in the overall model.

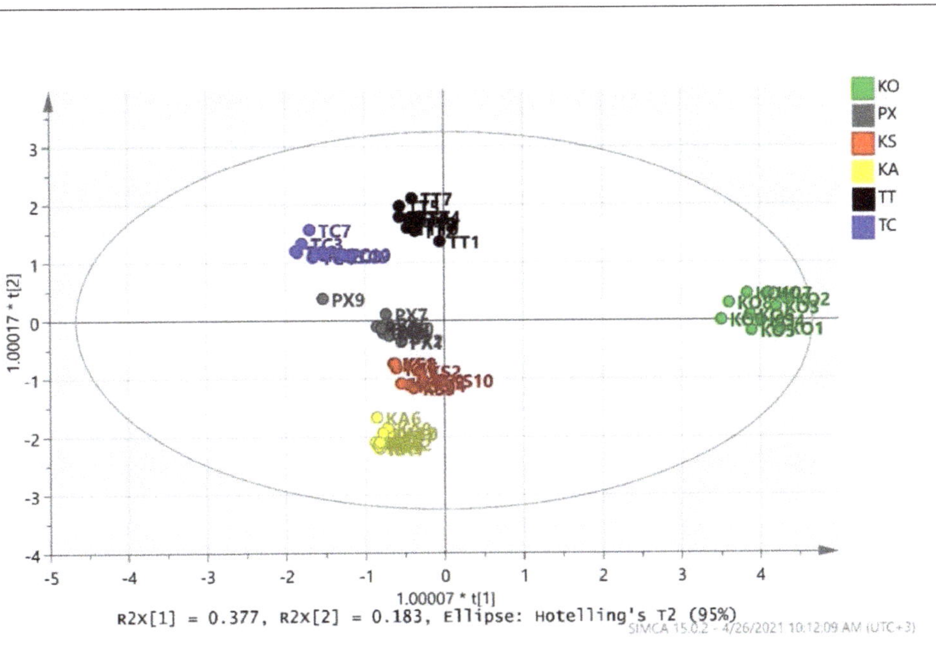

(a)

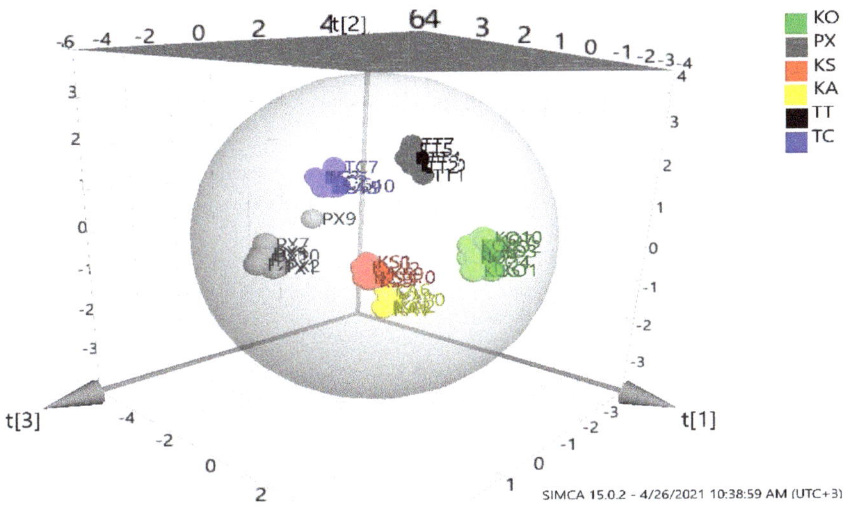

(b)

Figure 6. (**a**) The 2D and (**b**) 3D score scatter plots of the overall OPLS-DA model (R2X(cum) = 0.991, R2Y(cum) = 0.912 and Q2(cum) = 0.855.

To test the significance and adequacy of the model, the CV-ANOVA, which is considered as the most important test for the evaluation of significance of the developed model, was applied. The CV-ANOVA results show the value of the $F_{statistic}$ and *p*-value and are depicted in Table 4. The model is highly significant, due to the *p*-value of zero. Based on DF = 295, the null hypothesis should be rejected, and the alternate hypothesis is considered important due to that $F_{statistic} = 18.9 > F_{critical} = 2.24$ for probability level equal to 0.05.

Table 4. CV-ANOVA data obtained for the overall OPLS-DA model.

OPLS-DA	SS [1]	DF [2]	MS [3]	F [4]	p [5]	SD [6]
Total corr.	295	295	1			1
Regression	252.2	70	3.6	18.9	0	1.9
Residual	42.7	225	0.19			0.4

[1] SS = sum of squares, [2] DF = degree of freedom, [3] MS = mean squares, [4] F = F-test calculated value or $F_{statistic}$, [5] p = *p*-value of the test, [6] SD = standard deviation.

In addition, to validate further the goodness of fit and the predictability of these results, a random permutation test with 100 permutations was employed, as seen in Figure 7. Both R2 (original model) and Q2 (predictive model) located at right and permutated R2 (original model) and Q2 (predictive model) located left while all blue Q2 values to the left and right are lower than the green original R2 values. All the permuted models showed lower R2Y values if compared with the original model's R2Y value (0.912) and the majority of the Q2 regression lines showed negative intercepts (0.0, −0.688).

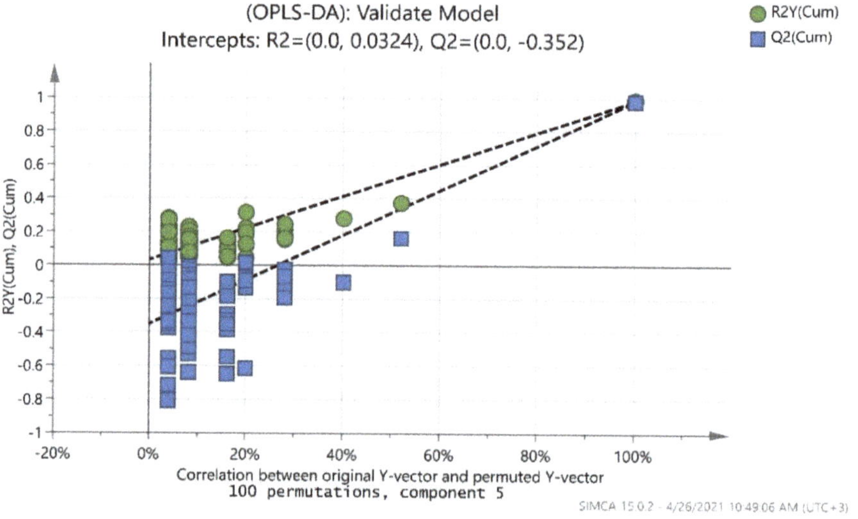

Figure 7. Permutation test of the overall OPLS-DA model took place with 100 permutations.

4. Discussion

The determination of olive variety with this method is very different from the use of precision instruments for material analysis. Not only does it reduce the cost of money and time consumption, but also is more efficient. The identification of the variety of table olives and especially the ones that have been characterized as PDO table olive varieties is required, as the PDO characterization products have higher prices. Variety is a major issue of authenticity and the use of the term PDO can lead to significant falsifications [42]. Since the final product of table olives is a fermented product and different types of table olives can be produced, the methods for determining the variety of origin of fresh olives are completely different from those of table olives, as many changes in the pulp of olives can occur [30]. It is well known that all table olives are fermented in sodium chloride brine [43], through a series of treatments that considerably vary depending on the region and variety [19]. Over the last decade, several studies have been published focusing primarily on reducing sodium chloride content. In this context, modified fermentation brines have been used, in order to satisfy consumers' demand for healthier table olives, with less sodium chloride [43–49]. Therefore, different physicochemical characteristics, and sensory and nutritional properties, may arise from the various fermentation procedures, which makes table olive classification difficult.

The study of morphological features of stones and weight measurements of examined Greek varieties constitutes an alternative method that permit us to determine each variety according to their different size, aspect and weight. Besides, using a high-resolution color camera the examined stones are presented in more details, than those, that can be measured with the human naked eye. In addition, the varietal identification achieved with the machine vision system in combination with the chemometric analysis allows fast classification, without the need for human observation and the subsequent errors. In many studies, in addition to the morphological characteristics of the stones, the morphological characteristics of the leaves and fruits have also been used to identify the olive variety. Olive stone information is the most valuable, among other morphological features of a variety, as they are little affected by environmental conditions. Therefore, olive stone characteristics tend to appear similar to olives belonging to the same variety and tend to differ in the opposite case. Martínez et al. approached the problem of varietal identification by feature extraction from the analysis of endocarp images, and then using partial least square-discriminant classifier [33].

This is the first time that the research group studied the authenticity of Greek varieties of table olives, although similar research has been conducted in different countries and varieties by other authors who combined imaging and chemometrics [19,20,30–32,34,35]. Esteves da Silva demonstrated the great usefulness of chemometrics in the classification of olive varieties. The morphological characteristics of the endocarp among other characteristics (for example olive fruits, trees, branches, leaves and flowers) were used to classify 22 Portuguese olive varieties. He also managed to demonstrate the similarities between the varieties studied and to show that some characteristics have a greater power of distinction than others [19].

Vanloot et al. achieved the discrimination of five French varieties, namely *Aglandau, Bouteillan, Lucques, Picholine, Tanche*, through artificial vision and chemometric analysis of olive stones with 100% of correct classification, working with the data obtained from front and profile pictures [30]. Even if the front and profile parameters are different for the discrimination of the varieties only the picture of profile was sufficient. Image processing techniques of olive fruit, olive leaves, and olive stones, were used for the identification of Turkish olive varieties namely *Sarı ulak, Gemlik, Edincik su, Memecik, Eşek zeytini, Ayvalık, Kilis yağlık, Uslu, Çilli,* and *Domat* [20], while in another study, image processing techniques with data obtained from the fruits and stones were used for the classification of Spanish olive cultivars, namely *Lechin De Granada, Arbequina, Picual, Verdial De V-M, Picudo, Hojiblanca* and *Empeltre* [31]. Seven Greek olive varieties, namely *Kalamon, Karidolia-Chalkidikis, Koroneiki, Lianomanako-Tyrou, Mastoidis, Megaron* and *Throumbolia*, were distinguished according to the morphological parameters of the olive fruit, olive leaves, and olive stones [34]. The study of biometric characteristics of the olive stone was also used to determine the relationships between wild and farmed olives [35].

The analysis regarding geographic origin of the Greek PDO table olive varieties reveals that there is regional clustering. KS and KA were expected to be located closer than the other groups as the locations which have been harvested are nearby. In addition, it can be said that regarding variety species, KS and KA as well as TT and TC were expected to be located "in pairs", meaning next to each other on the score scatter plot and this is what was observed. These observations show that the varieties from a particular PDO variety can easily be discriminated using the fruit and the stones characteristics. The stones of the KO variety are very characteristic with their very elongated and pointed shape. These stones are very similar to the *Lucques* variety, one of the best and most popular French table olives [50].

Moreover, other authors also stated that image processing alone or coupled with chemometrics can be the best combination in regard to rapidness and ease. Puerto et al. presented a methodology for differentiating olives collected from the ground from those harvested directly from the trees, as the former impoverishes quality of the subsequently produced olive [51]. An automatic inspection system, based on computer vision, was used to classify automatically different batches of olives, before being processed for oil extraction, with a success ratio of 100%. Ponce et al. proposed a non-invasive methodology, in which the classification is carried out uniquely using the morphology of the olive-fruits as distinguishing feature [52]. For this purpose, 2800 fruits belonging to seven different olive varieties, were photographed. It was designed by a procedure, based on image processing and analysis and convolutional neural networks, for developing a set of image classifier. These image classifiers showed a remarkable behaviour in terms of performance, as high rates of accuracy were obtained in general for all of them.

A new methodology, based on computer vision and feature modelling, was proposed by Ponce et al. [53] for automatic counting and individual size and mass estimation of olive-fruits. For its development, a total of 3600 olive-fruits from nine varieties were photographed, stochastically distributing the individuals on the scene, using an ad-hoc designed an imaging chamber. The results from the study indicated relative errors below 0.80% and 1.05% for the estimation of the major and minor axis length for all varieties, respectively.

In a very recent study, an efficient methodology to estimate the maximum/minimum (polar/equatorial) diameter length and mass of olive fruits by means of image analysis was proposed [54]. Different sets of olives from the varieties *Picual* and *Arbequina* were photographed, and an original algorithm based on mathematical morphology and statistical thresholding was developed for segmenting the acquired images. The performance of the models was evaluated on external validation sets, giving relative errors of 0.86% for the major axis, 0.09% for the minor axis and 0.78% for the *Arbequina* variety; analogously, relative errors of 0.03%, 0.29% and 2.39% were annotated for *Picual*.

Diaz et al. dealt the classification of table olive in different quality categories depending on the defects in the surface of the fruits [55]. Learning algorithms that allow the extraction of quality information from batches previously classified by experts have been applied. A colorimetric characterization of the most common defects was performed. An image analysis system was used to segment the parameter set with the olive quality information. The results show that a neural network with a hidden layer can classify olives with more than 90% accuracy.

New effective techniques for automatic detection and classification of external olive fruits defects based on image processing techniques, was presented by Hassan et al. [56]. The proposed techniques can separate between the defected and the healthy olive fruits, and then detect and classify the actual defected area. The proposed techniques are based on texture analysis and the homogeneity texture measure. The results reveal that proposed techniques have the highest accuracy rate among other techniques.

A comparative analysis of the discrimination of pepper (*Capsicum annuum* L.) based on the cross-section and seed textures determined using image processing was developed by Ropelewska and Szwejda-Grzybowska [57]. An effective method based on hyper- spectral imaging combined with a group sparse representation (GSR) classifier for the geographic origin authentication of Yangshan region peaches and to interpret the hyperspectral fingerprint with physiological metabolism using high-performance liquid chromatography (HPLC) analysis was developed by Sun et al. [58].

OPLS-DA method is a very efficient method for discrimination purpose. The NMR-based metabolic profiling tool for the quality assessment of table olives, from the *Konservolia*, *Kalamon* and *Chalkidikis* cultivars from different areas of Greece was used by Beteinakis et al. [59]. Specific biomarkers, related to the classification of olives based on different treatments, cultivars and geographical origin, were identified and OPLS-DA models were built by taking groups in pairs, in order to identify certain markers responsible for the differentiation of cultivars. Moreover, the comparison of similar species in different countries can verify the high discrimination accuracy of OPLS-DA method low-field nuclear magnetic resonance (LF-NMR) in combination with multivariate statistical analysis was used to identify the adulterated Spanish extra virgin olive oil with different rations of soybean oil or corn oil. The multi-blended oil could be 100% classified by OPLS-DA when the adulteration ratio was above 30% [60]. In a very recent study, NMR analysis to avocado oil to differentiate it from other oils including olive oil, was applied by Tang et al. [61]. Avocado oil and olive oil were efficiently classified by OPLS-DA method with an R^2 of 0.97, and a Q^2 of 0.91, indicating a very significant model.

This method gave satisfactory results for other agricultural products proving its effectiveness. Becerra-Martınez et al. [62] used NMR spectroscopy supported by principal component analysis PCA or OPLS-DA to differentiate between two Mexican cultivars of chili based on the difference of their metabolites. The authors were able to differentiate the two cultivars using PCA with an R^2 of 0.936; to better observe differences between groups, OPLS-DA was successfully applied (R^2 = 0.923). Chung et al. [63] analyzed the multi-element profile of rice samples procured from six different Asian countries using ICP-MS to investigate geographical origin. Rice samples were clearly discriminated through PCA and OPLS-DA as different countries exhibited a different proportion of micro and macro elements.

This work is a pre-study that should be continued in order to increase the database on Greek table olive stones. In future, similar research studies must test more samples per variety and focus on the harvest period. The discrimination of olive varieties can be definitely benefited from the current development of image analysis technology and big data analysis.

5. Conclusions

This research study shows that the morphological features of olives (fruit and endocarp/stone) as well the weight of stones in combination with chemometrics can be discriminated. OPLS-DA proved to be good method for visualizing and interpreting the data. Morphological characteristics of olive stone have enough discrimination capacity to allow to classify the olives. Further research and assessment will take place related to the Greek PDO table olives, and more models can be developed for future predictions related to their quality and authenticity. Ongoing research in the particular field will enlighten the authenticity of the Greek PDO table olives. This preliminary study shows encouraging results and that this visual authentication analysis is easy to implement. It will be more efficient when the image analysis is computerised as planned. This will save time and allow the Greek varieties to be compared with varieties from different geographic origins.

Author Contributions: Conceptualization, S.A., M.T., T.V., S.R. and J.A.; methodology, S.A., M.T., T.V., S.R. and J.A.; software, M.T., C.R.T., T.V., S.R. and J.A.; validation, S.A., M.T., M.A.B.B., C.R., N.D., C.R.T., T.V., S.R. and J.A.; formal analysis, M.T., C.R.T.; investigation, S.A., M.T., M.A.B.B., N.D., S.R., T.V. and J.A.; resources, S.A., M.T., M.A.B.B., C.R., N.D., C.R.T., T.V., S.R. and J.A.; data curation, S.A., M.T., M.A.B.B., C.R., N.D., C.R.T., T.V., S.R. and J.A.; writing—original draft preparation, S.A., M.T.; writing—review and editing, S.A., M.T., M.A.B.B., C.R., N.D., C.R.T., T.V., S.R. and J.A.; visualization, M.T. and C.R.T.; supervision, M.T., C.R., C.R.T., T.V., N.D., S.R. and J.A.; project administration, S.A., T.V., S.R. and J.A. All authors have read and agreed to the published version of the manuscript.

Funding: This research received no external funding.

Institutional Review Board Statement: Not applicable.

Informed Consent Statement: Not applicable.

Data Availability Statement: Not applicable.

Acknowledgments: Agriopoulou Sofia is the beneficiary of a grant by Institute of France and IKY in the framework of "Scholarships of Greek-French Cooperation for Higher Education in France, Academic Year 2020–2021" for a post-doctoral fellowship. Moreover, the authors would also like to acknowledge Agelis Konstantinos, Stampelos Xenofontas, Kaplanis Leonidas, Dimitriou Dimitrios, Kottis Georgios and Stamatelopoulou Eugenia for their kind offer of the samples.

Conflicts of Interest: The authors declare no conflict of interest.

References

1. Russo, G.; Beritognolo, I.; Bufacchi, M.; Stanzione, V.; Pisanelli, A.; Ciolfi, M.; Lauteri, M.; Brush, S.B. Advances in biocultural geography of olive tree (*Olea europaea* L.) landscapes by merging biological and historical assays. *Sci. Rep.* **2020**, *10*, 7673. [CrossRef]
2. Valvez, S.; Maceiras, A.; Santos, P.; Reis, P.N.B. Olive Stones as Filler for Polymer-Based Composites: A Review. *Materials* **2021**, *14*, 845. [CrossRef] [PubMed]
3. Breton, C.; Terral, J.-F.; Pinatel, C.; Médail, F.; Bonhomme, F.; Bervillé, A. The origins of the domestication of the olive tree. *C. R. Biol.* **2009**, *332*, 1059–1064. [CrossRef] [PubMed]
4. Jurado-Campos, N.; García-Nicolás, M.; Pastor-Belda, M.; Bußmann, T.; Arroyo-Manzanares, N.; Jiménez, B.; Viñas, P.; Arce, L. Exploration of the potential of different analytical techniques to authenticate organic vs. conventional olives and olive oils from two varieties using untargeted fingerprinting approaches. *Food Control.* **2021**, *124*, 107828. [CrossRef]
5. Bianchi, G. Lipids and phenols in table olives. *Eur. J. Lipid Sci. Technol.* **2003**, *105*, 229–242. [CrossRef]
6. International Olive Oil Council (IOC). *Trade Standard Applying to Table Olives*; International Olive Oil Council: Madrid, Spain, 2004.
7. International Olive Oil Council. World Table Olive Figures. 2021. Available online: https://www.internationaloliveoil.org/what-we-do/economic-affairs-promotionunit/#figures (accessed on 14 April 2021).
8. Roussos, S.; Rahmani, M. Les olives de table fermentées, un aliment fonctionnel comme le yogourt. In *Le Compagnon de l'Olivier-XXVIIeme Annee-N° 42*; Acopa: Aix-en-Provence, French, 2018; pp. 12–14.
9. Ghanbari, R.; Anwar, F.; Alkharfy, K.M.; Gilani, A.-H.; Saari, N. Valuable nutrients and functional bioactives in different parts of olive (*Olea europaea* L.)—A review. *Int. J. Mol. Sci.* **2012**, *13*, 1291–1340. [CrossRef]
10. Zoidou, E.; Melliou, E.; Gikas, E.; Tsarbopoulos, A.; Magiatis, P.; Skaltsounis, A.L. Identification of Throuba Thassos, a Traditional Greek Table Olive Variety, as a Nutritional Rich Source of Oleuropein. *J. Agric. Food Chem.* **2010**, *58*, 46–50. [CrossRef]
11. Perpetuini, G.; Prete, R.; Garcia-Gonzalez, N.; Khairul Alam, M.; Corsetti, A. Table Olives More than a Fermented Food. *Foods* **2020**, *9*, 178. [CrossRef]
12. International Olive Council. *World Catalogue of Olive Varieties*; International Olive Council: Madrid, Spain, 2000.
13. Council Regulation (EC). No. 2081/92 of 14 July 1992 on the protection of geographical indications and designations of origin for agricultural products and foodstuffs. *Off. J. Eur. Union* **1992**, *L208*, 1–8.
14. Council Regulation (EC). No. 2082/92 of 14 July 1992 on certificates of specific character for agricultural products and foodstuffs. *Off. J. Eur. Union* **1992**, *L208*, 9–14.
15. Council Regulation (EC). No. 510/2006 of 20 March 2006 on the protection of geographical indications and designations of origin for agricultural products and foodstuffs. *Off. J. Eur. Union* **2006**, *L93*, 12–25.
16. Council Regulation (EC). No. 1898/2006 of 14 December 2006 laying down detailed rules of implementation of Council Regulation (EC) no. 510/2006 on the protection of geographical indications and designations of origin for agricultural products and foodstuffs. *Off. J. Eur. Union* **2006**, *L369*, 1–23.
17. Council Regulation (EC). No 1151/2012 of 21 November 20122012 on quality schemes for agricultural products and foodstuffs. *Off. J. Eur. Union* **2012**, *L343*, 1–29.
18. Skiada, V.; Tsarouhas, P.; Varzakas, T. Preliminary Study and Observation of "Kalamata PDO" Extra Virgin Olive Oil, in the Messinia Region, Southwest of Peloponnese (Greece). *Foods* **2019**, *8*, 610. [CrossRef] [PubMed]
19. Esteves da Silva, J.C.G. Chemometric classification of cultivars of olives: Perspectives on Portuguese olives. In *Olives and Olive Oil in Health and Disease Prevention*; Elsevier: Amsterdam, The Netherlands, 2010; pp. 33–42.
20. Beyaz, A.; Öztürk, R. Identification of olive cultivars using image processing techniques. *Turk. J. Agric.* **2016**, *40*, 671–683. [CrossRef]
21. Consonni, R.; Cagliani, L.R. NMR Studies on Italian PDO Olive Oils and their Potential in Olive-Tree-Derived Products Characterization. *Eur. J. Lipid Sci. Tech.* **2019**, *121*, 1800174. [CrossRef]
22. Ben Othman, N.; Roblain, D.; Thonart, P.; Hamdi, M. Tunisian table olive phenolic compounds and their antioxidant capacity. *J. Food Sci.* **2008**, *73*, C235–C240. [CrossRef]
23. Dağdelen, A.; Tümen, G.; Özcan, M.M.; Dündar, E. Phenolics profiles of olive fruits (*Olea europaea* L.) and oils from Ayvalık, Domat and Gemlik varieties at different ripening stages. *Food Chem.* **2013**, *136*, 41–45. [CrossRef]
24. Albuquerque, T.G.; Costa, H.S.; Oliveira, M.B.P.P. An overview of ortuguese olive oils and table olives with protected designation of origin. *Eur. J. Lipid Sci. Tech.* **2019**, *121*, 1800129. [CrossRef]
25. Kalogiouri, N.P.; Aalizadeh, R.; Dasenaki, M.E.; Thomaidis, N.S. Authentication of Greek PDO Kalamata Table Olives: A Novel Non-Target High Resolution Mass. *Molecules* **2020**, *25*, 2919. [CrossRef] [PubMed]
26. Sánchez, A.H.; López-López, A.; Cortés-Delgado, A.; de Castro, A.; Montaño, A. Aroma profile and volatile composition of black ripe olives (Manzanilla and Hojiblanca cultivars). *Food Res. Int.* **2020**, *127*, 108733. [CrossRef]
27. Concepción, R.; García, P.; Medina, E.; Brenes, M. The PDO and PGI Table Olives of Spain. *Eur. J. Lipid Sci. Tech.* **2019**, *121*, 1–17. [CrossRef]
28. Selli, S.; Kelebek, H.; Kesen, S.; Sonmezdag, A.S. GC-MS olfactometric and LC-DAD-ESI-MS/MS characterization of key odorants and phenolic compounds in black dry-salted olives. *J. Sci. Food. Agric.* **2018**, *98*, 4104–4111. [CrossRef] [PubMed]
29. Tarapoulouzi, M.; Kokkinofta, R.; Theocharis, C.R. Chemometric analysis combined with FTIR spectroscopy of milk and Halloumi cheese samples according to species' origin. *Food Sci. Nutr.* **2020**, *8*, 3262–3273. [CrossRef] [PubMed]

30. Vanloot, P.; Bertrand, D.; Pinatel, C.; Artaud, J.; Dupuy, N. Artificial vision and chemometrics analyses of olive stones for varietal identification of five French cultivars. *Comput. Electron. Agric.* **2014**, *102*, 98–105. [CrossRef]
31. Beyaz, A.; Özkaya, M.T.; Duygu, İ. Identification of some spanish olive cultivars using image processing techniques. *Sci. Hortic.* **2017**, *225*, 286–292. [CrossRef]
32. Tarapoulouzi, M.; Skiada, V.; Agriopoulou, S.; Psomiadis, D.; Rébufa, C.; Roussos, S.; Theocharis, C.R.; Katsaris, P.; Varzakas, T. Chemometric Discrimination of the Geographical Origin of Three Greek Cultivars of Olive Oils by Stable Isotope Ratio Analysis. *Foods* **2021**, *10*, 336. [CrossRef] [PubMed]
33. Martínez, S.S.; Gila, D.M.; Beyaz, A.; Ortega, J.G.; García, J.G. A computer vision approach based on endocarp features for the identification of olive cultivars. *Comput. Electron. Agric.* **2018**, *154*, 341–346. [CrossRef]
34. Blazakis, K.N.; Kosma, M.; Kostelenos, G.; Baldoni, L.; Bufacchi, M.; Kalaitzis, P. Description of olive morphological parameters by using open access software. *Plant Methods* **2017**, *13*, 1–15. [CrossRef]
35. Piras, F.; Grillo, O.; Venora, G.; Lovicu, G.; Campus, M.; Bacchetta, G. Effectiveness of a computer vision technique in the characterization of wild and farmed olives. *Comput. Electron. Agric.* **2016**, *122*, 86–93. [CrossRef]
36. Oliveri, P.; López, M.I.; Casolino, M.C.; Ruisánchez, I.; Callao, M.P.; Medini, L.; Lanteri, S. Analytica Chimica Acta Partial least squares density modeling (PLS-DM)—A new class-modeling strategy applied to the authentication of olives in brine by near-infrared spectroscopy. *Anal. Chim. Acta* **2014**, *851*, 30–36. [CrossRef]
37. Brereton, R.G. Pattern recognition in chemometrics. *Chemom. Intell. Lab. Syst.* **2015**, *149*, 90–96. [CrossRef]
38. de Santana, F.B.; Borges Neto, W.; Poppi, R.J. Random forest as one-class classifier and infrared spectroscopy for food adulteration detection. *Food Chem.* **2019**, *293*, 323–332. [CrossRef] [PubMed]
39. Cubero-Leon, E.; De Rudder, O.; Maquet, A. Metabolomics for organic food authentication: Results from a long-term field study in carrots. *Food Chem.* **2018**, *239*, 760–770. [CrossRef]
40. Eriksson, L.; Trygg, J.; Wold, S. CV-ANOVA for significance testing of PLS and OPLS 1 models. *J. Chem.* **2008**, *22*, 594–600. [CrossRef]
41. Rodríguez, G.; Lama, A.; Rodríguez, R.; Jiménez, A.; Guillén, R.; Fernández-Bolaños, J. Olive stone an attractive source of bioactive and valuable compounds. *Bioresour. Technol.* **2008**, *99*, 5261–5269. [CrossRef]
42. Danezis, G.P.; Tsagkaris, A.S.; Camin, F.; Brusic, V.; Georgiou, C.A. Food authentication: Techniques, trends & emerging approaches. *TrAC Trends Anal. Chem.* **2016**, *85*, 123–132.
43. Mantzouridou, F.T.; Mastralexi, A.; Filippidou, M.; Tsimidou, M.Z. Challenges in the Processing Line of Spanish Style cv. Chalkidiki Green Table Olives Spontaneously Fermented in Reduced NaCl Content Brines. *Eur. J. Lipid Sci. Technol.* **2020**, *122*, 1900453. [CrossRef]
44. Pino, A.; Vaccalluzzo, A.; Solieri, L.; Romeo, F.V.; Todaro, A.; Caggia, C.; Arroyo-López, F.N.; Bautista-Gallego, J.; Randazzo, C.L. Effect of Sequential Inoculum of Beta-Glucosidase Positive and Probiotic Strains on Brine Fermentation to Obtain Low Salt Sicilian Table Olives. *Front. Microbiol.* **2019**, *10*, 174. [CrossRef] [PubMed]
45. Bautista-Gallego, J.; Arroyo-Lopez, F.N.; Gil, V.R.; Gómez, F.R.; Garcia, P.G.; Fernández, A.G. Chloride salt mixtures affect Gordal cv. green Spanish-style table olive fermentation. *Food Microbiol.* **2011**, *28*, 1316–1325. [CrossRef]
46. Moreno-Baquero, J.; Bautista-Gallego, J.; Garrido-Fernández, A.; López-López, A. Mineral and sensory profile of seasoned cracked olives packed in diverse salt mixtures. *Food Chem.* **2013**, *138*, 1–8. [CrossRef]
47. Bautista-Gallego, J.; Arroyo-Lopez, F.N.; Romero-Gil, V.; Rodríguez-Gómez, F.; García-García, P.; Garrido-Fernandez, A. Fermentation profile of green Spanish-style Manzanilla olives according to NaCl content in brine. *Food Microbiol.* **2015**, *49*, 56–64. [CrossRef]
48. Mateus, T.; Santo, D.; Saúde, C.; Pires-Cabral, P.; Quintas, C. The effect of NaCl reduction in the microbiological quality of cracked green table olives of the Maçanilha Algarvia cultivar. *Int. J. Food Microbiol.* **2016**, *218*, 57–65. [CrossRef] [PubMed]
49. Saúde, C.; Barros, T.; Mateus, T.; Quintas, C.; Pires-Cabral, P. Effect of chloride salts on the sensory and nutritional properties of cracked table olives of the Maçanilha Algarvia cultivar. *Food Biosci.* **2017**, *19*, 73–79. [CrossRef]
50. Boskou, D.; Camposeo, S.; Clodoveo, M.L. Table Olives as Sources of Bioactive Compounds. In *Olive and Olive Oil Bioactive Constituents*; AOCS Press: Urbana, IL, USA, 2015; pp. 217–259.
51. Puerto, D.A.; Gila, D.M.M.; García, J.G.; Ortega, J.G. Sorting olive batches for the milling process using image processing. *Sensors* **2015**, *15*, 15738–15754. [CrossRef]
52. Ponce, J.M.; Aquino, A.; Andújar, J.M. Olive-fruit variety classification by means of image processing and convolutional neural networks. *IEEE Access* **2019**, *7*, 147629–147641. [CrossRef]
53. Ponce, J.M.; Aquino, A.; Millán, B.; Andújar, J.M. Automatic counting and individual size and mass estimation of olive-fruits through computer vision techniques. *IEEE Access* **2019**, *7*, 59451–59465. [CrossRef]
54. Ponce, J.M.; Aquino, A.; Millán, B.; Andújar, J.M. Olive-fruit mass and size estimation using image analysis and feature modeling. *Sensors* **2018**, *18*, 2930. [CrossRef]
55. Diaz, R.; Gil, L.; Serrano, C.; Blasco, M.; Moltó, E.; Blasco, J. Comparison of three algorithms in the classification of table olives by means of computer vision. *J. Food Eng.* **2004**, *61*, 101–107. [CrossRef]
56. Hassan, N.M.H.; Nashat, A.A. New effective techniques for automatic detection and classification of external olive fruits defects based on image processing techniques. *Multidim. Syst. Sign. Process.* **2019**, *30*, 571–589. [CrossRef]

57. Ropelewska, E.; Szwejda-Grzybowska, J. A comparative analysis of the discrimination of pepper (*Capsicum annuum* L.) based on the cross-section and seed textures determined using image processing. *J. Food Process. Eng.* **2021**, *44*, 13694. [CrossRef]
58. Sun, Y.; Li, Y.; Pan, L.; Abbas, A.; Jiang, Y.; Wang, X. Authentication of the geographic origin of Yangshan region peaches based on hyperspectral imaging. *Postharvest Biol. Technol.* **2021**, *171*, 111320. [CrossRef]
59. Beteinakis, S.; Papachristodoulou, A.; Gogou, G.; Katsikis, S.; Mikros, E.; Halabalaki, M. NMR-based metabolic profiling of edible olives—Determination of quality parameters. *Molecules* **2020**, *25*, 3339. [CrossRef] [PubMed]
60. Wang, S.; Lai, G.; Lin, J.; Xia, F.; Ding, Z.; Feng, J.; Xu, J.; Shen, G. Rapid detection of adulteration in extra virgin olive oil by Low-Field Nuclear Magnetic Resonance combined with pattern recognition. *Food Anal. Methods* **2021**, *14*, 1322–1335. [CrossRef]
61. Tang, F.; Green, H.S.; Wang, S.C.; Hatzakis, E. Analysis and authentication of avocado oil using High Resolution NMR Spectroscopy. *Molecules* **2021**, *26*, 310. [CrossRef] [PubMed]
62. Becerra-Martínez, E.; Florentino-Ramos, E.; Pérez-Hernández, N.; Zepeda-Vallejo, L.G.; Villa-Ruano, N.; Velázquez-Ponce, M.; García-Mendoza, F.; Bañuelos-Hernández, A.E. ^{1}H NMR-based metabolomic fingerprinting to determine metabolite levels in serrano peppers (*Capsicum annum* L.) grown in two different regions. *Food Res. Internat.* **2017**, *102*, 163–170. [CrossRef]
63. Chung, I.M.; Kim, J.K.; Lee, K.J.; Park, S.K.; Lee, J.H.; Son, N.Y.; Jin, Y.I.; Kim, S.H. Geographic authentication of Asian rice (*Oryza sativa* L.) using multi-elemental and stable isotopic data combined with multivariate analysis. *Food Chem.* **2018**, *240*, 840–849. [CrossRef]

foods

MDPI

Article

Chemical Analysis Combined with Multivariate Statistical Methods to Determine the Geographical Origin of Milk from Four Regions in China

Ruting Zhao [1,2], Meicheng Su [1,2], Yan Zhao [1,2,*], Gang Chen [1,2], Ailiang Chen [1,2] and Shuming Yang [1,2]

1 Institute of Quality Standard and Testing Technology for Agro-Products, Chinese Academy of Agricultural Sciences, Beijing 100081, China; zhaoruting1201@163.com (R.Z.); sumeicheng1995@outlook.com (M.S.); chengang01@caas.cn (G.C.); ailiang.chen@gmail.com (A.C.); yangshuming@caas.cn (S.Y.)
2 Key Laboratory of Agro-Product Quality and Safety, Ministry of Agriculture, Beijing 100081, China
* Correspondence: zhaoyan01@caas.cn

Abstract: Traceability of milk origin in China is conducive to the implementation of the protection of regional products. In order to distinguish milk from different geographical distances in China, we traced the milk of eight farms in four neighboring provinces of China (Inner Mongolia autonomous region, Hebei, Ningxia Hui autonomous and Shaanxi), and multivariate data analysis was applied to the data including elemental analysis, stable isotope analysis and fatty acid analysis. In addition, orthogonal partial least squares discriminant analysis (OPLS-DA) is used to determine the optimal classification model, and it is explored whether the combination of different technologies is better than a single technical analysis. It was confirmed that in the inter-provincial samples, the combination of the two techniques was better than the analysis using a single technique (fatty acids: $R^2 = 0.716$, $Q^2 = 0.614$; fatty acid-binding isotopes: $R^2 = 0.760$, $Q^2 = 0.635$). At the same time, milk produced by farms with different distances of less than 11 km in each province was discriminated, and the discriminant distance was successfully reduced to 0.7 km (Ningxia Hui Autonomous Region: the distance between the two farms was 0.7 km, $R^2 = 0.771$, $Q^2 = 0.631$). For short-distance samples, the combination multiple technologies are not completely superior to a single technique, and sometimes, it is easy to cause model over-fitting.

Keywords: milk; fatty acids; isotopes; mineral elements; geographical origin; multivariate statistics

Citation: Zhao, R.; Su, M.; Zhao, Y.; Chen, G.; Chen, A.; Yang, S. Chemical Analysis Combined with Multivariate Statistical Methods to Determine the Geographical Origin of Milk from Four Regions in China. *Foods* **2021**, *10*, 1119. https://doi.org/10.3390/foods10051119

Academic Editor: Theodoros Varzakas

Received: 16 April 2021
Accepted: 14 May 2021
Published: 18 May 2021

1. Introduction

With the great improvement of people's living standard, China's dairy farming industry has also greatly developed, and has now become the third largest producer in the world. Milk has high nutritional value, and its quality is considered to be related to the geographical location of pasture, forage, water source and other factors. Therefore, consumers pay increasingly more attention to the origin of milk, resulting in the economic value of origin information. Traceability of milk origin in China is conducive to the implementation of the protection of regional products. It can also effectively prevent the spread of food safety incidents and recall products. Therefore, the traceability of China is of high importance. Chemical fingerprinting techniques occupy an important position among all traceability methods due to its advantages of simple operation, accurate results and so on. Increasingly, the traceability of milk utilizes fatty acids, stable isotopes and mineral elements to identify the geographical origins of dairy products.

At present, many studies on the geographical origin of milk have been carried out by isotope, mineral element and fatty acid techniques. Stable isotopes are commonly used to characterize geographical origin information and to describe agricultural products' origin information [1], where δ^2H and $\delta^{18}O$ can be used to distinguish altitude, $\delta^{15}N$ can be used to determine the type of grazing vegetation and $\delta^{13}C$ can determine the type of animal feed. Thus, the stable isotope ratios can be used to distinguish milk [2–4] and dairy products [5–7] of different areas.

Mineral elements have been widely used in the traceability of animal-derived agricultural products such as beef [8], pork [9], lamb [10], poultry meat [11] and honey [12,13]. This technique is also increasingly being used to identify the types and origins of milk and dairy products. In 2008, Benincasa et al. used 16 mineral elements in milk and buffalo milk to distinguish two types of milk from the same pasture [14]. In 2015, Osorio et al. determined the mineral elements in goat milk, Halloumi cheese and grazing plants in three parts of Cyprus, which can be completely distinguished, and found some mineral elements (Mn and Sr) with good traceability [15].

There have been a few studies on whether fatty acids can be used as potential chemical parameters to identify milk and dairy products of different origins. It has been reported that the proportion of fatty acids in milk produced in pastures at different latitudes varied significantly. Among them, essential fatty acids (EFA) contents and the ratios of Conjugated linoleic acid (CLA) and Polyunsaturated fatty acid (PUFA) in milk produced in mountain areas were higher than those produced in indoor cows [16]. Similar conclusions have been drawn in the study of fatty acids in milk from lowlands, mountains and highlands in Switzerland [17]. Moreover, a study of nutrients in milk from four provinces in China reported that the fatty acid contents were influenced by the geographical location [18].

Further studies showed that a model combining isotopes with mineral elements had a good differentiation effect in the geographical origin of milk, and the differentiation rate was above 90% [19–21]. This advantage has also been confirmed in the identification of the origins of milk, dairy products and other foods, especially in the identification of the origins of PDO foods [22–25]. In addition, in recent years, there have been studies using other technologies, such as nuclear magnetic resonance, metabolomics, infrared spectroscopy and elemental analysis, to analyze food origin. These studies have also shown that when multiple analytical techniques are combined, the results are better than when using only a single technical analysis [26–32].

To our knowledge, most of the research on the identification of milk producing areas has been carried out in countries or regions with far-reaching distances, such as Australia and New Zealand [1], the United States, Germany, China and France [3,4,33], or northern and central Italy [23] and northern, northwestern and southwestern China [24]. Only a few studies have focused on near-field production; in previous studies in our laboratory, Xie et al. paid attention to the traceability of milk in small-scale districts of Inner Mongolia Autonomous Region in China. It was found that a model combining all three techniques could distinguish milk samples from 11 regions in the same province and improve the accuracy of classification of a small-scale region tracking model [34]. However, in the study, Xie et al. did not verify the PCA and OPLS-DA models, which may lead to over-fitting. Although the combination of three techniques improves model accuracy, model reliability is unknown and has an impact on subsequent traceability applications. Therefore, in this study, we use permutation test to verify the model to ensure the reliability of the model.

In order to distinguish milk from different geographical distances in China, we traced the milk of eight farms in four neighboring provinces of China (Inner Mongolia autonomous region, Hebei, Ningxia Hui autonomous and Shaanxi), of which two farms in four provinces were not more than 11 km apart. We used isotopes, mineral elements and fatty acids to characterize milk origin information. Moreover, we used principal component analysis (PCA) for preliminary clustering, and further used OPLS-DA to classify milk from four provinces and distinguish milk from two farms in the province.

2. Materials and Methods

2.1. Materials

Milk samples (n = 120) were collected from eight large commercial farms in four provinces of China (Table 1). Milk samples were divided into three parts. One was processed according to the methods reported and used for the determination of fatty acids [18].

Two were freeze-dried for 24 h and then pulverized. The sample was mixed with a chloroform/methanol (2:1, *v/v*) solution at 1:5, vortexed for 10 min and centrifuged at 5000 rpm for 5 min, and the supernatant was discarded [35]. Then the previous degreasing step was repeated twice, the supernatant was discarded, the solid was retained and lyophilized to obtain a defatted dry matter (DDM) for the determination of stable isotopes and mineral elements. These samples were stored at −20 °C for subsequent analysis.

Table 1. Information on dairy farms in four regions of China.

Origin	Number of Samples	Number of Farms	Distance between Farms (km)	North Latitude	East Longitude	Altitude (m)	Staple Feed Species
HB	30	2	10.7	38°	117°	7	Yellow corn silage, Alfalfa hay, straw
NMG	30	2	4.2	40°	111°	1030	Corn silage, Alfalfa hay, *Leymus chinensis*
SX	30	2	2.9	34°	108°	468	Corn silage, Alfalfa hay, straw
NX	30	2	0.7	37°	106°	1160	Corn silage, Alfalfa hay, cottonseed, *Leymus chinensis*

HB = Hebei Province; NMG = Inner Mongolia Autonomous Region; NX = Ningxia Hui Autonomous; Region; SX = Shaanxi Province.

2.2. Analytical Methods

2.2.1. Analysis of Fatty Acids

The samples were analyzed by an Agilent 7890A gas chromatograph with a flame ionization detector. The column is an SP-2560 (100 m × 0.25 mm × 0.20 μm; Supelco Inc., Santa Clara, CA, USA). The initial temperature is 100 °C, and raised by 5 °C min^{-1} to 210 °C, which was maintained for 25 min, then raised to 230 °C, which was held for two minutes. The injector and detector temperature were maintained at 260 °C. In total, 32 fatty acids were measured (C4:0; C6:0; C8:0; C10:0; C11:0; C12:0; C13:0; C14:0; C14:1 *cis*-9; C15:0; C15:1 *cis*-10; C16:0; C16:1 *cis*-9; C17:0; C17:1 *cis*-10; C18:0; C18:1 *trans*-9; C18:1 *cis*-9; C18:2 *cis*-6; C18:3 *cis*-6,9,12; C18:3 *cis*-9,12,15; C20:0; C20:1-*trans*-11; C20:2-*cis*11,14; C20:3-*cis*8,11,14; C20:3-*cis*11,14,17; C20:4-*cis*5,8,11,14; C22:0; C22:1-*cis*13; C22:2-*cis*13,16; C24:1-*cis*15 and CLA).

2.2.2. Analysis of Stable Isotopes

For the stable isotope analysis of $\delta^{13}C$ and $\delta^{15}N$, DDM and other international reference materials (USGS43, USGS40 and Sorghum Flour) were weighed into tin capsules (5 × 8 mm), and then introduced into an elemental analyzer (Flash 2000, Thermo, Waltham, MA, USA), converting the entire material into carbon dioxide and nitrogen gas analyzed by an isotope ratio mass spectrometer (Delta V Advantage of Thermo, Waltham, MA, USA). Two-point normalization of international standard materials was used. For the values of $\delta^{13}C$, USGS40 and Sorghum Flour were used for two-point normalization, and USGS43 was used for QC. For the values of $\delta^{15}N$, USGS43 and USGS40 were used for two-point normalization, and Sorghum Flour was used for QC. Blanks consisting of an empty tin capsule were included and corrections were applied to the results.

For the stable isotope ratio analysis of $\delta^{2}H$ and $\delta^{18}O$, DDM and international reference materials (Caribou Hoof, Kudu Horn and EMA P2) were weighed into silver capsules (4 × 6 mm) along with other international reference materials and introduced into elemental analyzers (Flash 2000, Thermo, Waltham, MA, USA). The reactor packing is a glassy carbon reactor and silver wool. The element hydrogen and oxygen in samples were converted into H_2 and CO at 1380 °C via pyrolysis with glass carbon. The gas was transferred to an isotope ratio mass spectrometer (Delta V Advantage, Thermo, Waltham, MA, USA). For the values of $\delta^{2}H$, Caribou Hoof and Kudu Horn were used for two-point normalization, and EMA P2 was used for QC.

2.2.3. Analysis of Mineral Elements

The content of the mineral elements in DDM were determined according to published methods in our lab [36]. DDM underwent microwave digestion in a Microwave-Assisted Reaction System (MARS) (CEM, Matthews, NC, USA). A total of 0.20 g of each sample was accurately weighed directly into the PTFE digestion tube (15 mL) in triplicate, followed by the addition of 10 mL 65% HNO_3 (analytical grade) and 1.0 mL 30% H_2O_2 (analytical grade) and digested for 40 min. After the sample digestion was complete, the objects in the PTFE digestion tube were transferred to a 50 mL volumetric flask, diluted with ultra-pure water, and the volume was constant to 50 mL Next, 12 elements (sodium (Na), magnesium (Mg), potassium (K), calcium (Ca), titanium (Ti), cadmium (Cr), manganese (Mn), iron (Fe), nickel (Ni), zinc (Zn), strontium (Sr) and molybdenum (Mo)) were determined by inductively coupled plasma mass spectrometry (X Series 2, Thermo Fisher, Waltham, MA, USA). Three analyses were performed for each sample and external standard analysis was performed for quantification. All results are expressed as the average of three measurements.

2.3. Data Processing

Multivariate statistical analysis (PCA, OPLS-DA and Permutation test) was performed on all data using SIMCA 14.1.0 software (Umetrics, Umea, Sweden). The raw data were scaled using unit variance (UV-scale), and analyzed using supervised OPLS-DA, which was used to obtain the classifying model and synchronously extract the variables with important contributions to the classification. Permutation tests were used to assess the reliability of the model.

3. Results and Discuss

3.1. Multivariate Statistical Analysis

3.1.1. Identification of Milk Produced in Four Provinces

PCA Results

PCA is used to reduce the dimension of high dimensional variable space under the principle of minimum data information loss. These comprehensive indexes are called main components. The principal component will retain as much information as possible about the variation of the original index. In the preliminary study, single or multiple chemical parameter data (fatty acids, stable isotopes and mineral elements) were analyzed by PCA to study any possible milk clustering based on origin. PCA results (Supplementary Materials Figure S1A,B) showed that there was no obvious grouping in the score plots for inter-provincial samples, whether a single chemical parameter or the analysis with a combination of chemical parameters; however, other PCA models had no obvious classification. Thus, we consider conducting a supervised OPLS-DA of the data to improve the classification of samples.

OPLS-DA Results

A slight sign of classification was observed on the PCA score plot. Next, a supervised discriminant analysis of milk samples between four provinces was carried out using OPLS-DA. Moreover, we used the measure of fit of the model (R^2) and the measure of predictive ability of the model (Q^2) to evaluate the models.

There are three OPLS-DA score plots of mineral elements, isotopes, fatty acids and a combination of the best and no over-fitting in Figure 1. Four groups of milk data were analyzed by OPLS-DA. It was found that the results of the isotope and mineral element chemical parameter analysis showed no signs of classification in the score plot (Figure 1A,B). However, to our surprise, the fatty acid chemical parameter analysis showed good classification on the score plot (Figure 1C). As Figure 1C shows, Ningxia and Inner Mongolia were the most distinguished, followed by Hebei and Inner Mongolia and finally Ningxia and Shaanxi. This was because the fatty acid content and composition are affected by dairy cow breeds, feed and environmental factors such as altitude. Larsen et al. investigated the influence of regional climatic conditions on milk composition, especially fatty acid composition, and the result shows that the content of short-chain fatty acid (C4-C14), C18:0 and C18:3 n-3 are higher in central Sweden than in southern Sweden and that this is most likely because maize growing is limited to southern Sweden [37]. Thus, environmental factors affect the fatty acid content and composition in milk by affecting local plant types. Staple feed species differences (Table 1) may be the main cause of milk differences in four provinces, even more important than geographical factors. Moreover, some studies have shown that lactation also affects the fatty acid composition of milk [18,38]. Among the single techniques, the fatty acid model had the best predictive ability (Figure 1A–C). To sum up, each region in this study had a characteristic fatty acid content fingerprint and that the fatty acid chemical parameter analysis was more effective than the mineral element and isotope analysis at identifying the milk samples in the four provinces.

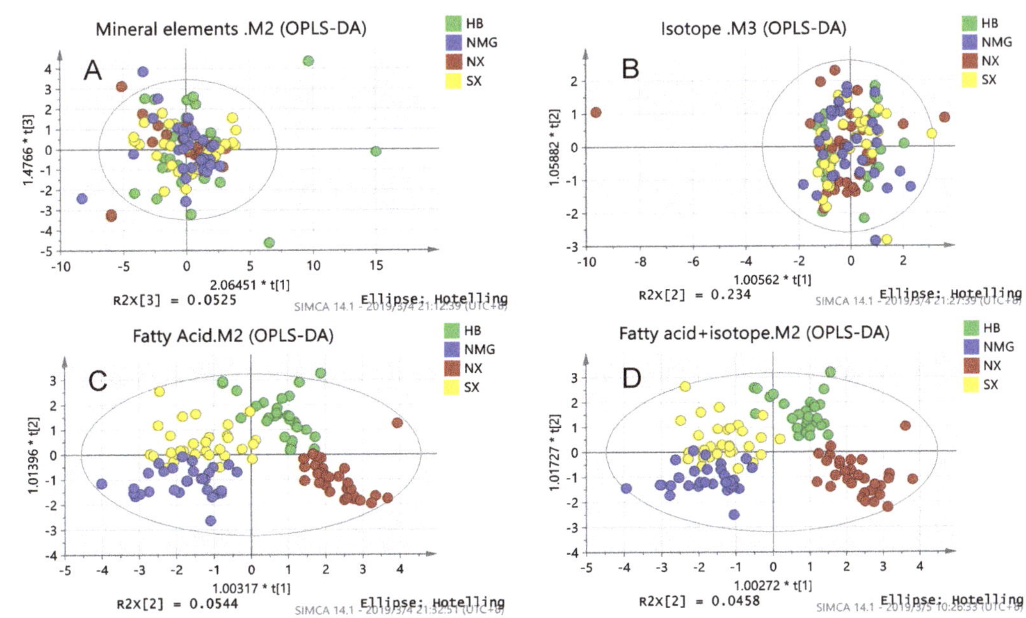

Figure 1. OPLS-DA score plots of inter-provincial samples obtained by a chemical analysis of: (**A**) Mineral elements; (**B**) Isotopes; (**C**) Fatty acids; (**D**) Fatty acids combined with isotopes.

As shown in Table 2, the R^2 of the isotope analysis, mineral element chemical parameter analysis and the combination of the two was less than 0.03, and the fitting degree of the model was extremely low, while the corresponding Q^2 was negative, which indicates that the prediction ability of the models is not good [39]. Except these three models (isotopes, mineral elements, isotopes + mineral elements), the fitting degree of other models was more than 71.60%, and the prediction ability of other models was more than 56.00%. It meant that these regression models are good. Among the single techniques, the fatty acid model had the well predictive ability ($R^2 = 0.716$, $Q^2 = 0.614$). Among the binding technologies, fatty acid technology is helpful to improve the model prediction ability, and the binding technologies including fatty acids have better model prediction ability (fatty acid and mineral element technologies: $R^2 = 0.717$, $Q^2 = 0.560$; fatty acid and isotope technologies: $R^2 = 0.760$, $Q^2 = 0.635$; three technologies: $R^2 = 0.754$, $Q^2 = 0.581$). This indicated that fatty acid chemical parameters play a major role in classification, while mineral element and isotope chemical parameters are less important for classification. The results show that it is the best that the fit and prediction ability of the model combines the fatty acid with isotope technologies ($R^2 = 0.760$, $Q^2 = 0.635$) in four provinces, even better than that of the three technologies ($R^2 = 0.754$, $Q^2 = 0.581$). Similar conclusions have been reported before [40].

When R^2 and Q^2 of each model are close to each other, we prefer to choose a combination of multiple technologies. To sum up, we chose the model combining the fatty acid and isotope chemical parameters as the best classification model for the milk samples from the four provinces. The classification between the provinces is consistent with the results of the K-fold cross validation (Supplementary Materials Table S1).

Table 2. Characteristics and evaluations of the OPLS-DA models for the inter-provincial milk samples.

Index	FA	ISO	ME	ISO/ME	FA/ME	FA/ISO	FA/ISO/ME
R^2	0.716	0.004	0.011	0.015	0.717	0.760	0.754
Q^2	0.614	−0.043	−0.109	−0.088	0.560	0.635	0.581
y-intercepts of R^2	0.112	0.031	0.061	0.045	0.179	0.143	0.211
y-intercepts of Q^2	−0.289	−0.035	−0.614	−0.054	−0.339	−0.348	−0.417

FA = Fatty acid; ISO = Isotope; ME = Mineral elements; R^2 = the measure of the fit of the model; Q^2 = the measure of the predictive ability of the model.

3.1.2. Identification of Milk Produced by Two Farms in the Same Province

We observed that milk samples at provincial geographic distances were differentiated significantly, so we will study the differentiation of milk samples within a smaller range. We suppose that the above methods can be used to identify the milk produced by two farms in the same province at a short distance. The analytical method used was the same as that of milk samples from different provinces.

PCA Analysis

For samples from two farms in Hebei, the PCA model of the isotope chemical parameter was completely divided into two categories (Supplementary Materials Figure S1C). However, no matter the other single chemical parameters or the combination of multiple chemical parameters, the score plots showed some trends of separation, though it was not completely separated. For two farm samples in Inner Mongolia and Shaanxi (Supplementary Materials Figure S1D,F), all PCA models showed a separation trend, but they were not completely separated. For two farm samples in Ningxia (Supplementary Materials Figure S1E), almost all milk samples in the model overlapped, and there was no classification trend. The above classification can be explained by the geographical distance of the two farms in the province (Table 1). The farther the geographical distance in the same province, the more obvious the classification of the samples; the closer the geographical distance in the same province, the less obvious the classification of the samples.

OPLS-DA Analysis

By using OPLS-DA, a good distinction between short-distance milk in the province was obtained. There are three OPLS-DA score plots of mineral elements, isotopes, fatty acids and a combination of the best and no over-fitting in Figure 2. From OPLS-DA score plots of Hebei samples, single chemical parameters or a combination of multiple chemical parameters could be used to separate the samples from the two farms in Hebei. In the mineral element and fatty acid chemical parameter model, there were some points that were confused, which affected the classification; however, the models of the isotope chemical parameter and the combination of isotopes and other chemical parameters were well classified. Among the single techniques, the isotope model had the well predictive ability (R^2 = 0.907, Q^2 = 0.876). Among the binding technologies, isotope technology is helpful to improve the model prediction ability, and the binding technologies including fatty acids have better model prediction ability (mineral element and isotope technologies: R^2 = 0.857, Q^2 = 0.678; fatty acid and isotope technologies: R^2 = 0.920, Q^2 = 0.814; three technologies: R^2 = 0.891, Q^2 = 0.707). This indicates that isotope chemical parameters play a major role in classification, while mineral element and fatty acid parameters are less important for classification. Milk samples from two dairy farms in Hebei were fingerprinted with isotope content, which is due to the geographical specificity of the isotopes in the local plants and water. The value of $\delta^{13}C$ in plants are affected by factors such as the type of plants, light, atmospheric CO_2 concentration, temperature, air pollution and soil moisture, salinity and nutritional status, showing geographical differentiation; the value of $\delta^{15}N$ in plants are influenced by parent material, soil types, topography, land use patterns and fertilization, showing geographical differentiation; the value of δ^2H and $\delta^{18}O$ of plants are related to the latitude, altitude and distance from the sea, showing geographical differentiation; the value of δ^2H and $\delta^{18}O$ of water content is affected by climate, season and precipitation, showing geographical differentiation [41–43]. Geographical differences in isotopes in plants and water are transferred to animals with breeding, distinguishing milk samples of different origin by determining the isotopes. As shown in Table 3, the Q^2 of the mineral element and fatty acid chemical parameter analysis and the combination of the two were less than 0.500, which indicates that the prediction ability of the models is not good. Except these three models, the fitting degree of the other models was more than 85.70%, and the prediction ability of other models was more than 67.80%. This model combines fatty acidwith isotope chemical parameters (R^2 = 0.920, Q^2 = 0.814), proving to be the best classification model for milk samples in Hebei. The classification in Hebei is consistent with the results of the K-fold cross validation (Supplementary Materials Table S1).

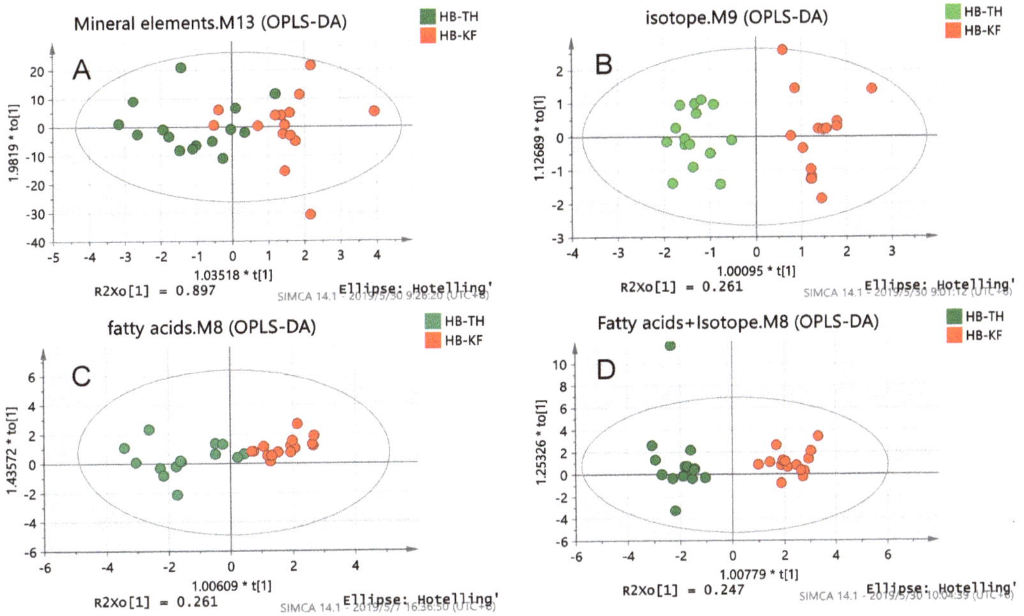

Figure 2. OPLS-DA score plots of Hebei samples obtained by the chemical analysis of: (**A**) Mineral elements; (**B**) Isotopes; (**C**) Fatty acids; (**D**) Fatty acids combined with isotopes.

Table 3. Characteristics of OPLS-DA models of milk in Hebei province.

Index	FA	ISO	ME	ISO/ME	FA/ME	FA/ISO	FA/ISO/ME
R^2	0.755	0.907	0.562	0.857	0.768	0.920	0.891
Q^2	0.469	0.876	−0.455	0.678	0.394	0.814	0.707
y-intercepts of R^2	0.269	−0.020	0.170	0.281	0.374	0.315	0.396
y-intercepts of Q^2	−0.884	−0.323	−0.182	−0.723	−0.819	−0.882	−0.826

FA = Fatty acid; ISO = Isotope; ME = Mineral elements; R^2 = the measure of the fit of the model; Q^2 = the measure of predictive ability of the model.

There are three OPLS-DA score plots of mineral elements, isotopes, fatty acids and a combination of the best and no over-fitting in Figure 3. For two farms samples in Inner Mongolia, among the single-index models, the fatty acid model was the best at separating the samples. Compared with the models in Hebei, the mineral element model in Inner Mongolia confused more points, and there were a few points in the isotope model that were not distinguished. This is probably because the geographical distance between the two dairy farms narrowed from 10.7 km to 4.2 km. After combining more methods, the samples were completely separated. Among them, the model that combines all three chemical parameters further aggregated the sample points. As shown in Figure 3, each farm had a characteristic fatty acid content fingerprint and the fatty acid chemical parameter analysis was more effective than the mineral element and isotope analysis at identifying the milk samples in Inner Mongolia. For two farms samples in Inner Mongolia (Table 4), the model that combines all three chemical parameters showed the best fit and prediction ability (R^2 = 0.985, Q^2 = 0.910), but its y-intercepts of R^2 was more than 0.40, which indicates that the model shows over-fitting. Thus, the model combining the fatty acid and isotope chemical parameters (R^2 = 0.954, Q^2 = 0.879) was determined as the best classification model for milk samples in Inner Mongolia. The classification in Inner Mongolia is consistent with the results of the K-fold cross validation (Supplementary Materials Table S1).

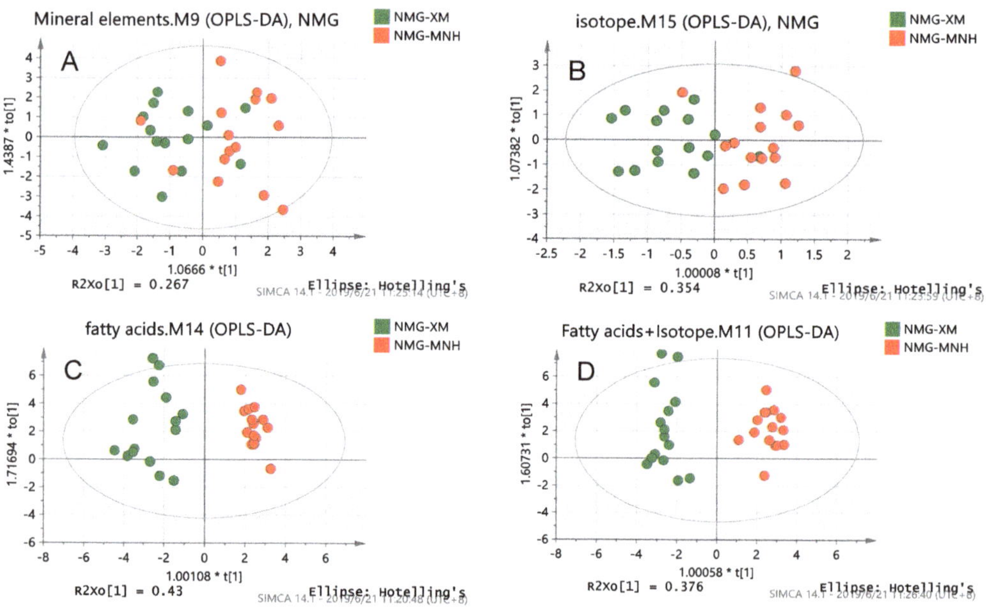

Figure 3. OPLS-DA score plots of Inner Mongolia samples obtained by a chemical analysis of: (**A**) Mineral elements; (**B**) Isotopes; (**C**) Fatty acids; (**D**) Fatty acids combined with isotopes.

Table 4. Characteristics of OPLS-DA models of milk in the Inner Mongolia autonomous region.

Index	FA	ISO	ME	ISO/ME	FA/ME	FA/ISO	FA/ISO/ME
R^2	0.919	0.599	0.410	0.763	0.955	0.954	0.985
Q^2	0.876	0.530	−0.243	0.432	0.813	0.879	0.910
y-intercepts of R^2	0.233	0.057	0.265	0.353	0.559	0.388	0.677
y-intercepts of Q^2	−0.680	−0.306	−0.339	−0.718	−1.130	−1.060	−1.560

FA = Fatty acid; ISO = Isotope; ME = Mineral elements; R^2 = the measure of the fit of the model; Q^2 = the measure of predictive ability of the model.

There are three OPLS-DA score plots of mineral elements, isotopes, fatty acids and a combination of the best and no over-fitting in Figure 4. For two farms samples in Shaanxi, among the single-index models, the fatty acid model was the best at separating the samples and the fatty acid and fatty acid-bound isotope models showed excellent separation abilities. The fitting degree of the fatty acid and fatty acids-binding isotope models of the samples in Shaanxi (Table 5) was 91.9% and 95.3%, respectively, and the prediction ability was 68.4% and 70.9%, respectively. However, the y-intercepts of R^2 of the fatty acid-binding isotope model was more than 0.40, which indicates that the model shows over-fitting. For the other models, there were some easily confused sampling points. As shown in Figure 4, each farm had a characteristic fatty acid content fingerprint and the fatty acid chemical parameter analysis was more effective than the mineral element and isotope analysis at identifying the milk samples in Inner Mongolia. For the two farms samples from Shaanxi (Table 5), the model of fatty acid chemical parameters was the best classification model (R^2 = 0.919, Q^2 = 0.684). The classification in Shaanxi is consistent with the results of the K-fold cross validation (Supplementary Materials Table S1).

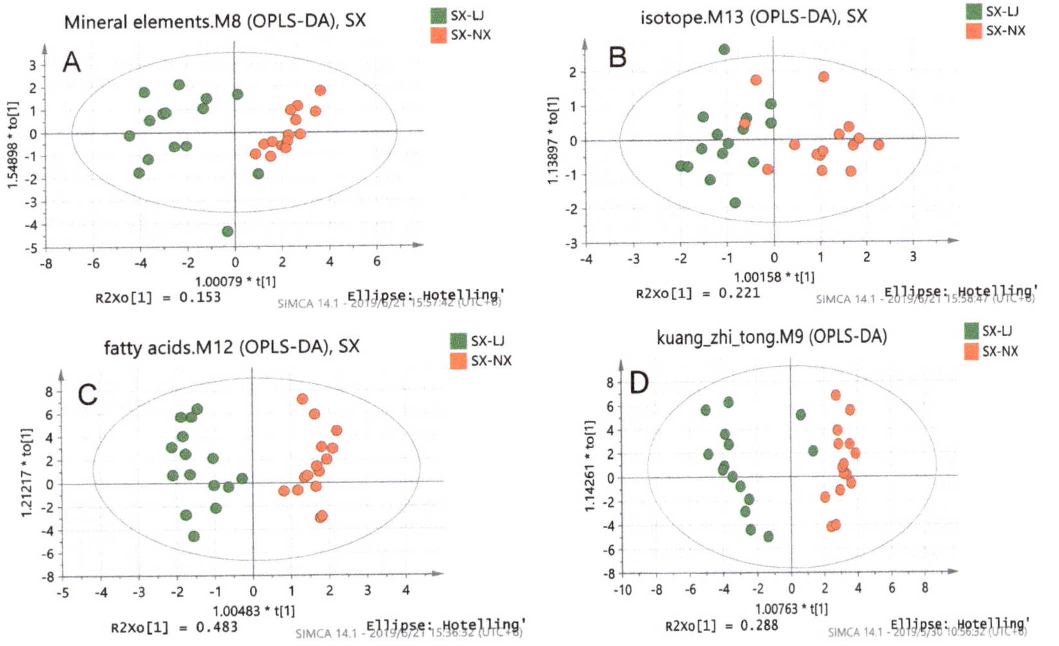

Figure 4. OPLS-DA score plots of the Shaanxi samples obtained by a chemical analysis of: (**A**) Mineral elements; (**B**) Isotopes; (**C**) Fatty acids; (**D**) a combination of three chemical parameters.

Table 5. Characteristics of OPLS-DA models of milk in Shaanxi province.

Index	FA	ISO	ME	ISO/ME	FA/ME	FA/ISO	FA/ISO/ME
R^2	0.919	0.673	0.773	0.725	0.810	0.953	0.839
Q^2	0.684	0.548	0.602	0.688	0.685	0.709	0.721
y-intercepts of R^2	0.371	0.058	0.134	0.035	0.240	0.417	0.301
y-intercepts of Q^2	−0.873	−0.300	−0.321	−0.303	−0.557	−1.020	−0.563

FA = Fatty acid; ISO = Isotope; ME = Mineral elements; R^2 = the measure of the fit of the model; Q^2 = the measure of predictive ability of the model.

As shown in Figure 5, there are a total of four OPLS-DA scores, of which three are of mineral elements, isotopes and fatty acids parameter models, and the remaining one is a combined parameter model with the best differentiation and no over-fitting..For two farm samples in Nngxia, among the single technology models, the fatty acid and mineral element models had very good predictive ability for milksamples, but their separation abilities were far less effective than in the other three provinces (Figures 2–4), because the geographical distance between the two dairy farms in Ningxia narrowed to 0.7 km. For the isotope model, there were no differentiation trends. For two farms samples in Ningxia (Table 6), only the models of fatty acid-bound element minerals (R^2 = 0.771, Q^2 = 0.631) showed great separation abilities. The classification in Ningxia is consistent with the results of the K-fold cross validation (Supplementary Materials Table S1).

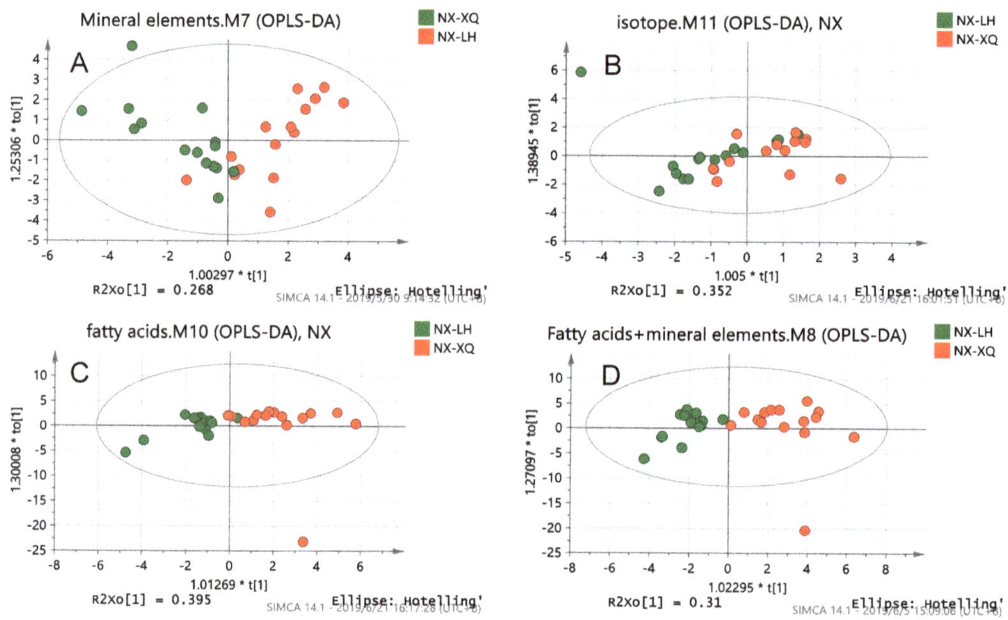

Figure 5. OPLS-DA score plots of the Ningxia samples obtained by a chemical analysis of: (**A**) Mineral elements; (**B**) Isotopes; (**C**) Fatty acids; (**D**) Fatty acids combined with mineral elements.

Table 6. Characteristics of OPLS-DA models of milk in the Ningxia Hui autonomous region.

Index	FA	ISO	ME	ISO/ME	FA/ME	FA/ISO	FA/ISO/ME
R^2	0.630	0.310	0.654	0.474	0.771	0.557	0.777
Q^2	0.434	−0.601	0.407	0.416	0.631	0.393	0.596
y-intercepts of R^2	0.139	0.011	0.176	0.137	0.328	0.232	0.434
y-intercepts of Q^2	−0.833	−0.217	−0.393	−0.257	−0.704	−0.453	−0.666

FA = Fatty acid; ISO = Isotope; ME = Mineral elements; R^2 = the measure of the fit of the model; Q^2 = the measure of predictive ability of the model.

3.1.3. Validation of the OPLS-DA Model

Generally, when using this type of supervised analysis, there is a risk of over-fitting the data. Therefore, validation is crucial to verify the reliability of the model. In order to check whether the model is over-fitting, we performed the permutation test. When using the permutation test, the order of the y-variable randomly permutes the specified number 200 times, and separate models are fitted to all the permuted y-variables. Then, the original y-variable and the permuted y-variable draw a regression line. Interception is a measure of over-fitting. Desirable values of y-intercepts should be less than 0.40 for R^2 intercept and less than 0.05 for Q^2 intercept, respectively [44], indicating that the model is effective and there is no over-fitting. The model test results are included in each table (Tables 2–6).

We found that the classification models, considered good in the previous section, showed over-fitting. In the identification of inter-provincial samples, the fatty acid and fatty acid-binding isotope models (y-intercepts of $R^2 = 0.143$, y-intercepts of $Q^2 = -0.348$) are applicable. Thus, we still chose the model combining the fatty acid and isotope chemical parameters as the best classification model for the milk samples from four provinces. Similarly, in the two farms samples in the same province, we chose the isotope-bound fatty acid model as the discriminant model for the Hebei and Inner Mongolia samples, the fatty acid chemical parameter was selected as the discriminant model for Shaanxi and the model combining fatty acid with mineral element chemical parameters was chosen as the discriminant model for Ningxia.

4. Conclusions

The above research shows that multivariate statistical analysis combined with chemical parameter analysis (fatty acids, isotopes and mineral elements) can distinguish milk from different geographical distances in China. The fatty acid-binding isotope model is the best for the classification of milk samples between provinces ($R^2 = 0.760$, $Q^2 = 0.635$). Moreover, the model combining fatty acid with isotope chemical parameters was the best classification model for milk samples within Hebei ($R^2 = 0.920$, $Q^2 = 0.814$) and Inner Mongolia ($R^2 = 0.954$, $Q^2 = 0.879$); the models of the fatty acid chemical parameter showed great separation abilities for milk samples in Shaanxi ($R^2 = 0.919$, $Q^2 = 0.684$); and the model combining fatty acid with element minerals chemical parameters showed the best separation abilities for two farms samples in Ningxia ($R^2 = 0.771$, $Q^2 = 0.631$). In this study, traceability technology reduced the geographical distance of identified milk samples to 0.7 km. Among the five OPLS-DA models of two farms in four provinces and within the provinces, the fatty acid chemical parameter analysis was more effective than the mineral element and isotope analysis at identifying the milk samples. In addition, we found that when the sample origin distance is relatively long, the combination of the two techniques is better than the analysis using a single technique, but using the three techniques together is not superior to the combination of two technologies or a single technique, and sometimes weakens the robustness of the model. When the sample origin distance is relatively close, the combination of various technologies is not always better than a single technique, and sometimes, it can easily cause model over-fitting. These findings may be used to improve the milk traceability in China. In a future study, we will collect unknown milk samples to verify the OPLS-DA model and judge the effect of its practical application.

Supplementary Materials: The following are available online at https://www.mdpi.com/article/10.3390/foods10051119/s1, Figure S1: PCA score plots for each technical model: (A, B) Isotope model and isotope-binding fatty acid model for interprovincial samples; (C–F) Isotope-bound fatty acid models of provincial samples, Table S1: Accuracy of OPLS-DA models for milk discrimination between provinces and within the same provinces.

Author Contributions: Conceptualization, Y.Z., G.C., A.C. and S.Y.; formal analysis, R.Z. and M.S.; resources, Y.Z.; visualization, M.S.; writing—original draft preparation, R.Z. and M.S.; writing—review and editing, R.Z. and Y.Z. All authors have read and agreed to the published version of the manuscript.

Funding: This research was funded by the National Natural Science Foundation of China (No. 32072314).

Institutional Review Board Statement: Not applicable.

Data Availability Statement: Data are contained within the article or Supplementary Materials.

Acknowledgments: The authors are very grateful to the National Natural Science Foundation of China for funding this study.

Conflicts of Interest: The authors declare no conflict of interest.

References

1. Crittenden, R.G.; Andrew, A.S.; LeFournour, M.; Young, M.D.; Middleton, H.; Stockmann, R. Determining the geographic origin of milk in Australasia using multi-element stable isotope ratio analysis. *Int. Dairy J.* **2007**, *17*, 421–428. [CrossRef]
2. Chesson, L.A.; Valenzuela, L.O.; O'Grady, S.P.; Cerling, T.E.; Ehleringer, J.R. Hydrogen and Oxygen Stable Isotope Ratios of Milk in the United States. *J. Agric. Food Chem.* **2010**, *58*, 2358–2363. [CrossRef] [PubMed]
3. Luo, D.H.; Dong, H.; Luo, H.Y.; Xian, Y.P.; Guo, X.D.; Wu, Y.L. Multi-Element (C, N, H, O) Stable Isotope Ratio Analysis for Determining the Geographical Origin of Pure Milk from Different Regions. *Food Anal. Methods* **2015**, *9*, 437–442. [CrossRef]
4. Dong, H.; Xiao, K.; Luo, D. Stability of carbon and nitrogen isotopic compositions of the protein extracted from milk and their potential as "fingerprints" of geographical origin. *RSC Adv.* **2017**, *7*, 18946–18952. [CrossRef]
5. Brescia, M.A.; Monfreda, M.; Buccolieri, A.; Carrino, C. Characterisation of the geographical origin of buffalo milk and mozzarella cheese by means of analytical and spectroscopic determinations. *Food Chem.* **2005**, *89*, 139–147. [CrossRef]
6. Manca, G.; Franco, M.A.; Versini, G.; Camin, F.; Rossmann, A.; Tola, A. Correlation between multielement stable isotope ratio and geographical origin in Peretta cows' milk cheese. *J. Dairy Sci.* **2006**, *89*, 831–839. [CrossRef]
7. Bontempo, L.; Lombardi, G.; Paoletti, R.; Ziller, L.; Camin, F. H, C, N and O stable isotope characteristics of alpine forage, milk and cheese. *Int. Dairy J.* **2012**, *23*, 99–104. [CrossRef]
8. Franke, B.M.; Haldimann, M.; Gremaud, G.; Bosset, J.O.; Hadorn, R.; Kreuzer, M. Element signature analysis: Its validation as a tool for geographic authentication of the origin of dried beef and poultry meat. *Eur. Food Res. Technol.* **2008**, *227*, 701–708. [CrossRef]
9. Kim, J.S.; Hwang, I.M.; Lee, G.H.; Park, Y.M.; Choi, J.Y.; Jamila, N.; Khan, N.; Kim, K.S. Geographical origin authentication of pork using multi-element and multivariate data analyses. *Meat Sci.* **2017**, *123*, 13–20. [CrossRef]
10. Sun, S.M.; Guo, B.L.; Wei, Y.M.; Fang, M.T. Geographical origin traceability of lamb based on mineral element fingerprints. *Trans. Chin. Soc. Agric. Eng.* **2012**, *28*, 237–243.
11. Bai, T.; Cai, H.Y.; Deng, Y.H.; Xu, X.F.; Zhang, G.J.; Zhou, Y.; Sun, Q. Study on origin of HeiShui Phoenix chicken based on trace element fingerprint. *China Meas. Test.* **2018**, *44*, 57–62, 74.
12. Silva, B.; Gonzaga, L.V.; Maltez, H.F.; Samochvalov, K.B.; Fett, R.; Costa, A.C.O. Elemental profiling by ICP-MS as a tool for geographical discrimination: The case of bracatinga honeydew honey. *J. Food Compos. Anal.* **2021**, *96*, 103727. [CrossRef]
13. Bontempo, L.; Camin, F.; Ziller, L.; Perini, M.; Nicolini, G.; Larcher, R. Isotopic and elemental composition of selected types of Italian honey. *Measurement* **2017**, *98*, 283–289. [CrossRef]
14. Benincasa, C.; Lewis, J.; Sindona, G.; Tagarelli, A. The use of multi element profiling to differentiate between cow and buffalo milk. *Food Chem.* **2008**, *110*, 257–262. [CrossRef] [PubMed]
15. Osorio, M.T.; Koidis, A.; Papademas, P. Major and trace elements in milk and Halloumi cheese as markers for authentication of goat feeding regimes and geographical origin. *Int. J. Dairy Technol.* **2015**, *68*, 573–581. [CrossRef]
16. Kraft, J.; Collomb, M.; Mockel, P.; Sieber, R.; Jahreis, G. Differences in CLA isomer distribution of cow's milk lipids. *Lipids* **2003**, *38*, 657–664. [CrossRef]
17. Collomb, M.; Butikofer, U.; Sieber, R.; Jeangros, B.; Bosset, J.O. Correlation between fatty acids in cows' milk fat produced in the Lowlands, Mountains and Highlands of Switzerland and botanical composition of the fodder. *Int. Dairy J.* **2002**, *12*, 661–666. [CrossRef]
18. Liang, K.H.; Zhao, Y.; Han, J.; Liu, P.; Qiu, J.; Zhu, D.Z.; Qin, Y.C.; Lu, L.G.; Wang, X.H. Fatty acid composition, vitamin A content and oxidative stability of milk in China. *J. Appl. Anim. Res.* **2018**, *46*, 566–571. [CrossRef]
19. Magdas, D.A.; Dehelean, A.; Feher, I.; Cristea, G.; Puscas, R.; Dan, S.D.; Cordea, D.V. Discrimination markers for the geographical and species origin of raw milk within Romania. *Int. Dairy J.* **2016**, *61*, 135–141. [CrossRef]
20. Nečemer, M.; Potočnik, D.; Ogrinc, N. Discrimination between Slovenian cow, goat and sheep milk and cheese according to geographical origin using a combination of elemental content and stable isotope data. *J. Food Compos. Anal.* **2016**, *52*, 16–23. [CrossRef]
21. Griboff, J.; Baroni, M.V.; Horacek, M.; Wunderlin, D.A.; Monferran, M.V. Multielemental plus isotopic fingerprint enables linking soil, water, forage and milk composition, assessing the geographical origin of Argentinean milk. *Food Chem.* **2019**, *283*, 549–558. [CrossRef]
22. Sola-Larranaga, C.; Navarro-Blasco, I. Chemometric analysis of minerals and trace elements in raw cow milk from the community of Navarra, Spain. *Food Chem.* **2009**, *112*, 189–196. [CrossRef]
23. Bontempo, L.; Barbero, A.; Bertoldi, D.; Camin, F.; Larcher, R.; Perini, M.; Sepulcri, A.; Zicarelli, L.; Piasentier, E. Isotopic and elemental profiles of Mediterranean buffalo milk and cheese and authentication of Mozzarella di Bufala Campana PDO: An initial exploratory study. *Food Chem.* **2019**, *285*, 316–323. [CrossRef]
24. Liu, H.Y.; Zhao, Q.Y.; Guo, X.Q.; Tang, C.H.; Yu, X.N.; Zhan, T.F.; Qin, Y.C.; Zhang, J.M. Application of isotopic and elemental fingerprints in identifying the geographical origin of goat milk in China. *Food Chem.* **2019**, *277*, 448–454. [CrossRef] [PubMed]

25. Zhao, Y.; Tu, T.; Tang, X.Y.; Zhao, S.S.; Yang, S.M. Authentication of organic pork and identification of geographical origins of pork in four regions of China by combined analysis of stable isotopes and multi-elements. *Meat Sci.* **2020**, *165*, 108129. [CrossRef]
26. Renou, J.P.; Bielicki, G.; Deponge, C.; Gachon, P.; Micol, D.; Ritz, P. Characterization of animal products according to geographic origin and feeding diet using nuclear magnetic resonance and isotope ratio mass spectrometry. Part II: Beef meat. *Food Chem.* **2004**, *86*, 251–256. [CrossRef]
27. Camin, F.; Pavone, A.; Bontempo, L.; Wehrens, R.; Paolini, M.; Faberi, A.; Marianella, R.M.; Capitani, D.; Vista, S.; Mannina, L. The use of IRMS, H-1 NMR and chemical analysis to characterise Italian and imported Tunisian olive oils. *Food Chem.* **2016**, *196*, 98–105. [CrossRef]
28. Jandrić, Z.; Frew, R.D.; Fernandez-Cedi, L.N.; Cannavan, A. An investigative study on discrimination of honey of various floral and geographical origins using UPLC-QToF MS and multivariate data analysis. *Food Control* **2017**, *72*, 189–197. [CrossRef]
29. Circi, S.; Ingallina, C.; Vista, S.; Capitani, D.; Di Vecchia, A.; Leonardi, G.; D'Achille, G.; Centauri, L.; Camin, F.; Mannina, L. A Multi-Methodological Protocol to Characterize PDO Olive Oils. *Metabolites* **2018**, *8*, 43. [CrossRef] [PubMed]
30. Perini, M.; Nardin, T.; Camin, F.; Malacarne, M.; Larcher, R. Combination of sugar and stable isotopes analyses to detect the use of nongrape sugars in balsamic vinegar must. *J. Mass Spectrom.* **2018**, *53*, 772–780. [CrossRef] [PubMed]
31. Perini, M.; Paolini, M.; Camin, F.; Appendino, G.; Vitulo, F.; De Combarieu, E.; Sardone, N.; Martinelli, E.M.; Pace, R. Combined use of isotopic fingerprint and metabolomics analysis for the authentication of saw palmetto (*Serenoa repens*) extracts. *Fitoterapia* **2018**, *127*, 15–19. [CrossRef] [PubMed]
32. Yao, S.; Li, J.Q.; Li, T.; Duan, Z.L.; Wang, Y.Z. Geographical traceability of Boletaceae mushrooms using data fusion of FT-IR, UV, and ICP-AES combined with SVM. *Int. J. Food Prop.* **2019**, *22*, 414–426. [CrossRef]
33. Ehtesham, E.; Baisden, W.T.; Keller, E.D.; Hayman, A.R.; Van Hale, R.; Frew, R.D. Correlation between precipitation and geographical location of the δ2H values of the fatty acids in milk and bulk milk powder. *Geochim. Cosmochim. Acta* **2013**, *111*, 105–116. [CrossRef]
34. Xie, L.N.; Zhao, S.S.; Rogers, K.M.; Xia, Y.N.; Zhang, B.; Suo, R.; Zhao, Y. A case of milk traceability in small-scale districts-Inner Mongolia of China by nutritional and geographical parameters. *Food Chem.* **2020**, *316*, 126332. [CrossRef]
35. Zhao, S.S.; Zhang, H.B.; Zhang, B.; Xu, Z.Z.; Chen, A.L.; Zhao, Y. A rapid sample preparation method for the analysis of stable isotope ratios of beef samples from different countries. *Rapid Commun. Mass Spectrom.* **2020**, *34*, e8795. [CrossRef] [PubMed]
36. Zhao, Y.; Zhang, B.; Chen, G.; Chen, A.L.; Yang, S.M.; Ye, Z.H. Tracing the Geographic Origin of Beef in China on the Basis of the Combination of Stable Isotopes and Multielement Analysis. *J. Agric. Food Chem.* **2013**, *61*, 7055–7060. [CrossRef]
37. Larsen, M.K.; Nielsen, J.H.; Butler, G.; Leifert, C.; Slots, T.; Kristiansen, G.H.; Gustafsson, A.H. Milk quality as affected by feeding regimens in a country with climatic variation. *J. Dairy Sci.* **2010**, *93*, 2863–2873. [CrossRef]
38. Kgwatalala, P.M.; Ibeagha-Awemu, E.M.; Mustafa, A.F.; Zhao, X. Influence of stearoyl-coenzyme a desaturase 1 genotype and stage of lactation on fatty acid composition of Canadian Jersey cows. *J. Dairy Sci.* **2009**, *92*, 1220–1228. [CrossRef]
39. Triba, M.N.; Le Moyec, L.; Amathieu, R.; Goossens, C.; Bouchemal, N.; Nahon, P.; Rutledge, D.N.; Savarin, P. PLS/OPLS models in metabolomics: The impact of permutation of dataset rows on the K-fold cross-validation quality parameters. *Mol. Biosyst.* **2015**, *11*, 13–19. [CrossRef]
40. Jandric, Z.; Haughey, S.A.; Frew, R.D.; McComb, K.; Galvin-King, P.; Elliott, C.T.; Cannavan, A. Discrimination of honey of different floral origins by a combination of various chemical parameters. *Food Chem.* **2015**, *189*, 52–59. [CrossRef]
41. Franke, B.M.; Gremaud, G.; Hadorn, R.; Kreuzer, M. Geographic origin of meat—elements of an analytical approach to its authentication. *Eur. Food Res. Technol.* **2005**, *221*, 493–503. [CrossRef]
42. Monahan, F.J.; Schmidt, O.; Moloney, A.P. Meat provenance: Authentication of geographical origin and dietary background of meat. *Meat Sci.* **2018**, *144*, 2–14. [CrossRef] [PubMed]
43. Zhao, R.T.; Yang, S.M.; Zhao, Y. Research progress in traceability of agricultural products using stable isotope. *J. Nucl. Agric. Sci.* **2020**, *34* (Suppl. S1), 120–128.
44. Belmonte-Sanchez, J.R.; Romero-Gonzalez, R.; Arrebola, F.J.; Vidal, J.L.M.; Garrido Frenich, A. An innovative metabolomic approach for Golden Rum classification combining ultrahigh-performance liquid chromatography-orbitrap mass spectrometry and chemometric Strategies. *J. Agric. Food Chem.* **2019**, *67*, 1302–1311. [CrossRef] [PubMed]

foods

MDPI

Article

Chemometric Discrimination of the Geographical Origin of Three Greek Cultivars of Olive Oils by Stable Isotope Ratio Analysis

Maria Tarapoulouzi [1], Vasiliki Skiada [2,3], Sofia Agriopoulou [2], David Psomiadis [4], Catherine Rébufa [5], Sevastianos Roussos [5], Charis R. Theocharis [1], Panagiotis Katsaris [3] and Theodoros Varzakas [2,*]

1 Department of Chemistry, Faculty of Pure and Applied Science, University of Cyprus, P.O. Box 20537, CY-1678 Nicosia, Cyprus; tarapoulouzi.maria@ucy.ac.cy (M.T.); charis@ucy.ac.cy (C.R.T.)
2 Department of Food Science and Technology, Faculty of Agriculture and Food, University of the Peloponnese, Antikalamos, 24100 Kalamata, Greece; vpskiada@yahoo.gr (V.S.); sagriopoulou@gmail.com (S.A.)
3 Department of Olive and Horticultural Plants, Hellenic Agricultural Organization—DEMETER, 24100 Kalamata, Greece; pankatsaris@yahoo.gr
4 Imprint Analytics GmbH, Werner von Siemens Straße 1, 7343 Neutal, Austria; psomiadis@imprint-analytics.at
5 Aix Marseille Univ, Avignon Université, CNRS, IRD, IMBE, Campus Saint Jerome, 13013 Marseille, France; c.rebufa@univ-amu.fr (C.R.); sevastianos.roussos@imbe.fr (S.R.)
* Correspondence: t.varzakas@uop.gr

Abstract: A stable isotope ratio mass spectrometer was used for stable isotope ratio (i.e., $\delta^{13}C$, $\delta^{18}O$, and $\delta^{2}H$) measurements, achieving geographical discrimination using orthogonal projections to latent structures discriminant analysis. A total of 100 Greek monovarietal olive oil samples from three different olive cultivars (cv. Koroneiki, cv. Lianolia Kerkyras, and cv. Maurolia), derived from Central Greece and Peloponnese, were collected during the 2019–2020 harvest year aiming to investigate the effect of botanical and geographical origin on their discrimination through isotopic data. The selection of these samples was made from traditionally olive-growing areas in which no significant research has been done so far. Samples were discriminated mainly by olive cultivar and, partially, by geographical origin, which is congruent with other authors. Based on this model, correct recognition of 93.75% in the training samples and correct prediction of 100% in the test set were achieved. The overall correct classification of the model was 91%. The predictability based on the externally validated method of discrimination was good (Q^2 (cum) = 0.681) and illustrated that $\delta^{18}O$ and $\delta^{2}H$ were the most important isotope markers for the discrimination of olive oil samples. The authenticity of olive oil based on the examined olive varieties can be determined using this technique.

Keywords: stable isotope ratios; Greek olive oils; chemometric analysis; OPLS-DA; discrimination; fraud; authenticity; adulteration; geographical origin; quality

Citation: Tarapoulouzi, M.; Skiada, V.; Agriopoulou, S.; Psomiadis, D.; Rébufa, C.; Roussos, S.; Theocharis, C.R.; Katsaris, P.; Varzakas, T. Chemometric Discrimination of the Geographical Origin of Three Greek Cultivars of Olive Oils by Stable Isotope Ratio Analysis. *Foods* **2021**, *10*, 336. https://doi.org/10.3390/foods10020336

Academic Editors: Seung-Hyun Kim and Enrico Valli
Received: 27 December 2020
Accepted: 29 January 2021
Published: 4 February 2021

1. Introduction

Olive oil plays an important role in the diet in Greece as well as in other Mediterranean countries. The various health benefits of olive oil consumption are well known, increasing its reputation [1–5]. The high price of extra virgin olive oil makes it susceptible to fraudulent activities. For instance, deliberate mislabeling of lower commercial-grade olive oils or even mislabeling by a false declaration of origin may occur [6]. For these reasons, a series of criteria and standards associated with the genetic variety, the geographical origin, and the quality grade have been established by the European Union (EU) so as to offer both fair trade in the olive sector as well as safety and protection guarantees for consumers [7,8]. Two of the most well-known denominations related to foodstuff authenticity are the Protected Designation of Origin (PDO) and the Protected Geographical Indication (PGI) [9–13].

The continuous evaluation of olive oil is a key priority as it holds great importance in the Mediterranean diet. Recently, specific attention has been focused on geographical and cultivar traceability by using several advanced analytical techniques, for instance, proton-nuclear magnetic resonance [14], liquid chromatography [15], luminescence [16], Fourier transform infrared spectroscopy [17], near-and- mid-infrared spectroscopy [18–20], as well as triacylglycerol, fatty acid, squalene, and tocopherol determinations [19,21–23].

Simple and efficient cultivar and geographical discrimination are of crucial importance to judge labeling compliance. Agricultural practices applied in specific areas, as well as geoclimatic characteristics, are reflected in the content of stable isotopes of bioelements (H, C, N, O, and S) and determined experimentally with isotope ratio mass spectrometry (IRMS) [24]. The IRMS technique is a well-established method for determining the authenticity of olive oil [14,25–29]. The stable isotope ratios of $^{13}C/^{12}C$, $^{18}O/^{16}O$, and $^{2}H/^{1}H$ give many promises for proof of food authenticity, as this analytical technique offers more accuracy and sensitivity compared to other less expensive and equally effective techniques [30]. Latitude and altitude, distance from the sea, and environmental and climate-related conditions are some of the factors that influence $^{13}C/^{12}C$, $^{18}O/^{16}O$, and $^{2}H/^{1}H$ ratios [31].

Although indigenous Greek monocultivar olive oils are more than 40 [32], with cv. Koroneiki being the most systematically cultivated variety, detailed investigation has not been carried out for local autochthonous monocultivars. For example, the local variety of Lianolia Kerkyras is cultivated exclusively on the coasts of northwestern Greece, while cv. Maurolia, which according to our knowledge has never before been characterized, is cultivated exclusively in northern Peloponnese, and more precisely in a small area of the regional unit of Messinia [33,34]. Hence, the combination of the cultivar species (cv. Koroneiki, cv. Lianolia Kerkyras, and cv. Maurolia) that have been chosen for study have only been partly evaluated for genotyping and molecular characterization [35,36], as well as for antioxidant content [1,37] and chemical composition [38].

Chemometrics and discriminant analysis are important tools for food authenticity. The supervised method orthogonal projections to latent structures discriminant analysis (OPLS-DA) is very useful for the discrimination of samples in food analysis [39–44]. Recently, OPLS-DA has been used extensively by several authors in the olive oil authentication field with great success [15,45–49].

The first goal of this research study was the achievement of a preliminary discrimination of the above-mentioned Greek monocultivar olive oils according to olive cultivar by applying a multivariate analysis to the IRMS measurements. The second goal was to produce a robust model capable of identifying any adulteration of unknown olive oils in Greece regarding cultivar and geographical origin. One of the most innovative points of this study was that the examined three monocultivar olive oils (cv. Koroneiki, cv. Lianolia Kerkyras, and cv. Maurolia) had never before been evaluated by IRMS in combination with chemometrics. By measuring the O, H, and C isotope ratios, the goal of this study has been successfully reached. Chemometric analysis with combination of SIMCA software was used to interpret the measurements. The OPLS-DA chemometric method was implemented. External validation was achieved by dividing the dataset into two groups. The first group (training set) was used for the adjustment of the model parameters and the second group (test set) was used for the estimation of generalization error [50].

2. Materials and Methods

2.1. Sampling

A set of 100 monovarietal olive oil samples were gathered from various olive mills in Central Greece and Peloponnese, located in different geographical areas. Particularly, 38, 29, and 33 samples of the three cultivars cv. Maurolia from Messinia (37.25° N, 21.95° E) and both cv. Lianolia Kerkyras and cv. Koroneiki from Preveza (38.95° N, 20.75° E) were collected, respectively. All samples were derived from the 2019–2020 harvesting period. This was a significant factor in the evaluation of core geographical information from the dataset in order to avoid differences due to seasonal effects. Additionally, long storage effects of olive oils (e.g., through oxidation) had to be eliminated [51]. To this end, only fresh, recently harvested olive oils were used in this study. Since stable isotope ratios are related to parameters like latitude, mean annual temperature, and average relative humidity at the collection area, the samples were collected from coastal areas, i.e., Messinia and Preveza.

2.2. IRMS Analysis

Stable isotope ratios of carbon ($^{13}C/^{12}C$) were measured using a horizon isotope ratio mass spectrometer (Nu Instruments Limited, Wrexham, UK) following total combustion in a Euro EA-CHNSO 2 dual elemental analyzer (EuroEA3000, EuroVector Srl, Pavia, Italy). Stable isotope ratios of oxygen and hydrogen ($^{18}O/^{16}O$, $^{2}H/^{1}H$) were measured with a HTCEA (high temperature conversion elemental analyzer, Hekatech Gmbh, Wegberg, Germany) connected to a horizon isotope ratio mass spectrometer (Nu Instruments Limited, Wrexham, UK). Approximately 1 mL sample volume was used for each measurement.

The measured isotope ratios of each sample were normalized to a pulse of respective reference gases (CO_2, CO, H_2). Every sample was measured in duplicate. Each measurement sequence included two reference materials with known isotope signatures (2-point referencing) and multiple quality control samples to monitor sequence precision and accuracy.

Carbon, hydrogen, and oxygen stable isotope ratio analysis was performed at Imprint Analytics GmbH (Neutal, Austria) following the requirements of EN ISO17025:2018 accreditation standard. The method validation reported repeatability of 0.2‰, 1.9‰, and 0.2‰ for carbon, hydrogen, and oxygen isotope ratios, respectively. The accuracy of the reference materials (international reference materials, inhouse reference materials, and quality control samples) was controlled by the SD of replicate analysis during the runs, as well as quality control charts, and was under 0.1‰, 1.3‰, and 0.1‰ for carbon, hydrogen, and oxygen isotope ratios, respectively.

The values were denoted in delta (δ) in relation to the international VPDB (Vienna-Pee Dee Belemnite) and VSMOW (Vienna-Standard Mean Ocean Water) standards for $\delta^{13}C$, $\delta^{18}O$, and $\delta^{2}H$, respectively, according to the following general equation:

$$\delta^{13}Cs = (R_s/R_{std}) - 1 \quad (1)$$

where R is the $^{13}C/^{12}C$ ratio of the sample (s) and of the standard (std) [52] (similarly for $^{18}O/^{16}O$ and $^{2}H/^{1}H$) expressed in per mil (‰).

2.3. Statistical Analysis

SIMCA version 15.0.2 (Umetrics, Umeå, 907 29, Sweden) was used for multivariate statistical analysis. Initially, by applying chemometrics, the unsupervised (principal component analysis) PCA method was performed (not shown here). The supervised OPLS-DA procedure was then applied to discriminate and classify the olive oil samples. OPLS-DA was applied to distinguish the wrongly classified samples and to test the robustness of the model. Scaling to unit variance (UV) and mean-centering were the two settings before chemometric analysis. The application of classification methods was done after dividing the olive oil samples into three classes: cv. Maurolia (Messinia): class 1, cv. Lianolia Kerkyras (Preveza): class 2, and cv. Koroneiki (Preveza): class 3.

The 100-sample dataset was divided into training and test sets to apply external validation. Additionally, 80 and 20 samples were randomly selected to represent the training and the test set, respectively. The symbol A was used for the number of important selected components. Hotelling's T2 confidence region is defined by the ellipse in the score scatter plots, providing a multivariate generalization of Student's *t*-test, and a 95% confidence interval for the observations. Determination coefficient R^2 was used for evaluation of the internal validation. R^2 reflects the goodness of fit, while Q^2 reflects the predictive ability of the model. The 7-fold leave out procedure (default setting in SIMCA 15.0.2) was used for Q^2 measurements. R^2X is the amount of variation in X that is uncorrelated to Y with systematic variation. It shows whether data can be well interpreted. R^2X(cum) is the total sum of variation in X that is uncorrelated to Y. R^2Y is the proportion of the variance of the response variable that is explained by the model. R^2Y(cum) is the total sum of variation in Y explained by the model. Q^2(cum) reflects the goodness of prediction calculated by full cross validation. R^2X, R^2Y, and Q^2values (not less than 0.5) recommended a powerful model with predictive reliability [53]. R^2X(cum) and Q^2(cum) values must be less than 0.2–0.3 [54,55]. In addition, misclassification tables were produced for the constructed models, as well as a permutation test (repeated 100 times) for the overall model. Referring to the latter, the criteria for validity of both R^2 (original model) and Q^2 (predictive model), located to the right, and permutated R^2 (original model) and Q^2 (predictive model), located to the left, and all Q^2 values to the left and right, are lower than the original R^2 values.

3. Results and Discussion

3.1. Stable Isotope Analysis of Olive Oils

Hydrogen and oxygen isotope composition are generally affected by climatic and environmental conditions [56]. More specifically, factors such as temperature [57], precipitation [58], air humidity, soil and plant evapotranspiration [59], and water stress [60] are related with isotopic composition of hydrogen and oxygen. Moreover, carbon isotope composition is influenced by humidity [61], ground and rain water, temperature [60], sea distance, longitude, and latitude [57]. The climatic parameters of the examined olive oil cultivation areas in Greece according to the data of EMY (the Hellenic National Meteorological Service) are presented in Table 1.

Table 1. The climatic parameters of the examined olive oil cultivation areas in Greece for the harvest year 2019.

	Precipitation (mm)/Month		Temperature (°C)		Relative Humidity (%)	
	Messinia	**Preveza**	**Messinia**	**Preveza**	**Messinia**	**Preveza**
Average	64	91	17	17.2	69	67
Minimum	6	13.4	9.8	8.7	57.7	59.2
Maximum	141.7	199.8	26.5	26.5	75	74.1

The stable isotope values act as ecophysiological tracers of natural processes with very good discriminatory power [62]. The traceability that can be achieved by measuring stable isotope values is based on the assumption that the isotope values of plants reflect the characteristics of the specific environment [62]. In (Table 2), the mean, standard deviation, minimum, and maximum values of δ^2H, $\delta^{18}O$, and $\delta^{13}C$ of three monocultivars, i.e., Maurolia (Messinia), Lianolia Kerkyras (Preveza), and Koroneiki (Preveza), thus from two geographic areas, i.e., Messinia and Preveza, are presented.

The stable isotope values of our samples varied between −152.1 and 23.4. Carbon isotope values varied between −31.4 and −27.8, hydrogen isotope values varied between −152.1 and −130.4, and oxygen isotope values varied between 16.7 and 23.4. It is important to note that both cultivars from Preveza (i.e., Lianolia Kerkyras and Koroneiki) had more similar isotope values compared to the Maurolia cultivar from Messinia, as shown in Table 2. Relatively lower values of oxygen isotopes are correlated with areas of high elevation, inland location, and cool climate, while higher values are associated with low elevation, coastal location, and warmer climate [62]. The very specific climatic data that differentiates the two areas from which the samples were taken refers to rainfall, with Preveza having higher values of precipitation as presented in (Table 1). This is related to the lower values of carbon isotopes in Preveza compared to Messinia. In general, samples from the southern region (Messinia) had higher values for carbon and hydrogen isotopes compared to the northern region (Preveza). Moreover, the oxygen isotope ratio reflects water-related processes in plants. Comparing the climatic data (Table 1) with the isotopic composition of examined samples (Table 2), cv. Lianolia Kerkyras and cv.

Koroneiki, both from Preveza, had higher oxygen isotope compositions due to higher values of precipitation.

Table 2. Stable isotope ratios ($^{13}C/^{12}C$, $^{2}H/^{1}H$, $^{18}O/^{16}O$) of Greek olive oils from three monocultivars, i.e., Maurolia (Messinia), Lianolia Kerkyras (Preveza), and Koroneiki (Preveza), thus from two geographic areas, i.e., Messinia and Preveza.

Cultivar (Area)	Stable Isotopes	Mean	Standard Deviation	Minimum	Maximum
Maurolia (Messinia)	$\delta^{2}H/‰$	−134.7	2.4	−140.4	−130.4
	$\delta^{13}C/‰$	−29.3	0.5	−30.4	−28.4
	$\delta^{18}O/‰$	19.7	1.1	16.7	21.4
Lianolia Kerkyras (Preveza)	$\delta^{2}H/‰$	−138.4	1.9	−141.7	−135.8
	$\delta^{13}C/‰$	−30.1	0.5	−30.8	−29.2
	$\delta^{18}O/‰$	21.8	0.6	20.0	22.4
Koroneiki (Preveza)	$\delta^{2}H/‰$	−137.4	3.3	−152.1	−139.2
	$\delta^{13}C/‰$	−29.7	0.7	−31.4	−27.8
	$\delta^{18}O/‰$	21.7	0.8	20.2	23.4

3.2. *Chemometric Discrimination of Olive Oils*

The training set in Figure 1 indicates that good discrimination of the samples in the three classes was achieved. Difference of microclimate and soil could be a possible explanation for the dispersal of samples of the same cultivar [63]. The validation values $R^2X(cum) = 0.998$, $R^2Y(cum) = 0.723$, and $Q^2(cum) = 0.708$ showed a good fit and prediction ability of the training set. Moreover, the misclassification table for the OPLS-DA modeling is presented in Table 3, and it also showed 93.75% correct classification of the 80 samples in the three classes regarding cultivar. A low Fischer value of $p < 0.05$ emphasized the statistical importance of the training set. In more detail, Maurolia cultivar samples (class 1) were all correctly classified (100% correct classification), but Lianolia Kerkyras (class 2) and Koroneiki (class 3) cultivar samples reduced their class' correct classification to 83.33 and 96.15%, respectively. Classes 2 and 3, both originating from Preveza, were well separated.

It is important to note that sample 68, which was labelled as cv. Koroneiki from Preveza (class 3), fell into cv. Lianolia Kerkyras from Preveza (class 2). The first reason for this discrepancy is that a labeling mistake might have occurred during sampling. It is very common in fields for the majority of olive trees to have one main variety with some other varieties scattered amongst them. Particularly, in our case, cv. Lianolia Kerkyras and cv. Koroneiki were harvested at the same time in the Preveza area. The outliers found in this study are justified by the fact that the specific samples were indeed blended (i.e., from olive fruits that came from both varieties due to co-cultivation in the same field). Co-cultivation essentially reinforces the fact that this model is capable of detecting samples that are not pure cv. Lianolia Kerkyras or pure cv. Koroneiki, thus not purely monovarietal. There are oils sold in the market with an indication of monovarietal, but since this is not always the truth, this indication clearly has an economic impact. Monovarietal olive oils are more expensive, and so this is consumer deception, which points to fraud. The model presented here can detect this type of adulteration and consequently detect an important type of fraud. The second reason is due to sample contamination during olive oil extraction in the olive mill. Contamination may be related to impurities, such as traces of different cultivar residues.

The test set in (Figure 2) indicates that the 20 randomly selected samples from the three classes were able to present a very good classification. All 20 samples were 100% correctly classified in the three classes, as also shown by (Table 4). $R^2X(cum) = 0.894$, $R^2Y(cum) = 0.706$, and $Q^2(cum) = 0.641$ were acceptable and the test set was valid.

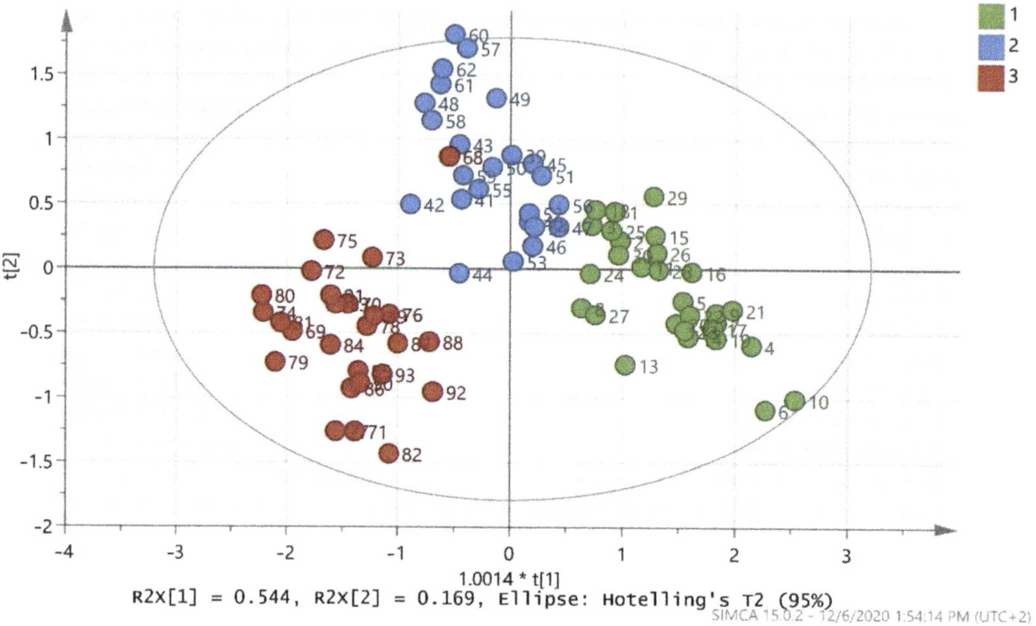

Figure 1. Score scatter plot (t2/t1) from OPLS-DA modeling of the training set where 1 = Maurolia (Messinia), 2 = Lianolia Kerkyras (Preveza), and 3 = Koroneiki (Preveza). A = 2 + 1 components, $R^2X(cum) = 0.998$, $R^2Y(cum) = 0.723$, and $Q^2(cum) = 0.708$ and 1.0014 * t[1] means 1.0014 × t[1].

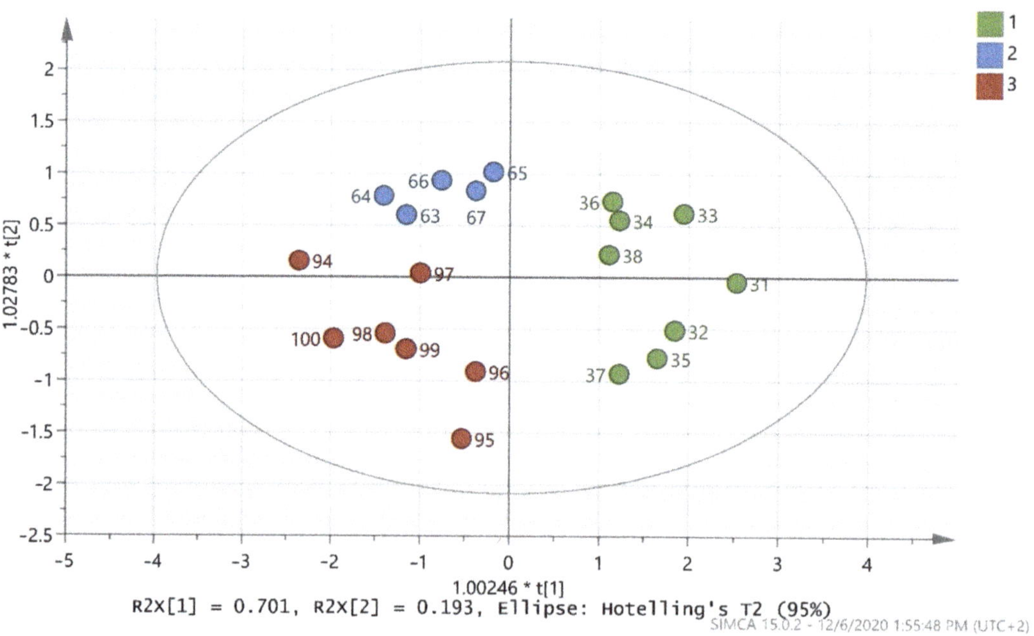

Figure 2. Score scatter plot (t2/t1) from OPLS-DA modeling of the test set where 1 = Maurolia (Messinia), 2 = Lianolia Kerkyras (Preveza), and 3 = Koroneiki (Preveza). A = 2 + 0 + 0 components, $R^2X(cum) = 0.894$, $R^2Y(cum) = 0.706$, and $Q^2(cum) = 0.641$ and 1.00246 * t[1] means 1.00246 × t[1] as well as 1.02783 * t[2] means 1.02783 × t[2].

Table 3. Misclassification table from OPLS-DA modeling of the training set, where 1 = Maurolia (Messinia), 2 = Lianolia Kerkyras (Preveza), and 3 = Koroneiki (Preveza).

	Members	Correct	1	2	3
1	30	100%	30	0	0
2	24	83.33%	3	20	1
3	26	96.15%	0	1	25
Total	80	93.75%	33	21	26
Fisher's prob.	4.3×10^{-11}				

Table 4. Misclassification table from OPLS-DA modeling of the test set where 1 = Maurolia (Messinia), 2 = Lianolia Kerkyras (Preveza), and 3 = Koroneiki (Preveza).

	Members	Correct	1	2	3
1	8	100%	8	0	0
2	5	100%	0	5	0
3	7	100%	0	0	7
Total	20	100%	8	5	7
Fisher's prob.	1×10^{-8}				

After merging the training and test sets, the overall model (whole database) was constructed, which is presented in (Figure 3). The validation values $R^2X(cum) = 0.999$, $R^2Y(cum) = 0.695$, and $Q^2(cum) = 0.681$ showed a good fit and prediction ability of the model. In addition, (Figure 4) presents the three-dimensional illustration of the score scatter plot in (Figure 3). Furthermore, misclassification table for the OPLS-DA modeling is presented in (Table 5), and it shows 91% correct classification of the 100 samples in the three classes regarding cultivar. The low Fischer value of $p < 0.05$ emphasized the statistical importance of the model. In more detail, Maurolia (Messinia) class had only 1 sample out of 38 wrongly classified and 97.37% correct classification. Lianolia Kerkyras (Preveza) class gave 75.86% correct classification and it had 7 samples out of 29 whose distance from the center of class 2 was longer than what was expected; further study needs to take place for this class. However, a possible explanation of the low percentage of classification in cv. Lianolia Kerkyras samples (75.86%) could be explained by the fact that those 7 wrongly classified samples (out of 29) could be derived from an olive grove where both cv. Lianolia Kerkyras and cv. Koroneiki olive trees were co-cultivated. From Koroneiki (Preveza) class, only 1 sample out of 33 was wrongly classified, and the class correct classification was 96.97%. In addition, sample 68 was indicated as a misclassified sample in (Figures 3 and 4) as well as in (Figure 1), confirming the robustness of both training and overall sets. Figure 5 shows a random permutation test with 100 permutations used for the validation of goodness of fit and the predictability of these results. The R^2Y values of all permuted models were lower than the original model's R^2Y value (0.695); most of the Q^2 regression lines showed negative intercepts (0.0−0.101).

Further chemometric analysis showed that O and H stable isotopes were more important variables for the constructed model than C isotopes. Stable isotope ratio analysis variations of $^{13}C/^{12}C$ proved to be a useful tool for characterizing samples from different regions with very different climatological and geographic characteristics [26,27,64–69] as isotope ratios are affected by latitude, which indicates the distance from the sea, and environmental conditions during the growing of trees (water stress, atmospheric humidity, and temperature) as co-factors of variability. Since all the samples of this study were cultivated in coastal locations, this explains why C stable isotopes were not as important as O and H. Samples were discriminated mainly by olive cultivar and, partially, by geographical origin. This is congruent with the recent study of Alves de Carvalho et al. (2020) [70].

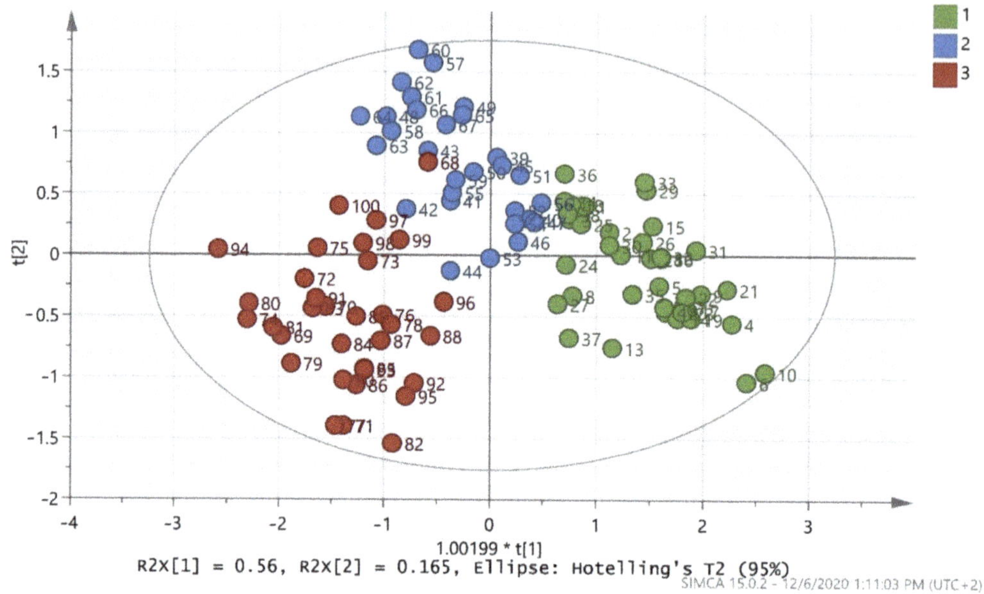

Figure 3. Score scatter plot (t2/t1) from OPLS-DA modeling of the overall model (whole database) where 1 = Maurolia (Messinia), 2 = Lianolia Kerkyras (Preveza), and 3 = Koroneiki (Preveza). A = 2 + 1 + 0 components, $R^2X(cum) = 0.999$, $R^2Y(cum) = 0.695$, and $Q^2(cum) = 0.681$ and 1.00199 * t[1] means 1.00199 × t[1].

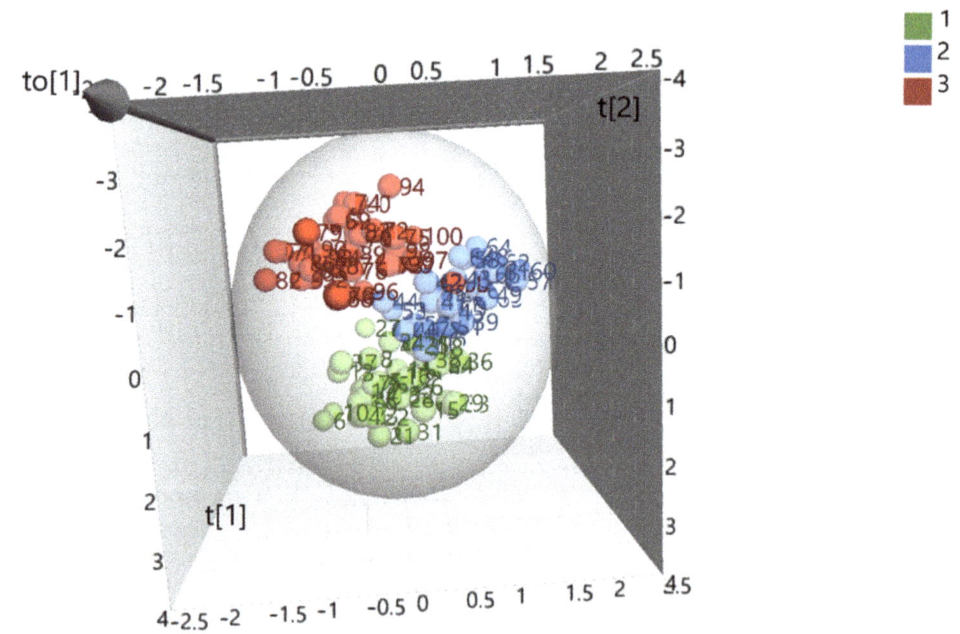

Figure 4. Three-dimensional illustration of the score scatter plot in Figure 3 where 1 = Maurolia (Messinia), 2 = Lianolia Kerkyras (Preveza), and 3 = Koroneiki (Preveza).

Table 5. Misclassification table from OPLS-DA modeling of the overall model (whole database) where 1 = Maurolia (Messinia), 2 = Lianolia Kerkyras (Preveza), and 3 = Koroneiki (Preveza).

	Members	Correct	1	2	3
1	38	97.37%	37	1	0
2	29	75.86%	6	22	1
3	33	96.97%	0	1	32
Total	100	91%	43	24	33
Fisher's prob.	2.3×10^{-8}				

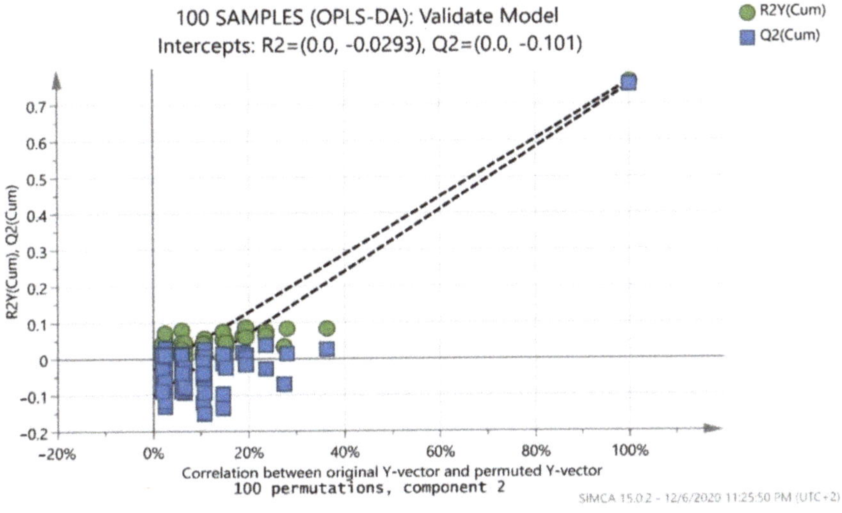

Figure 5. Permutation test of the overall model (whole database) with 100 permutations where both R^2 (original model) and Q^2 (predictive model) located at right and permutated R^2 (original model) and Q^2 (predictive model) located left.

The choice for the study of these varieties was made based on specific parameters. Specifically, the Maurolia variety, along with cv. Koroneiki and cv. Athinolia, are the main varieties grown in southeastern Peloponnese, one of the most important olive oil-producing areas. Compared to the other two varieties, the Maurolia variety matures and is harvested first at the beginning of the olive harvesting period. This is very important, because increased profit could arise from the very first harvest. Moreover, as monovarietal Maurolia olive oil is characterized by balanced qualitative characteristics, manufacturers prefer it in the case that they want produce blends with more bitter or more spicy oils [37]. Although Maurolia olive oil is of high quality, similar to Koroneiki olive oil, [37], this variety has not been included in any catalogue of denominations related to foodstuff authenticity, in contrast to the Koroneiki and Lianolia Kerkyras olive oils, which have been characterized as PDO and PGI varieties, respectively. Instead, Koroneiki cultivar is the most widespread all over the Greece, and the most well-known Greek olive cultivar. It is cultivated in many areas of the country, mainly in Peloponnese, Crete, and in the northwestern part of country [38,71]. Regarding Lianolia Kerkyras, this cultivar is mainly cultivated in the Ionian Islands and in the geographical region of Epirus. As the Koroneiki variety has the greatest reputation both in Greece and abroad, and thus achieves the best prices, it could possibly be a blend of the lesser known and less widespread varieties of Maurolia and Lianolia Kerkyras.

This is the first qualitative attempt to study three important Greek cultivars and extract a robust chemometric model capable of discriminating olive oils based on geographic origin. The qualitative results of this study answer the important question of which geographic area the samples of the three cultivars come from. The method proposed here can be enriched in the future by creating synthetic samples from different geographic and cultivar origins. The big challenge involves discriminating adulterant mixtures of olive oils from the market. By adding the synthetic samples in the present chemometric model, a possible fourth class will be generated. Unknown adulterated samples may be detected in the future, as the adulterant samples will be classified in the fourth class of synthetic samples.

Moreover, a quantitative method is considered necessary for the next attempt, which will be answering the significant question of how much adulteration the unknown samples contain. Those synthetic samples could be mixtures made of different proportions (adulteration levels, $\%v/v$) of olive oils, as Tsopelas et al. proposed [72]. In the future, the more similar the adulteration level of every unknown adulterated and synthetic sample, the closer the distance between them on a score scatter plot. Subsequently, the adulteration level of the spiked samples will show the adulteration level of every unknown sample in question by observing their positions after classification. Future predictions are important, and this study is the beginning of a bigger model which will be developed in order to classify unknown olive oils from the market.

4. Conclusions

By combining IRMS data and OPLS-DA through a multivariate statistical approach, a statistical model able to discriminate olive oils based on geographical origin was obtained with a successful discrimination ability at around 91%. Both $\delta^{18}O$ and $\delta^{2}H$ were the most important isotope markers for discrimination of olive oils. The investigation carried out in the present work can be used as a reliable and powerful tool for the characterization and authentication of Greek olive oils. Future determinations of unknown samples can easily occur based on the model depicted here.

Author Contributions: Conceptualization, M.T., V.S., S.A. and T.V.; methodology, M.T., V.S., S.A., D.P. and T.V.; software, M.T., D.P., C.R.T. and T.V.; validation, M.T., V.S., S.A., D.P., C.R., S.R., C.R.T., P.K. and T.V.; formal analysis, M.T., C.R.T.; investigation, M.T., V.S., S.A., D.P. and T.V.; resources, V.S., S.A., D.P., C.R.T., P.K. and T.V.; data curation, M.T., V.S., S.A., D.P. and T.V.; writing—original draft preparation, M.T.; writing—review and editing, M.T., V.S., S.A., D.P., C.R., S.R., C.R.T., P.K. and T.V.; visualization, M.T. and C.R.T.; supervision, V.S., S.A., D.P., C.R., S.R., C.R.T. and T.V.; project administration, V.S., S.A., D.P. and T.V. All authors have read and agreed to the published version of the manuscript.

Funding: This research received no external funding.

Institutional Review Board Statement: Not applicable.

Informed Consent Statement: Not applicable.

Data Availability Statement: Not applicable.

Acknowledgments: Part of the collected olive oil samples were derived in the framework of the European program "Interreg Greece Italy" MIS Code 5003145, under the project "Authentic Olive Net".

Conflicts of Interest: The authors declare no conflict of interest.

References

1. Kalogeropoulos, N.; Tsimidou, M.Z. Antioxidants in Greek virgin olive oils. *Antioxidants* **2014**, *3*, 387–413. [CrossRef]
2. Preedy, V.R.; Watson, R.R. (Eds.) *Olives and Olive Oil in Health and Disease Prevention*; Academic Press: Cambridge, MA, USA, 2010; p. 1520.
3. Tripoli, E.; Giammanco, M.; Tabacchi, G.; Di Majo, D.; Giammanco, S.; La Guardia, M. The phenolic compounds of olive oil: Structure, biological activity and beneficial effects on human health. *Nutr. Res. Rev.* **2005**, *18*, 98–112. [CrossRef]
4. Yubero-Serrano, E.M.; Lopez-Moreno, J.; Gomez-Delgado, F.; Lopez-Miranda, J. Extra virgin olive oil: More than a healthy fat. *Eur. J. Clin. Nutr.* **2019**, *72*, 8–17. [CrossRef] [PubMed]
5. Aparicio, R.; Morales, M.T.; Aparicio-Ruiz, R.; Tena, N.; García-González, D.L. Authenticity of olive oil: Mapping and comparing official methods and promising alternatives. *Food Res. Int.* **2013**, *54*, 2025–2038. [CrossRef]

6. European Commission. Commission Regulation (EEC) No 2568/91, on the characteristics of olive oil and olive-residue oil and on the relevant methods of analysis. *Off. J. Eur. Union* **1991**, *248*, 1–128.
7. Council Regulation (EC). No. 182/2009 of 6 March 2009 amending Regulation (EC) no. 1019/2002 on marketing standards for olive oil. *Off. J. Eur. Union* **2009**, *63*, 6–8.
8. Likudis, Z. Olive oils with protected designation of origin (PDO) and protected geographical indication (PGI). In *Products from Olive Tree*; Boskou, D., Clodoveo, M.L., Eds.; IntechOpen: London, UK, 2016; pp. 175–190.
9. Council Regulation (EC). No. 2081/92 of 14 July 1992 on the protection of geographical indications and designations of origin for agricultural products and foodstuffs. *Off. J. Eur. Union* **1992**, *208*, 1–8.
10. Council Regulation (EC). No. 2082/92 of 14 July 1992 on certificates of specific character for agricultural products and foodstuffs. *Off. J. Eur. Union* **1992**, *208*, 9–14.
11. Council Regulation (EC). No. 510/2006 of 20 March 2006 on the protection of geographical indications and designations of origin for agricultural products and foodstuffs. *Off. J. Eur. Union* **2006**, *93*, 12–25.
12. Council Regulation (EC). No. 1898/2006 of 14 December 2006 laying down detailed rules of implementation of Council Regulation (EC) no. 510/2006 on the protection of geographical indications and designations of origin for agricultural products and foodstuffs. *Off. J. Eur. Union* **2006**, *369*, 1–23.
13. Skiada, V.; Tsarouhas, P.; Varzakas, T. Preliminary Study and Observation of "Kalamata PDO" Extra Virgin Olive Oil, in the Messinia Region, Southwest of Peloponnese (Greece). *Foods* **2019**, *8*, 610. [CrossRef]
14. Alonso-Salces, R.M.; Segebarth, N.; Garmón-Lobato, S.; Holland, M.V.; Moreno-Rojas, J.M.; Fernández-Pierna, J.A.; Baeten, V.; Fuselli, S.R.; Gallo, B.; Berrueta, L.A. 1H-NMR and isotopic fingerprinting of olive oil and its unsaponifiable fraction: Geographical origin of virgin olive oils by pattern recognition. *Eur. J. Lipid Sci. Technol.* **2015**, *117*, 1991–2006. [CrossRef]
15. Gil-Solsona, R.; Raro, M.; Sales, C.; Lacalle, L.; Díaz, R.; Ibáñez, M.; Beltran, J.; Sancho, J.V.; Hernández, F.J. Metabolomic approach for Extra virgin olive oil origin discrimination making use of ultra-high performance liquid chromatography-Quadrupole time-of-flight mass spectrometry. *Food Control* **2016**, *70*, 350–359. [CrossRef]
16. González-Domínguez, R.; Sayago, A.; Morales, M.T.; Fernández-Recamales, Á. Assessment of virgin olive oil adulteration by a rapid luminescent method. *Foods* **2019**, *8*, 287. [CrossRef] [PubMed]
17. Gouvinhas, I.; de Almeida, J.M.; Carvalho, T.; Machado, N.; Barros, A.I. Discrimination and characterisation of extra virgin olive oils from three cultivars in different maturation stages using Fourier transform infrared spectroscopy in tandem with chemometrics. *Food Chem.* **2015**, *174*, 226–232. [CrossRef] [PubMed]
18. Bevilacqua, M.; Bucci, R.; Magrì, A.D.; Magrì, A.L.; Marini, F. Tracing the origin of extra virgin olive oils by infrared spectroscopy and chemometrics: A case study. *Anal. Chim. Acta* **2012**, *717*, 39–51. [CrossRef] [PubMed]
19. Laroussi-Mezghani, S.; Vanloot, P.; Molinet, J.; Dupuy, N.; Hammami, M.; Grati-Kamoun, N.; Artaud, J. Authentication of Tunisian virgin olive oils by chemometric analysis of fatty acid compositions and NIR spectra. Comparison with Maghrebian and French virgin olive oils. *Food Chem.* **2015**, *173*, 122–132. [CrossRef]
20. Sinelli, N.; Casale, M.; Di Egidio, V.; Oliveri, P.; Bassi, D.; Tura, D.; Casiraghi, E. Varietal discrimination of extra virgin olive oils by near and mid infrared spectroscopy. *Food Res. Int.* **2010**, *43*, 2126–2131. [CrossRef]
21. Guissous, M.; Le Dréau, Y.; Boulkhroune, H.; Madani, T.; Artaud, J. Chemometric characterization of eight Monovarietal Algerian virgin olive oils. *J. Am. Oil Chem. Soc.* **2018**, *95*, 267–281. [CrossRef]
22. Ollivier, D.; Artaud, J.; Pinatel, C.; Durbec, J.P.; Guérère, M. Triacylglycerol and fatty acid compositions of French virgin olive oils. Characterization by chemometrics. *J. Agric. Food Chem.* **2003**, *51*, 5723–5731. [CrossRef]
23. Ollivier, D.; Artaud, J.; Pinatel, C.; Durbec, J.; Guérère, M. Differentiation of French virgin olive oil RDOs by sensory characteristics, fatty acid and triacylglycerol compositions and chemometrics. *Food Chem.* **2006**, *97*, 382–393. [CrossRef]
24. Laursen, K.H.; Bontempo, L.; Camin, F.; Roßmann, A. Advances in isotopic analysis for food authenticity testing. In *Advances in Food Authenticity Testing*; Elsevier: Amsterdam, The Netherlands, 2016; pp. 227–252.
25. Bontempo, L.; Camin, F.; Larcher, R.; Nicolini, G.; Perini, M.; Rossmann, A. Coast and year effect on H, O and C stable isotope ratios of Tyrrhenian and Adriatic italian olive oils. *Rapid Commun. Mass Spectrom.* **2009**, *23*, 1043–1048. [CrossRef]
26. Camin, F.; Larcher, R.; Perini, M.; Bontempo, L.; Bertoldi, D.; Gagliano, G.; Nicolini, G.; Versini, G. Characterisation of authentic Italian extra-virgin olive oils by stable isotope ratios of C, O and H and mineral composition. *Food Chem.* **2010**, *118*, 901–909. [CrossRef]
27. Camin, F.; Pavone, A.; Bontempo, L.; Wehrens, R.; Paolini, M.; Faberi, A.; Marianella, R.M.; Capitani, D.; Vista, S.; Mannina, L. The use of IRMS, 1H NMR and chemical analysis to characterise Italian and imported Tunisian olive oils. *Food Chem.* **2016**, *196*, 98–105. [CrossRef]
28. Iacumin, P.; Bernini, L.; Boschetti, T. Climatic factors influencing the isotope composition of Italian olive oils and geographic characterisation. *Rapid Commun. Mass Spectrom.* **2009**, *23*, 448–454. [CrossRef]
29. Portarena, S.; Gavrichkova, O.; Lauteri, M.; Brugnoli, E. Authentication and traceability of Italian extra-virgin olive oils by means of stable isotopes techniques. *Food Chem.* **2014**, *164*, 12–16. [CrossRef] [PubMed]
30. Carter, J.F.; Chesson, L.A. (Eds.) *Food Forensics: Staple Isotopes as a Guide to Authenticity and Origin*; CRC Press: Boca Raton, FL, USA, 2017; pp. 303–331.
31. Bontempo, L.; Paolini, M.; Franceschi, P.; Ziller, L.; García-González, D.L.; Camin, F. Characterisation and attempted differentiation of European and extra-European olive oils using stable isotope ratio analysis. *Food Chem.* **2019**, *276*, 782–789. [CrossRef] [PubMed]

32. Hagidimitriou, M.; Katsiotis, A.; Menexes, G.; Pontikis, C.; Loukas, M. Genetic diversity of major Greek olive cultivars using molecular (AFLPs and RAPDs) markers and morphological traits. *J. Am. Soc. Hort. Sci.* **2005**, *130*, 211–217. [CrossRef]
33. Kostelenos, G. *Olive Fruit Data, History, Description and Geographical Distribution of Olive Varieties in Greece*, 1st ed.; Kostelenos Georgios: Poros Island, Greece, 2011; pp. 1–436.
34. Linos, A.; Nikoloudakis, N.; Katsiotis, A.; Hagidimitriou, M. Genetic structure of the Greek olive germplasm revealed by RAPD, ISSR and SSR markers. *Sci. Hortic.* **2014**, *175*, 33–43. [CrossRef]
35. Xanthopoulou, A.; Ganopoulos, I.; Koubouris, G.; Tsaftaris, A.; Sergendani, C.; Kalivas, A.; Madesis, P. Microsatellite high-resolution melting (SSR-HRM) analysis for genotyping and molecular characterization of an Olea europaea germplasm collection. *Plant. Genet. Resour.* **2014**, *1*, 5. [CrossRef]
36. Skiada, V.; Tsarouhas, P.; Varzakas, T. Comparison and Discrimination of Two Major Monocultivar Extra Virgin Olive Oils in the Southern Region of Peloponnese, According to Specific Compositional/Traceability Markers. *Foods* **2020**, *9*, 155. [CrossRef] [PubMed]
37. Anastasopoulos, E.; Kalogeropoulos, N.; Kaliora, A.C.; Falirea, A.; Kamvissis, V.N.; Andrikopoulos, N.K. Quality characteristics and antioxidants of mavrolia cv. virgin olive oil. *J. Am. Oil Chem. Soc.* **2012**, *89*, 253–259. [CrossRef]
38. Skiada, V.; Agriopoulou, S.; Tsarouhas, P.; Katsaris, P.; Stamatelopoulou, E.; Varzakas, T. Evaluation and Origin Discrimination of Two Monocultivar Extra Virgin Olive Oils, Cultivated in the Coastline Part of North-Western Greece. *Appl. Sci.* **2020**, *10*, 6733. [CrossRef]
39. Black, C.; Haughey, S.A.; Chevallier, O.P.; Galvin-King, P.; Elliott, C.T. A comprehensive strategy to detect the fraudulent adulteration of herbs: The oregano approach. *Food Chem.* **2016**, *210*, 551–557. [CrossRef] [PubMed]
40. Consonni, R.; Cagliani, L.R.; Cogliati, C. Geographical discrimination of honeys by saccharides analysis. *Food Control* **2013**, *32*, 543–548. [CrossRef]
41. Hrbek, V.; Rektorisova, M.; Chmelarova, H.; Ovesna, J.; Hajslova, J. Authenticity assessment of garlic using a metabolomic approach based on high resolution mass spectrometry. *J. Food Composit. Anal.* **2018**, *67*, 19–28. [CrossRef]
42. Tarapoulouzi, M.; Kokkinofta, R.; Theocharis, C.R. Chemometric analysis combined with FTIR spectroscopy of milk and Halloumi cheese samples according to species' origin. *Food Sci. Nutr.* **2020**, *8*, 3262–3273. [CrossRef]
43. Yahagi, T.; Masada, S.; Oshima, N.; Suzuki, R.; Matsufuji, H.; Takahashi, Y.; Watanabe, M.; Yahara, S.; Iida, O.; Kawahara, N. Determination and identification of a specific marker compound for discriminating Shrub Chaste Tree Fruit from Agnus Castus Fruit based on LC/MS metabolic analysis. *Chem. Pharm. Bull.* **2016**, *64*, 305–310. [CrossRef]
44. Tarapoulouzi, M.; Theocharis, C.R. Discrimination of Cheddar and Kefalotyri Cheese Samples: Analysis by Chemometrics of Proton-NMR and FTIR Spectra. *J. Agric. Sci. Technol.* **2019**, *9*, 347–355. [CrossRef]
45. Rongai, D.; Sabatini, N.; Del Coco, L.; Perri, E.; Del Re, P.; Simone, N.; Marchegiani, D.; Fanizzi, F.P. 1H NMR and multivariate analysis for geographic characterization of commercial extra virgin olive oil: A possible correlation with climate data. *Foods* **2017**, *6*, 96. [CrossRef]
46. Del Coco, L.; De Pascali, S.A.; Iacovelli, V.; Cesari, G.; Schena, F.P.; Fanizzi, F.P. Following the olive oil production chain: 1D and 2D NMR study of olive paste, pomace, and oil. *Eur. J. Lipid Sci. Technol.* **2014**, *116*, 1513–1521. [CrossRef]
47. del Mar Contreras, M.; Jurado-Campos, N.; Arce, L.; Arroyo-Manzanares, N. A robustness study of calibration models for olive oil classification: Targeted and non-targeted fingerprint approaches based on GC-IMS. *Food Chem.* **2019**, *288*, 315–324. [CrossRef]
48. Ghisoni, S.; Lucini, L.; Angilletta, F.; Rocchetti, G.; Farinelli, D.; Tombesi, S.; Trevisan, M. Discrimination of extra-virgin-olive oils from different cultivars and geographical origins by untargeted metabolomics. *Food Res. Int.* **2019**, *121*, 746–753. [CrossRef] [PubMed]
49. Mohamed, M.B.; Rocchetti, G.; Montesano, D.; Ali, S.B.; Guasmi, F.; Grati-Kamoun, N.; Lucini, L. Discrimination of Tunisian and Italian extra-virgin olive oils according to their phenolic and sterolic fingerprints. *Food Res. Int.* **2018**, *106*, 920–927. [CrossRef]
50. Marini, F.; Balestrieri, F.; Bucci, R.; Magrì, A.D.; Magrì, A.L.; Marini, D. Supervised pattern recognition to authenticate Italian extra virgin olive oil varieties. *Chemom. Intellig. Lab. Syst.* **2004**, *73*, 85–93. [CrossRef]
51. Aykas, D.P.; Karaman, A.D.; Keser, B.; Rodriguez-Saona, L. Non-Targeted Authentication Approach for Extra Virgin Olive Oil. *Foods* **2020**, *9*, 221. [CrossRef]
52. Coplen, T.B. Guidelines and recommended terms for expression of stable-isotope-ratio and gas-ratio measurement results. *Rapid Commun. Mass Spectrom.* **2011**, *25*, 2538–2560. [CrossRef] [PubMed]
53. Yang, Y.; Zheng, N.; Zhao, X.; Zhang, Y.; Han, R.; Yang, J.; Zhao, S.; Li, S.; Guo, T.; Zang, C.; et al. Metabolomic biomarkers identify differences in milk produced by Holstein cows and other minor dairy animals. *J. Proteom.* **2016**, *136*, 174–182. [CrossRef] [PubMed]
54. Eriksson, L.; Johansson, E.; Kettaneh-Wold, N.; Trygg, J.; Wikström, C.; Wold, S. *Multi-and Megavariate Data Analysis*; Umetrics: Umea, Sweden, 2006; Volume 1.
55. Fotakis, C.; Zoga, M.; Baskakis, C.; Tsiaka, T.; Boutsikou, T.; Briana, D.; Dendrinou, K.; Malamitsi-Puchner, A.; Zoumpoulakis, P. Investigating the metabolic fingerprint of term infants with normal and increased fetal growth. *RSC Adv.* **2016**, *6*, 79325–79334. [CrossRef]
56. Mihailova, A.; Abbado, D.; Kelly, S.D.; Pedentchouk, N. The impact of environmental factors on molecular and stable isotope compositions of n-alkanes in Mediterranean extra virgin olive oils. *Food Chem.* **2015**, *173*, 114–121. [CrossRef]

57. Jiménez-Morillo, N.T.; Palma, V.; Garcia, R.; Pereira, J.A.; Barrocas-Dias, C.; Cabrita, M.J. Multivariate geostatistical analysis of stable isotopes in Portuguese varietal extra virgin olive oils. *Microchem. J.* **2020**, *157*, 105044. [CrossRef]
58. Spangenber, J.E. Bulk C, H, O, and fatty acid C stable isotope analyses for purity assessment of vegetable oils from the southern and northern hemispheres. *Rapid Commun. Mass Spectrom.* **2016**, *30*, 2447–2461. [CrossRef]
59. Jiménez-Morillo, N.T.; Palma, V.; Garcia, R.; Dias, C.B.; Cabrita, M.J. Combination of Stable Isotope Analysis and Chemometrics to Discriminate Geoclimatically and Temporally the Virgin Olive Oils from Three Mediterranean Countries. *Foods* **2020**, *9*, 1855. [CrossRef]
60. Karalis, P.; Poutouki, A.E.; Nikou, T.; Halabalaki, M.; Proestos, C.; Tsakalidou, E.; Gougoura, S.; Diamantopoulos, G.; Tassi, M.; Dotsika, E. Isotopic Traceability (^{13}C and ^{18}O) of Greek Olive Oil. *Molecules* **2020**, *25*, 5816. [CrossRef]
61. O'Leary, M.H. Environmental Effects on Carbon Isotope Fractionation in Terrestrial Plants. In *Stable Isotopes in the Biosphere*; Kyoto University Press: Kyoto, Japan, 1995; pp. 78–91. Available online: https://ci.nii.ac.jp/naid/10014710615/#cit (accessed on 23 January 2021).
62. Chiocchini, F.; Portarena, S.; Ciolfi, M.; Brugnoli, E.; Lauteri, M. Isoscapes of carbon and oxygen stable isotope compositions in tracing authenticity and geographical origin of Italian extra-virgin olive oils. *Food Chem.* **2016**, *202*, 291–301. [CrossRef] [PubMed]
63. Portarena, S.; Farinelli, D.; Lauteri, M.; Famiani, F.; Esti, M.; Brugnoli, E. Stable isotope and fatty acid compositions of monovarietal olive oils: Implications of ripening stage and climate effects as determinants in traceability studies. *Food Control* **2015**, *57*, 129–135. [CrossRef]
64. Angerosa, F.; Basti, C.; Vito, R. Virgin Olive Oil Volatile Compounds from Lipoxygenase Pathway and Characterization of Some Italian Cultivars. *J. Agric. Food Chem.* **1999**, *47*, 836–839. [CrossRef]
65. Aramendía, M.A.; Marinas, A.; Marinas, J.M.; Moreno, J.M.; Moalem, M.; Rallo, L.; Urbano, F.J. Oxygen-18 measurement of Andalusian olive oils by continuous flow pyrolysis/isotope ratio mass spectrometry. *Rapid Commun. Mass Spectrom.* **2007**, *21*, 487–496. [CrossRef]
66. Baum, B.A.; Lu, Y.; Muccio, Z.; Jackson, G.P.; Harrington, P.B. Differentiation Between Origins of Extra Virgin Olive Oils by GC-C-IRMS Using Differentiation Between Origins of Extra Virgin Olive Oils by GC-C-IRMS Using Principal Component Analysis, Linear Discriminant Analysis, and Hierarchical Cluster Analysis. *Spectroscopy* **2010**, *25*, 40.
67. Bréas, O.; Guillou, C.; Reniero, F.; Sada, E.; Angerosa, F. Oxygen-18 Measurement by Continuous Flow Pyrolysis/Isotope Ratio Mass Spectrometry of Vegetable Oils. *Rapid Commun. Mass Spectrom.* **1998**, *12*, 188–192. [CrossRef]
68. Royer, A.; Naulet, N.; Mabon, F.; Lees, M.; Martin, G.J. Stable Isotope Characterization of Olive Oils: II-Deuterium Distribution in Fatty Acids Studied by Nuclear Magnetic Resonance (SNIF-NMR). *J. Am. Oil Chem. Soc.* **1999**, *76*, 1–9. [CrossRef]
69. Gumus, Z.P.; Celenk, V.U.; Tekin, S.; Yurdakul, O.; Ertas, H. Determination of trace elements and stable carbon isotope ratios in virgin olive oils from Western Turkey to authenticate geographical origin with a chemometric approach. *Eur. Food Res. Technol.* **2017**, *243*, 1719–1727. [CrossRef]
70. Alves de Carvalho, A.G.; Olmo-García, L.; Gaspar, B.R.A.; Carrasco-Pancorbo, A.; Castelo-Branco, V.N.; Torres, A.G. Preliminary discrimination of commercial extra virgin olive oils from Brazil by geographical origin and olives' cultivar: A call for broader investigations. In Proceedings of the 1st International Electronic Conference on Food Science and Functional Foods, Basel, Switzerland, 10–25 November 2020.
71. Mikrou, T.; Pantelidou, E.; Parasyri, N.; Papaioannou, A.; Kapsokefalou, M.; Gardeli, C.; Mallouchos, A. Varietal and Geographical Discrimination of Greek Monovarietal Extra Virgin Olive Oils Based on Squalene, Tocopherol, and Fatty Acid Composition. *Molecules* **2020**, *25*, 3818. [CrossRef] [PubMed]
72. Tsopelas, F.; Konstantopoulos, D.; Kakoulidou, A.T. Voltammetric fingerprinting of oils and its combination with chemometrics for the detection of extra virgin olive oil adulteration. *Anal. Chim. Acta* **2018**, *1015*, 8–19. [CrossRef] [PubMed]

MDPI
St. Alban-Anlage 66
4052 Basel
Switzerland
Tel. +41 61 683 77 34
Fax +41 61 302 89 18
www.mdpi.com

Foods Editorial Office
E-mail: foods@mdpi.com
www.mdpi.com/journal/foods

www.ingramcontent.com/pod-product-compliance
Lightning Source LLC
Chambersburg PA
CBHW061441210926
51850CB00008B/1572
9783036566610